Dan Lindley

2. 1-5, 8
3. 2, 3, 4, 7
4. 1, 2, 3, 4, 5, 6, 8, 10

# Matter in Motion

## The Spirit and Evolution of Physics

**Ernest S. Abers**
Professor of Physics
*University of California at Los Angeles*

**Charles F. Kennel**
Professor of Physics
*University of California at Los Angeles*

Allyn and Bacon, Inc.
Boston · London · Sydney · Toronto

*Ernest Abers dedicates his share of this book to Geoffrey and Rebecca Abers, hoping they will learn something from it.*

*Charles Kennel dedicates his share of this book to Deborah Bochner Kennel, who awakened in him a love of history, and to Salomon Bochner, her father, whose presence and career, both scientific and historical, have been an inspiration.*

Copyright © 1977 by Allyn and Bacon, Inc.
470 Atlantic Avenue, Boston, Massachusetts 02210

*All rights reserved. Printed in the United States of America. No part of the material protected by this copyright notice may be reproduced or utilized in any form or by any means, electronic or mechanical, including photocopying, recording, or by any information storage and retrieval system, without written permission from the copyright owner.*

**Library of Congress Cataloging in Publication Data**

Abers, Ernest S   1936-
   Matter in motion.

   Includes bibliographical references and index.
   1. Physics—History.  2. Astrophysics—History.
I. Kennel, Charles F., 1939-   joint author.
II. Title.
QC7.A24     530'.09     77-23185
ISBN 0-205-05790-X

# Contents

*Preface*     xv
*Introduction*     xix

## 1. How the Heavenly Bodies Move    3

Here we assemble, describe, and summarize the basic observations of the heavens that can be made with the naked eye. To explain these observations was the central preoccupation of physics and astronomy from ancient times to that of Newton. The facts we first present here will appear and reappear throughout our story.

### *Ancient Astronomy*     3

Introductory remarks; Babylonian astronomy; definition of units of angular measure; how big is a degree?

### *The Motions of the Stars*     6

Simplicity of patterns of motions in the heavens; constellations; appearance of individual stars; apparent rotation of the heavens about the north star; celestial sphere model; sun, moon, and planets exceptions to simple model.

### *The Apparent Motion of the Sun*     11

Sun seems unlike stars; sun's diurnal motion; motion of the sun around the ecliptic.

### *The Motions of the Moon and Planets*     14

Moon and planets seem unlike sun and stars; diurnal motion of planets; the zodiac; motion of the moon; moon's phases correlated with the position of the sun; retrograde motion; Venus' motion; relation of Venus to sun; Mercury's motion; motions of Jupiter and Saturn; Mars, Jupiter, and Saturn retrogress at opposition, Mercury and Venus at conjunction; list of scientific questions posed by naked eye observations.

### *Summary*     20

### *Appendix to Chapter 1*     21

The precession of the equinoxes and the astrological calendar.

### *Questions*     21

## 2. The Greek Cosmos, The Beginning of Science    25

The Ancient Greeks first posed many scientific questions still asked today. The science they created lasted for two-thousand years, longer than modern science. No thinker has been more influential than Aristotle, and his physics, created in the context of the Greek picture of the heavens, defined the nature of modern science's, and particularly Galileo's, rebellion against it.

### *Pythagoras*    27
Introductory remarks; Pythagoras' theorem.

### *The Shape of the Earth*    28
What is the geometrical shape of the earth and heavens? The earth a sphere; three proofs that earth is round: the way ships disappear below the horizon, altitude of celestial pole above the horizon, and shape of the earth's shadow on the moon during eclipses; Greek explanation of eclipses.

### *Eratosthenes and the Size of the Earth*    34
Calculation of earth's radius, using Pythagoras' theorem; calculating using altitude of pole; Eratosthenes' ancient calculation.

### *The Two-Sphere Universe*    36
The spherical earth at the very center of the celestial sphere; what the model seemed to explain; support from Plato's philosophy; Platonic perfection and uniform circular motion.

### *The Universe of Eudoxus*    38
"Save the appearances"; Eudoxus' research program; adding a sphere for the sun; correction for the different lengths of the seasons; the moon's sphere; a system of spheres for the retrograde motions of the planets; difficulties with Eudoxus' model.

### *Aristotle's Synthesis of Physics and Cosmology*    43
Aristotle's physics created in a framework of Eudoxan cosmology; common sense; biological analogies in Aristotle's thought; Aristotle's four elements; definition of natural motion; speed of falling bodies; a vacuum impossible; unnatural motions require a force; Aristotle's cosmology; the fifth element—weightless celestial matter; reasons why earth is round.

### *Heraklides, Aristarchus, and the Sun-Ordered Universe*    48
Could the sun, and not the earth, be at the center of the universe? Perhaps the earth and planets orbit the sun; parallax and the distance to the stars; perhaps the stars are so far away that their distance is not measurable by parallax. Aristotelian arguments against the earth's motion.

### *Summary*    52

*Appendix to Chapter 2*   53

Pythagoras' Theorem. Calculation of the Earth's Size, Algebraic Details. Aristarchus' Method for Measuring the Distance to the Moon. Stellar Parallax and the Distances of the Stars.

*Questions*   59

*References*   60

3. **Ptolemy and Copernicus, Technical Astronomy in the Greek Tradition**   63

In this chapter, we adopt the unusual expedient of discussing Ptolemy and Copernicus together. Both had the same aim—accurate calculation of planetary positions—and similar calculational techniques. Ptolemy's system was earth-centered; Copernicus' sun-centered. The juxtaposition helps us compare the two systems, but, unfortunately, prevents us from discussing medieval science, both Arab and Christian, created in the centuries separating them.

*Ptolemaic Astronomy*   64

Ptolemy's system is two-dimensional; epicycles; Hipparchus' epicycle theory of different lengths of seasons; Ptolemy's use of epicycles for retrograde motion; epicycle for Saturn; epicycles for Jupiter and Mars; epicycles for Venus and Mercury; detailed description of theory for Venus; simplicity and complexity of Ptolemaic system.

*Copernicus*   74

Was Copernicus the last great Ptolemaic astronomer, or the first great modern astronomer? Biographical sketch; *On the Revolutions of the Heavenly Spheres*; the Copernican revolution.

*On the Revolutions of the Heavenly Spheres*   75

Why did Copernicus reformulate planetary astronomy? Copernicus' philosophical and physical reasoning; suppose the earth rotated; rotation of earth and rotation of celestial sphere explain astronomical observations equally well. Is there any astronomical evidence against the earth moving around the sun? Retrograde motion explained by earth's motion; Copernican model of Saturn's retrograde motion; Mars' retrogression of opposition a natural feature of his theory; calculation of Venus' distance from sun; calculation of sidereal periods of planets.

*Comparison of Ptolemaic and Copernican Astronomy*   90

Harmony of theories; Copernican calculations not much simpler; Copernican theory explains more with fewer assumptions; phases of Venus would be a proof of Copernicus; Copernican model apparently violated laws of physics; Copernicus required universe to be frighteningly large.

*Summary*   94

*Appendix to Chapter 3*   95
The Construction of Venus' Epicycle. The Precession of the Equinoxes and the Calendar. The Sizes of the Planets' Orbits. Calculation of the Sidereal Periods of the Planets.

*Questions*   103

*References*   103

4. **Tycho Brahe and Johannes Kepler, Reform of Planetary Astronomy**   105

We come now to the first two great astronomers to follow in Copernicus' footsteps. Brahe, freed from the preconceptions of ancient theory, measured the planets' positions with unprecedented accuracy. Kepler used Brahe's data to put the Copernican system on a more solid physical and mathematical foundation.

*Tycho Brahe*   106
Biographical sketch; Uraniborg castle; the nova of 1572; the nova was further than the moon, contrary to Aristotle; the comet of 1577; Tycho's systematic methods; Tycho's arguments against Copernican model; Tycho's model of solar system.

*Johannes Kepler and the Cosmographic Mystery*   113
Biographical sketch; Kepler's teacher, Mästlin; Kepler's first book, the *Cosmographic Mystery*, asserts complete reality of Copernican system; the five regular solids fit between the six planetary orbits.

*Kepler's War on Mars*   118
Brahe and Kepler's stormy relationship; the *Astronomia Nova;* Kepler refers all planetary distances to the sun; Kepler rejects uniform speeds; Mars' orbit lies in a plane intersecting the ecliptic plane at the sun; Kepler's circular theory for Mars—its failure; shape of the earth's orbit; how Kepler calculated the earth's orbit; Mars' orbit an ellipse; conic sections; foci of ellipses; Equal Areas law.

*Kepler's Laws*   128
Orbits of all planets are ellipses, with sun at one focus; each planet sweeps out equal areas in equal times; the square of the period of revolution divided by the cube of the orbital radius is same for all planets.

*Summary*   132

*Appendix to Chapter 4*   133
The Five Regular Solids. More on Kepler's Speed Rule.

*Questions*   138

*References*   140

# Contents

## 5. Galileo Galilei — 143

Galileo was the first to turn a telescope to the heavens. What he saw created a sensation. Galileo was the first abstract theoretician. He formed quantitative laws for the motion of falling bodies. To explain the nature of Galileo's work on mechanics, we insert into our historical narrative a careful modern-style discussion of speed and acceleration.

### *Galileo's Astronomical Discoveries* — 145

First astronomical use of telescope; five great discoveries; the *Starry Messenger;* the moon has mountains; a vast crowd of stars; Jupiter has four satellites; they obey Kepler's laws; reception of *Starry Messenger;* criticisms of Galileo; stars remain points of light in the telescope; phases of Venus; Galileo's work still consistent with Tycho's model; sunspots; *Dialogue on the Two Great World Systems;* Galileo's abjuration.

### *The Science of Motion—Inertia* — 159

*Dialogues Concerning the Two New Sciences;* medieval developments and criticisms of Aristotelian mechanics; Galileo's realization that criticisms were condemnation; inertia and importance of relative motion.

### *Speed and Acceleration* — 161

Definition of speed; formula for constant speed; changing speed; average speed; instantaneous speed; definition of acceleration; acceleration as rate of change of speed or of direction.

### *Falling Bodies* — 167

Bodies fall with constant acceleration; inclined planes; distance an object falls in one second given by successively increasing odd numbers; $d = 16t^2$; proof that acceleration is constant.

### *Compound Motion* — 172

Separation of horizontal and vertical projectile motion; impetus; horizontal acceleration is zero and vertical acceleration is constant.

### *Galileo's Principle of Relativity* — 175

Motion has absolute meaning in Aristotelian mechanics; relativity principle derived from law of compound motion.

### *Relativity and the Earth's Motion* — 177

Refutation of arguments against earth's motion; inertia is straight line on small scale, but is it circular on astronomical scale? Galileo's contributions to the style of science; logical consistency and experimental agreement only scientific authorities; Galileo first abstract theoretician.

### *Summary* — 179

*Appendix to Chapter 5*   181

The Height of the Mountains on the Moon. Kepler's Third Law Applied to Jupiter's Moons. Another Proof That $d = \frac{1}{2} gt^2$ Is the Same As $v = gt$. The General Motion of Projectiles.

*Questions*   187

*References*   189

## 6  Heaven and Earth in one Framework: Sir Isaac Newton   191

Isaac Newton created a logically consistent and complete mathematical mechanics. With it he proved that gravitation is a universal force, and calculated the planets orbits with high precision. While we avoid mathematical complexity, we do not evade the logical content of Newton's arguments.

*The Seventeenth Century in Science*   191

Great increase in activity; scientific societies.

*Seventeenth Century Astronomy and Mechanics*   192

Aristotelian thought fades away; mechanistic philosophy; what forces keep planets in orbit? Mathematical difficulties with mechanics; Isaac Newton, biographical sketch.

*Newton's Miraculous Year*   194

Plague in London, Newton goes home; Galileo's laws of mechanics must also apply to heavens; gravity, apples, and the moon; acceleration of circular motion; calculation of $a = v^2/R$; calculation of moon's acceleration; acceleration of planets from Kepler's third law; gravity controls orbits of planets and moons; artificial satellites; problems remaining after Newton's miraculous year.

*The* Principia Mathematica   203

Newton refuses to publish; Halley reawakens Newton's interest in mechanics; the incredible effort of writing the *Principia*.

*Newton's Three Laws*   205

Collisions; Newton's First Law defines when forces are needed; Newton's Second Law defines force as time rate of change of momentum. Mass and Weight—separate experiment needed to measure weight; weight equals mass times acceleration of gravity; mass is proportional to weight; if mass is constant, force equals mass times acceleration. Newton's Third Law—Descartes' momentum conservation; conservation of momentum in collisions; to every action, there is an equal and opposite reaction; when we drop a ball, why do we not feel the earth move?

*Newton's System of the World*    214
Summary of Books I and II of the *Principia*; Book III discusses motion under gravity; gravity provides unified, detailed understanding of motions of heavenly bodies.

*Universal Gravitation*    215
How do we know that all bodies exert gravitational force? Is gravity exerted only between objects of planetary size or larger, or between small bodies too? Proof urging Newton's third law; mathematical description of gravitational force; Newton's constant of gravitation; Why do we not feel mutual force between small bodies?

*An Important Mathematical Theorem*    217
Earth's total mass must act as if it were concentrated at the earth's center; this would be true if earth were exactly spherical; separate experiment needed to measure $G$.

*Inertial and Gravitational Mass*    220
Galileo's law of falling bodies would be violated if they were unequal; Newton's experiments to test equality of inertial and gravitational mass much more accurate than Galileo's.

*Precession of the Equinoxes*    221
Since earth rotates, it is not exactly spherical; moon and sun's gravity pull slightly differently at equator and poles; spinning earth wobbles.

*Proof of Kepler's Laws*    222
Proof that equal areas law applies to all central forces; most general orbits under gravity are conic sections; Kepler's laws are an approximation; planetary perturbations; Laplace, Lagrange, and the stability of the solar system.

*The Tides, Comets, and New Planets*    225
Lunar theory of the tides; how far does gravitational force extend? At least as far as the comets; Herschel's discovery of Uranus; discovery of Neptune from perturbations of Uranus; Mercury does not obey Newton's law exactly; discovery of Pluto; star clusters and galaxies.

*The Mystery of Gravitation*    228
Newton's method does not penetrate "ultimate" reality; "I frame no hypotheses."

*Philosophical Impact of Newtonian Physics*    230
Letters to Richard Bentley; Newton's cosmology; order of planets' orbits evidence for Divine Providence.

*On Absolute Space and Time*    231
Newtonian time absolute; universe as "God's Sensorium."

*Newton's Later Life*   232

Newton leaves Cambridge; master of the mint; president of the Royal Society; summary of Newton's genius.

*Summary*   235

*Appendix to Chapter 6*   237

More about the Moon. Weighing the Earth, Sun, and Planets.

*Questions*   241

*References*   242

7. **Light**   245

The science of motion should explain not only the motions of ordinary matter but also of light. We return back to Greek physics to trace the evolution of our ideas about light. Many names now familiar to us—Aristotle, Kepler, Galileo, and Newton—reappear, together with some new ones: Alhazen, Huygens, Young, Fresnel, and Fizeau.

*Ancient Models of Light*   245

Euclidean optics; atomist theory of light; Aristotle's objection to atomist theory; difficulty in separating physical and physiological aspects of light and vision.

*The Geometry of Light*   248

Light moves in straight lines; shadows; reflection and refraction; Ptolemy's quantitative rule for refraction.

*Alhazen*   253

Flourishing science in medieval Islam; Alhazen, and his *camera obscura*; angle of incidence equals the angle of refraction; Alhazen separated physics and physiology.

*Kepler*   255

Kepler carefully analyzed *camera obscura*; Kepler understands that the eye has a lens; Kepler corrects Ptolemy's law of refraction; Snell's law of refraction.

*Diffraction*   261

The bending of light rays around sharp obstacles; "Light propagates not only in a straight line, reflected or refracted, but also in a certain other direction, diffracted."

*Color*   261

Problem of chromatic aberration in telescopes; Newton's prism experiment; different colors refract differently; white light a mixture of all the colors; colors of objects; first reflecting telescope.

### The Speed of Light    266
Aristotle believed speed finite. Kepler infinite; Galileo unable to measure speed; Roemer measures speed using eclipses of Galilean satellites of Jupiter; summary of basic facts about light that first theories strove to explain.

### Newton and the Particle Model    270
Light not a wave since waves could not propagate in a vacuum; support for particle model from law of reflection; particle model requires faster speed of light in denser media.

### The Wave Model    273
Huygens; geometry of wave propagation can explain straight-line propagation, reflection and refraction; details of wave model explanations; wave model requires speed to be slower in denser media.

### Thomas Young and the Revival of the Wave Theory    278
Young; definition of frequency and wavelength; interference; explanation of diffraction; analysis of Young's experiment; calculation of visible light wavelength; Fresnel's mathematical description of diffraction; Fizeau's measurement of the speed of light in water.

### Summary    286

### Appendix to Chapter 7    287
More on Lenses. The Reflecting Telescope. The Wavelength of Light. Diffraction through a Slit.

### Questions    292
### References    293

## 8. Magnetism and Electricity    295
It came as a surprise that to understand light, we also had to understand electricity and magnetism. We retrace our steps again, this time to trace the evolution of our knowledge of electromagnetism. We encounter, for the first time, the great 19th century of science, and its two greatest practitioners, Michael Faraday and James Clerk Maxwell. When we finish, we will be at the threshold of our own twentieth century, and the theory of relativity, which unifies what we have learned.

### Magnets    295
Lodestones; north and south poles; superstition surrounding magnets; the first compasses; Christopher Columbus averts a near disaster; compasses do not point to north celestial pole exactly; magnetic dip; William Gilbert; the earth is a giant magnet; the earth's magnetism extends into space; permanent and induced magnetism; Gilbert's Copernicanism.

### Electricity   302

Amber; frictional electricity; Gilbert's electrics and nonelectrics; Von Guericke and his electric machine; conduction of electricity; Dufay's two fluid theory of electricity; Benjamin Franklin's single fluid theory.

### Coulomb's Law   308

Coulomb and electrical charge; Coulomb's torsion balance; inverse-square law for electrostatic forces; difference between gravitational and electrostatic forces.

### Moving Electricity   312

Volta, "animal electricity" and the first batteries; Humphrey Davy and the electrolysis of water; Oersted, and magnetism from an electrical current; explosion of activity at the French Academy of Sciences; Ampere's laws.

### Michael Faraday   322

His life, a classic success story; lines of force, changing magnetism induces an electrical force; reality of lines of force; Maxwell's interest in Faraday's idea.

### James Clerk Maxwell   331

As a boy he was unusually precocious; Saturn's rings, foundations of the physics of gases; physical content of Maxwell's equations; lines of force surrounding charges; concept of field; changing electric fields produce magnetic fields; electromagnetic waves; light is an electromagnetic wave; Hertz's detection of the electromagnetic wave; Marconi and the invention of the radio.

### Foundations of Electrical Civilization   342

Henry, Wheatstone, Morse, and the invention of the telegraph; generation of large quantities of electricity; the electrical motor.

### Summary   344

### Appendix to Chapter 8   346

The Electroscope. Franklin and Lightning. More on Microscopic Magnetism. Lines of Force and the Inverse-Square Law. Some Quantitative Rules for Fields and Forces. Faraday and Electrochemistry. A Charging Capacitor.

### Questions   353

### References   354

## 9. Einstein and Relativity  357

In this chapter, we complete our revision of our view of man's place in the universe, which began with Copernicus. Einstein's special theory of relativity unites light, electricity and magnetism, and mechanics into one framework; his general theory adds in gravity. With Copernicus, we had to give up the idea that the earth is stationary; with Einstein, we must renounce our common-sense notions of space and time.

### *The Aether Returns*  357

Is the speed of light defined with respect to the aether? A. A. Michelson, his early life; Michelson's first measurement of the speed of light; the Michelson interferometer; the Michelson-Morley experiment; "saving the appearances" again; Lorentz contraction.

### *Albert Einstein*  363

His early life; influence of Ernst Mach; Brownian motion; the photoelectric effect.

### *The Special Theory of Relativity*  368

Basic observations; speed of light constant in all reference frames; Newtonian time; Einstein's arguments concerning simultaneity; light clocks; time dilation; length contraction; addition of velocity; relativistic mass increase.

### *A New View of Gravitation*  383

Einstein rises in the world of physics; need to extend relativity theory in accelerating frames; principal of equivalence; thought experiments indicating its significance; does light fall? Deflection of light by gravitating bodies; formulation of relativity theory for non-uniform gravitational fields; curved space; precession of the perihelion of Mercury; deflection of light by the sun; reception of Einstein by the general public; Einstein advocates building the atomic bomb; the expanding universe.

### *Summary*  397

### *Appendix to Chapter 9*  399

Time-Difference in the Michelson-Morley Experiment. Contraction of Moving Yardsticks. Relativistic Addition of Speeds.

### *References*  407

## Index  409

# Preface

This text is one of the results of our having taught a one-term course titled "Physics for Non-Science Majors" for several years at the University of California at Los Angeles. For most students the course was not required, but was one option they could choose in order to fulfill the college's "breadth" or "distribution" requirements.

The students were a mixture of freshmen, sophomores, juniors, and seniors with the broadest imaginable range of interests, talents, and facility with mathematics. As anyone who has taught such a disparate group of students knows, the greatest difficulty in designing such a course is trying to make it possible for all to derive something from it.

We believe that a simplified version of a standard physics text is not the appropriate format for such a course, and we do not attempt to teach the techniques a scientist must learn (e.g., through extensive problem solving). Rather, it is our purpose to explain some of the theoretical ideas and experimental facts that are included in the body of knowledge generally called "physics", and to communicate something of the elegance that physicists see in their subject.

Finally, we feel that knowledge of the physical world is, and always has been, an integral part of our culture, and that the human drama of scientific discovery is often as exciting as the results themselves. Therefore, we explain not only the "facts", or the systems of organizing them, but also how they came to be known. What were the arguments for and against accepted physical ideas, and why did these ideas prevail? We have found it useful to adopt a historical approach to our subject, and explain how physics developed.

## Preface

All our students know that the earth is a round planet that orbits the sun, but few know how this improbable explanation of everyday experience came to be generally accepted. A historical presentation that raises questions of this sort can be interesting for its own sake, and it can also provide the motivation for the non-scientist to take an interest in physics, which he or she never had before. While this text may be a useful adjunct in beginning history of science courses—because it teaches some physics—our pedagogical method prevents it from being a true history of science text. For example, we have limited our discussion to the truly key scientific personalities, and have not discussed the roles played by their lesser contemporaries.

A great barrier to teaching physics to non-scientists is that even its simplest concepts are couched in quantitative language. We have tried to use mathematics as little as possible; and, when we do, have kept in mind that mathematical argument is quite foreign to most readers. No more mathematics is required beyond a year each of high school algebra and geometry, which are usually prerequisites for admission to a university. We completely avoid calculus, trigonometry, and vector algebra, even when using them could shorten an explanation. Nevertheless, some mathematics is unavoidable, especially in discussing the foundations of mechanics. Science is quantitative, and always has been. At the very least, we can explain why physicists use mathematical language.

Most of the chapters are followed by an appendix. These contain material that is either peripheral to the main subject or an extension of it. Some of the appendices are mathematically too complicated to be worthwhile for most students. Those with a good background or talent for mathematical argument may find them interesting. Parts of Chapters 5 and 6 contain algebraic arguments that are rather more involved than those in the rest of the text. This is true also for parts of Chapter 9. We feel that this is unavoidable, since the formulation of mechanics by Galileo and Newton, which is the cornerstone of all physics, was achieved by conceiving the problem in a mathematical way. Students who find these sections difficult should first understand what are the hypotheses and what are the conclusions; i.e., what goes in and what comes out, and only then try to follow the algebraic argument in detail.

Our book does not include many twentieth century discoveries about the nature of matter, such as atoms and nuclei. These subjects are at least as important and interesting as those we cover, but we found that it is necessary to develop first the basic concepts of mechanics and the properties of light, magnetism, and electricity. The topics we have selected make a unified subject—roughly the science of motion—and our material is more than can reasonably be covered in one term. For those students who wish to learn more, we have a second term at UCLA following the same historical methods as in the present text, and perhaps shall one day finish a companion volume.

We acknowledge with gratitude the assistance and advice of many friends and colleagues: Our students, whose responses and criticisms of this material have been our single most important guide; the many historians who have been patient in answering our simple-minded questions, including John Burke, Robert Frank, Amos Funkenstein, Robert Westman, and Curtis Wilson; our teaching assistants, including Clinton Ar and Victor Decyk, who helped organize the material in its early stages; Professor Donald Weinshank for his constructive criticisms; our colleagues Ferdinand Coroniti, Kenneth Mackenzie, and Kimball Milton among others who have taken an active interest in this project. Betty Rae Brown, Joan Kaufman, and Gilda Reyes suffered through the production of many draft typescripts with unfailing good humor. Our list is certainly incomplete.

Ernest S. Abers<br>
Charles F. Kennel

# Introduction

Most modern physics books start with the work of Isaac Newton. He unified the achievements of the scientific revolution during the sixteenth centuries into a mathematical system for describing the physical world. Newton was the first to fuse a long tradition of patient observation of the planets with a rigorous mathematical theory for their motions, derived from a general theory of motion based upon logically consistent philosophical statements. While many before Newton had used mathematics in science, never before had there been a theory capable of explaining so many phenomena with complete accuracy. The subsequent development of all other branches of science are patterned after Newton's achievement. He showed how science ought to be done.

To understand modern physics, we must understand what Newton achieved, and to understand that, we must understand what came before Newton. Newton remarked that if he had seen far, it was because he had "stood on the shoulders of giants." The fascinating story of the development of the science of motion is the substance of our book. Since before Newton the mathematics used to describe motion was elementary, all the essential arguments concerning the development of Newtonian mechanics can be presented using only elementary algebra and geometry. Just because pre-Newtonian science was from a modern point of view elementary does not mean it was trivial, and we are able to present one very deep scientific problem as it was treated at the time. In this way, we hope to communicate not only the flavor of scientific activity as it was years ago but also as it is now, not only the nature of scientific reasoning but also the mutual influences between science and philosophy, religion, and politics; for science is, after all, inseparable from other human activities.

There was a time before most people dared think the earth moves through the heavens. If anything is obviously stationary, it is the solid earth beneath our feet; we should not imagine without good reason that it is whirling on its axis and moving around the sun at frightening speeds. Our first few chapters will be concerned with how we realized that the earth moves. The ancient Greeks had a plan of the universe with the solid, immovably heavy earth at its center. Consistent with this plan was a particular science of motion—Aristotle's physics. When Copernicus revised the plan of the solar system and universe, a new science of motion, Newton's mechanics, was invented to explain it. And when Faraday's and Maxwell's discoveries concerning electromagnetism and light forced Einstein to revise Newton's mechanics, our view of the universe changed again.

Therefore, we begin with the first great question of physics: "How do things move?" We start with the motions of the celestial bodies as they appear to any naked-eye observer and the Greek concept of the structure of the universe derived from these observations. We will present the Greek view of the universe as convincingly as we can without reference to our modern view. We will discuss the physics that explains the Greek universe. We shall see the Greek universe and physics prevail for two thousand years only to dissolve and be reformed by the work of Copernicus, Kepler, Galileo, and Newton. We shall see their new scientific method applied to electricity, magnetism, and light, as well as to mechanics and astronomy. We shall see the development of Einstein's theory of relativity in response to discoveries about electromagnetism and light. Just as Copernicus taught us that the earth moves even when it seems stationary, Einstein taught us that space and time are not absolute even though they seem so. We finish then with these twentieth century discoveries about the nature of space and time. Only after we had learned to answer "how do things move", could we begin to answer the second great question of physics that dominates today's research: "What is everything made of?"

**STAR MOTION**

*A Star Field Showing Circumpolar Motion of the Stars. (Courtesy of Harvard College Observatory and by permission of the Houghton Library, Harvard University)*

# CHAPTER 1

# HOW THE HEAVENLY BODIES MOVE

### ANCIENT ASTRONOMY

People have always observed nature carefully, and have modified it to suit themselves. For primitive people, intimate familiarity with the natural world around them was a matter of survival. The marvelously detailed drawings of animals on the walls of caves illustrated how seriously nature was observed. Yet, the first observational science, which emerged at the dawning of literate civilization, was not concerned with how plants grow or what makes the weather, but with the heavens. It was astronomy. And for good reason: the ancient peoples were more concerned with the stars in their day-to-day life than we are today. Their sky was unobscured by city lights or tall buildings or atmospheric pollution, and the bowl of stars overhead was their constant companion. Events in the sky affected their lives. Before clocks were invented, people told time by the sun. The sun seemed to control the seasons. When the sun was high in the sky at noon, it was summer, days were long and plants grew; when the sun was low in the sky at noon, it was winter, days were short and snow fell. The full moon lit the way for travelers at night. Eclipses of the sun and moon seemed as frightening then as do earthquakes today. And those who could interpret, or better yet, predict eclipses, were regarded with awe.

Nearly all civilizations, including some surprisingly primitive ones, collected a systematic body of astronomical observations. Stonehenge, a temple each of whose stones probably marks the position in the sky of an important astronomical event, was constructed by a barbarian tribe in England that had not learned to write. The Chinese in Asia compiled centuries of observations, as did the Mayans in North America. Over four millenia ago, the Egyptians and Babylonians began to accumulate observations of planetary and stellar positions. The Babylonian king, Hammurabi, "the Law-Giver," is thought to have inspired the production of their first star tables about 1800 B.C. In the past century, modern archaeology has uncovered direct evidence of the Babylonians' proficiency in astronomy. Cuneiform tablets from the ancient cities of Ur, Uruk, and Babylon contain, among the administrative and harvest records, astronomical tables. At first, their record of star positions was primitive: A planet was "three fingers from the Great Bear constellation". Later, the Babylonians developed more accurate ways to measure the stars' and planets' positions. Beginning about 750 B.C., there were uniform dated records so accurate that today we can place all their astronomical events within one day or less. Using such records, the Babylonian Kidinnu made predictions of planetary positions in the 4th century B.C. whose accuracy was unsurpassed until the nineteenth century. The ancient Greeks used these many centuries of Babylonian observations, together with their own, to formulate theories of the universe and its laws that are direct antecedents of our own. Our astronomical tradition and nomenclature come from ancient Greece, and the Greeks relied upon Babylonian astronomy.

While very little Babylonian astronomy has come directly to us, one thing has, our present system of angular measure. The Babylonians counted in sixties. Since we will be using their units often, we introduce here our modern names for them. We divide a circle into 360 degrees, each degree into 60 minutes of arc, and each minute of arc into 60 seconds of arc. We use the phrases, "minutes of arc" and "seconds of arc," to distinguish these units of angular measure from the units of the same name used to measure time, which were also invented by the Babylonians. We often use the shorthand notation below:

$$1 \; degree = 1°$$

$$1 \; minute \; of \; arc = 1'$$

$$1 \; second \; of \; arc = 1''$$

so that,

$$360° = 6 \times 60° = 1 \; full \; circle$$

$$60' = 1°$$

$$60'' = 1'$$

If we use this notation, we will not confuse minutes of arc and minutes of time.

When we say that two stars are five degrees apart, we mean that if imaginary lines are drawn from each star to our eye, the angle between the two lines at our eye is 5°. Equivalently, we can imagine drawing in the sky a circle that passes through both stars and whose center is at our eye. The fraction of the circle between the two stars is 5/360 of its circumference, or 5°. How big is a degree? The sun and moon are both about 1/2° in diameter. To get an idea of sizes in degree measure, you might go outside one night, extend your arm as far as it will go with your thumb up, and observe your thumb against the background of stars, first with your left eye closed and then with your right eye closed. The apparent position of your thumb against the star background shifts by roughly 5°. (This apparent change in position of an object viewed against the stars from two separated points is called "parallax", a concept we will encounter again and again.) You might also look for the Big Dipper, a constellation of seven prominent stars that are sketched in Fig. 1-1. The stars labelled 1 and 2 are almost exactly 5° apart. What is the smallest angle the human eye can distinguish? Star number 6 is actually a double star whose two components are

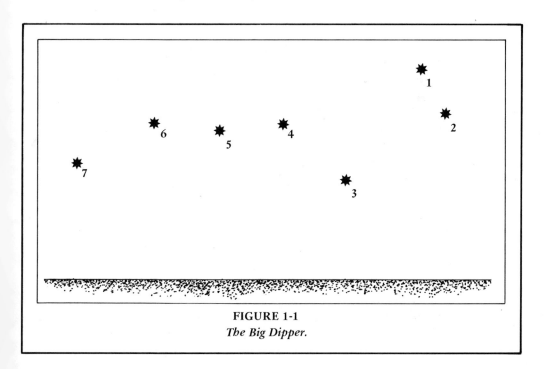

**FIGURE 1-1**
*The Big Dipper.*

about 11' (eleven minutes of arc) apart. The Roman army used to test their recruits' eyesight by asking them to distinguish this double star. Ten minutes of arc is about the smallest angle the average human eye can resolve. Until 1609, when Galileo first used the telescope to observe the stars, this was the maximum accuracy of any astronomical observation; this natural limitation was built into the tables of stellar and planetary positions, which served as the basis for the first scientific theories. The tables were not accurate to better than ten minutes of arc.

## THE MOTIONS OF THE STARS

In retrospect, we can see another reason why astronomy became the first science rather than, for example, meteorology. The events immediately around us are complicated, and it is hard for us to perceive a pattern. On the other hand, the motions of the stars and planets through the sky are relatively simple. We can perceive the patterns more easily. Moreover, the patterns repeat themselves over and over again so that we can check and refine our ideas. We mentioned that the Greeks used Babylonian observations, as well as their own, to construct theories of the motions of stars and planets. In the next chapter we will discuss Greek astronomy and physics; but, before we do so, it is important to understand the information available to them. We must put ourselves in their place and try to construct in our minds a picture of what they knew, putting aside all that we have learned about the sun and stars since childhood. Of course, it is not profitable for us to read ancient astronomical tables and texts directly, but we can reconstruct in modern language the basic ideas that emerged from the ancient, naked-eye observations of the heavens. Our discussion of the motions of the stars and planets is necessarily brief, but we will return again and again to these basic facts in the next few chapters, so that eventually they will become familiar to us.

To begin, perhaps we should imagine ourselves shepherds in the fields, alone, with the stars for company. Gradually, the stars become familiar and we begin to imagine they fall into patterns that seem to be pictures of familiar things: that is, animals, gods, and so forth. We then give them names: Great Bear, Big Dipper, and so on. We call such imaginative groupings of stars *constellations*; different cultures, seeing different things in the stars, have invented different constellations. Imagining and naming constellations is a useful way of remembering the map of the heavens. When we look up at the sky, we can orient ourselves by identifying constellations familiar to us, as ancient peoples did, and as modern navigators still do.

Constellations would not have been useful for so many centuries if the individuals stars in them had moved relative to one another. If the stars had moved, the constellations would have distorted with time until they became unrecognizable. However, centuries of naked-eye observation never revealed a change in the angular separations between any pair of stars in the sky, whether belonging to a constellation or not. In time the relative positions of the stars came to be thought of as fixed.

What else was fixed about the stars? Different stars had different brightnesses, but the individual stars never changed their brightness, neither from night to night nor from year to year nor from century to century. Not only that but the apparent sizes of the stars never changed. The fixed brightness and size of the stars seemed to say that the stars all remained at a constant distance from the earth; for if a star had been moving towards the earth, would it not have become noticeably bigger and brighter? Moreover, since the apparent sizes of the stars were about the same, were not all the stars at the *same* unchanging distance from the earth?

While the stars were fixed relative to one another, the sky was not fixed. It moved with respect to an observer's horizon. We may observe this today by watching the change in position relative to our horizon of a particular constellation or star throughout the night. Some stars remain above our horizon throughout the night, some rise during the night, and still others set. Even though a given star may set during the night, we will find it again the next night almost where we found it at first. In fact, the entire pattern of motion repeats itself every 23 hours and 56 minutes. There is a gradual change in this pattern throughout the year; we can see different constellations in different seasons, but the difference from night to night is slight. At the end of one year, the sky returns to the configuration it had a year ago, and the cycle begins again.

One star, the North Star, hardly moves at all. It is always the same distance above our horizon night after night throughout the year—year after year. Should you move to another latitude on earth, you would find the North Star at a different angle above your new horizon, but it would remain there throughout the night. In fact, the North Star seems to organize the motions of the stars around it. These all move around the North Star. Some, which are close to the North Star, move in circles that never dip below the horizon. They are always visible and neither rise nor set during the night. They are called *circumpolar* stars. The others, which do rise and set, nonetheless move in segments of circles about the pole star when they are visible. Facing south, an observer in the northern hemisphere sees the stars also move in circular arcs throughout the night. The farther south the star, the smaller the portion of the circle that is above the horizon. During the day, the sun traces out a circular arc through the sky similar to that of a southern star. Figures 1-2 and 1-3 sketch the paths of northern and southern stars, respectively.

It is easy for us to see that other stars move about the North Star by following a few bright stars in the northern sky on a clear night. However, it required much more systematic observations for the ancient astronomers to establish that each and every star in the northern sky moved on a *perfect* circle about the North Star. The angle between the North Star and any other *never* changed as the star moved around the North Star. (Of course, this was consistent with the observation that the angle between *any* two stars never changed.) Each northern star moved in a circle with the North Star at its center; the center of the different northern stars' circles was the North Star. Moreover, ancient astronomers, most of whom lived in the northern hemisphere,

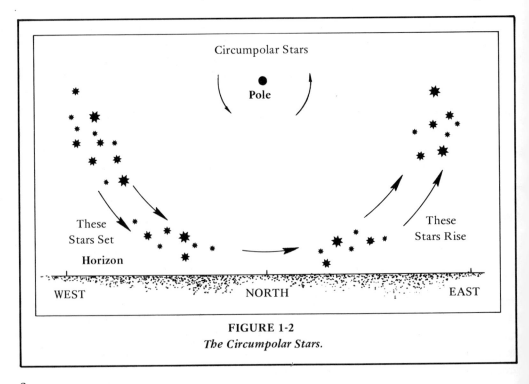

**FIGURE 1-2**
*The Circumpolar Stars.*

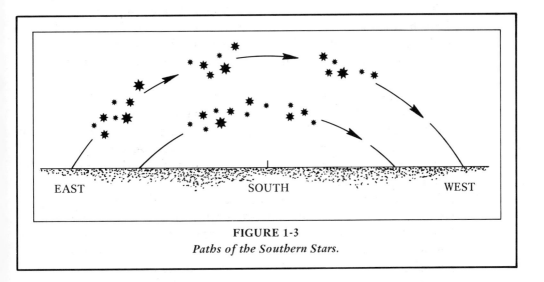

**FIGURE 1-3**
*Paths of the Southern Stars.*

established that the southern stars moved in segments of perfect circles and that all the circles had a common center below the horizon, separated by 180° from the North Star. Finally, they found that all the northern stars rotated about the North Star at the same rate. The stars nearer the North Star did not complete their circle ahead of the ones farther away, or *vice versa;* every star traced out the same fraction of its circle in a given time. The stars did not speed up or slow down. The same rules were true for the circular motions of the southern stars about their common center. They moved about their common center at the same rate northern stars moved about the North Star.

This, then, was the basic information about the stars accumulated from millenia of ancient astronomical observations. The stars, which were fixed in brightness and size and whose angular separation from one another was also fixed, moved around perfect concentric circles at an absolutely uniform rate. The center of these circles was either the North Star or a point half a circle away below the southern horizon.

We will now construct, using modern terminology, a conceptual model first posed by the ancient Greeks and still useful today, which summarizes the apparent motions of the stars in the heavens. By model we mean a scheme, a set of ideas, which is easier to remember than all the observations of the stars and their motions, but still describes all the motions. This model will also make it easier for us to describe the motions of the sun, moon, and planets through the stars, as it did for the Greeks. Let us imagine that the earth is at the center of a giant hollow sphere and that the stars are attached to the inner surface of the sphere. If the idea that the stars are actually attached to a sphere is difficult to accept, it is sufficient to imagine that a star map is drawn on the inside surface of the sphere, which has been given the name, *celestial sphere.*

An observer on the earth can see that portion of the celestial sphere that is above the horizon. All the apparent motions of the stars can then be reproduced if we further imagine that the celestial sphere rotates steadily on its axis, from east to west, once every 23 hours and 56 minutes, at a constant rate. As seen from the interior of the celestial sphere, all the stars will appear to move in circles about the axis of rotation. Since the North Star remains almost fixed in the sky, we must imagine that the axis of rotation passes almost through the North Star. We call the point in the northern sky about which the stars appear to rotate the *North Celestial Pole.* Since the North Star is very near the North Celestial pole, it hardly appears to move. The opposite point in the southern sky, where the axis of rotation again intersects the celestial sphere, about which the stars in the southern sky appear to move, is the *South Celestial Pole.* There is no "south star," which happens to lie near the South Celestial Pole. The north pole is above the horizon of observers in the earth's northern hemisphere; the South Celestial Pole is always below it. The great circle on the celestial sphere that lies exactly halfway between the North and South Celestial Poles is, by analogy with maps of the earth, the *celestial equator*\*. (A plane that passes through the center of a sphere cuts the surface of that sphere in a great circle; planes that do not pass through the sphere's center produce smaller circles.)

Figure 1-4 is a schematic drawing of the celestial sphere. It obviously is easier to remember this one picture than it is to remember our description of the motions of the stars. This makes the celestial sphere a convenient model, one which is used today for navigation. However, the fact that the celestial sphere is a convenient aid to our thought processes does not necessarily mean that the stars are really on a celestial sphere. That step requires further scientific justification. We can ask, nevertheless, whether the model is complete. Does the celestial sphere model describe the motions of *all* the heavenly bodies? The answer is no. It was known in antiquity that the motions of the sun, moon, and the five planets observable by the naked eye—Mercury, Venus, Mars, Jupiter, and Saturn—did not easily fit into the model. We now describe their motions, since it was the central preoccupation of

---

\*A "great circle" has its center at the center of the sphere; examples on the earth are the equator or a meridian of longitude. The centers of the latitude circles are on the polar axis, but are not at the center of the sphere; they are "small circles."

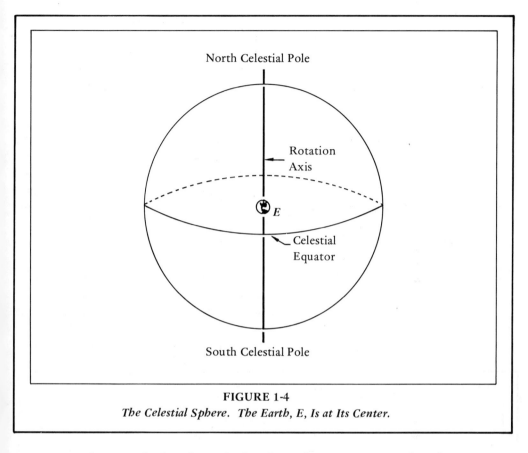

**FIGURE 1-4**
*The Celestial Sphere. The Earth, E, Is at Its Center.*

astronomy from antiquity through the time of Newton to resolve the puzzle of their unusual behavior.

## THE APPARENT MOTION OF THE SUN

Surely the sun is not a star. It warms us, and the stars do not. Its position in the sky controls the seasons, the weather, and the harvest. Whereas the stars are mere points of light in the sky, the sun has a big round image. The stars twinkle and the sun does not. The sun is immensely brighter than the stars, so bright that we can not see stars during the day. Here is the key to understanding the sun's motion. While we can not see stars during the day, we must imagine that they are still there, on that half of the celestial sphere that is "up"; they are invisible because of the sun's glare. On rare occasions this idea is dramatically verified. During a total eclipse of the sun, the moon blots out the sun's light and we can see the stars around the sun directly. Therefore, we can always locate the sun on the celestial sphere, the star map. For example, the sun might be near the star "Spica" or on the "celestial equator", even though Spica is invisible in the daytime. Astronomers can accurately fix the sun's position on the celestial sphere by doing a few simple calculations.

The sun's motion through the sky proves to be more complex than the stars', but it shares one feature with them. On any given day, the sun traces out a circular arc like those of the stars. This daily motion of the sun and stars is called *diurnal* motion. Figure 1-5 shows the path of the sun as seen by an observer in the earth's northern hemisphere facing south. The sun's path is similar to the paths of the stars that would be observed the following night ( see Fig. 1-3). Moreover, the sun traces out its circular arc through the sky at almost the same rate as the stars at night.

Is the sun always in the same place relative to the background stars? The answer is no. If at noon one day we measured the position of the sun relative to the background stars on the celestial sphere, we would find at noon on the next day that the sun had moved roughly 1° eastward. The sun moves relative to the stars! As time passes, the sun

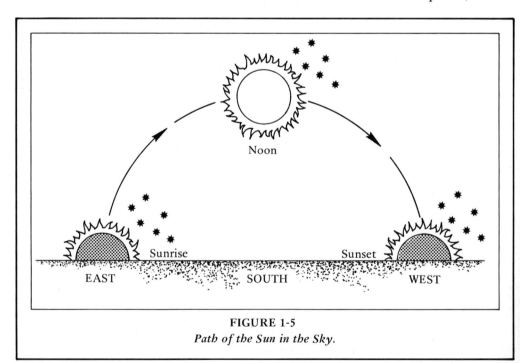

**FIGURE 1-5**
*Path of the Sun in the Sky.*

# The Apparent Motion of the Sun

gradually traces out a path, the *ecliptic*, through the constellations. After exactly one year, the sun returns to the same position on the celestial sphere and begins to retrace the same path around the ecliptic. The sun's motion around the ecliptic is called its *annual* motion. Since there are about 365 days in a year, the sun travels about 1/365th of its way around the ecliptic each day. This is not far from the 1/360th of the circumference of a circle, or one degree. Because the sun moves about one degree eastward relative to the stars each day, the *solar* day—the time between noon one day and noon on the next—is not exactly the same length as the *sidereal* day—the time it takes a star fixed on the celestial sphere to return to the same position in the sky. Since the sun's slow annual motion is eastward, opposite the westward diurnal motion of the stars, the sun lags behind the stars and makes its daily trip in 24 hours rather than 23 hours 56 minutes. If the sun is right on top of a star at noon one day, the star will be a degree west of the sun at noon the next day.

Shown in Fig. 1-6 is the celestial sphere, which rotates westward about its axis—the straight line drawn between the north and south celestial poles (*NCP* and *SCP*). The celestial equator is a great circle half-way between the *NCP* and *SCP*. The ecliptic is a great circle on the celestial sphere that crosses the celestial equator twice, at an angle of 23 1/2°. At its northernmost point, the ecliptic is 23 1/2° north of the celestial equator and it is 23 1/2° south at its southernmost point*. Each day the sun makes a circular motion about the axis that is the same as for the stars behind the sun on that particular day. The paths marked "*W*" and "*S*" (Fig. 1-6) are the sun's daily motion on the first days of northern winter and summer, December 21 and June 21. Between December and June, the sun's daily circles move along the ecliptic, in the direction opposite its daily rotation.

The sun's motion may be summarized almost completely by saying that, superimposed on its rapid westward diurnal motion shared with the stars, is a much slower eastward annual motion through the celestial sphere along the ecliptic. Unfortunately, this simple picture is not quite right. The sun's annual motion around the ecliptic, unlike the daily motion of the sun and stars, is not always exactly at the same rate. The sun moves along the ecliptic a little more quickly in the winter than in summer. Winter is, consequently, three days shorter than summer. That is why we have one short winter month, February.

---

*It is not immediately obvious that if two great circles cross at a certain angle, this angle equals the maximum number of degrees of arc between them. Nevertheless, it is true. The ecliptic and the equator intersect at an angle of 23 1/2°. Therefore, the farthest north (or south) of the equator the ecliptic gets is also 23 1/2°. This fact is an elementary theorem of solid geometry.

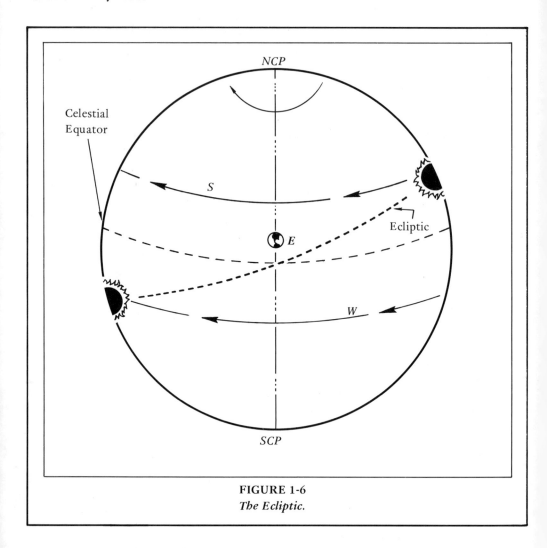

FIGURE 1-6
*The Ecliptic.*

## THE MOTIONS OF THE MOON AND THE PLANETS

The moon and the planets appear different from both sun and stars. While the sun and moon have disks, the five planets visible to the unaided eye do not, but look like bright stars. Venus and Jupiter are brighter than any fixed star, while Mercury, Mars, and Saturn are about as bright as any one of the brightest ten or twenty fixed stars. The planets do not twinkle like stars, but shine with a steady light. And, completely unlike the sun and stars, the brightness of the planets changes with time, Mars' changing most noticeably. Usually the faintest of the five visible planets, Mars increases in brightness every two years and outshines Jupiter for several months.

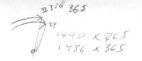

# The Motions of the Moon and the Planets

The planets do move in diurnal circles through the night. In fact, *all* objects in the heavens—sun, moon, planets, and stars—have the same westward diurnal motion described earlier for the stars. Since this is so, we will focus our attention on the motions of the moon and planets relative to the stars, and not on their diurnal motions. By observing the position of the moon or a planet relative to the stars at intervals of 23 hours and 56 minutes, when the stars return to the same position, we find that the moon and planets move. By making a succession of such observations, we may trace out their paths through the stars. Qualitatively, the moon and planets move in a way similar to the sun. While their most rapid motion is the daily westward motion of the stars, they also have a gradual eastward motion through the stars. The moon and planets each have separate paths that do not exactly follow the ecliptic, but they stay within seven degrees north or south of it. Thus, there is a region on the celestial sphere in which all irregular activity is located, within which the sun, moon, and planets are always found. This band, which circles the sky about seven degrees on either side of the ecliptic, is called the *zodiac*. Outside the zodiac all motions are perfectly simple and regular, and only stars are located there. The zodiac crosses twelve constellations in the sky: Aries, "the Ram"; Taurus, "the Bull"; Gemini, "the Twins"; Cancer, "the Crab"; Leo, "the Lion"; Virgo, "the Virgin"; Libra, "the Scales"; Scorpio, "the Scorpion"; Sagittarius, "the Archer"; Capricorn, "the Goat"; Aquarius, "the Water-Bearer"; and Pisces, " the Fish". The sun is "in" each of these constellations about 1/12 of each year, approximately one month each.

Only one object moves eastward through the zodiac faster than the sun: the moon. The moon's motion can be discerned over the course of an hour or two when it is near a bright star. The moon circles the zodiac once a month; it moves on the celestial sphere on a great circle that crosses the ecliptic at an angle of about six degrees. The sun and moon do not trace out exactly the same path through the constellations of the zodiac. Occasionally, however, the two paths intersect when the moon crosses the ecliptic. If at the same time the moon is directly in front of the sun, the sun will be eclipsed. Thus, eclipses can only occur when the moon crosses the sun's path through the sky. Our modern word for the sun's path, the ecliptic, comes from the word *eclipse*.

The moon's phase is correlated in a very simple way with its position relative to the sun. When the moon is in the same part of the zodiac as the sun, it is new (dark). Gradually, as it moves eastward more rapidly than the sun, a thin crescent appears on the side towards the sun. Seven days after the new moon, the moon is a full quarter-circle (90°

around the zodiac) from the sun, and, therefore, sets at midnight. It is a first quarter moon. In another week the moon is exactly opposite the sun, being highest in the sky at midnight; it is a full moon. As the month continues, the moon approaches the sun from the west, rising later and later at night, and repeats the above system of phases in reverse order, the lighted part of the moon always being toward the sun.

Like the sun and moon, the five visible planets generally move eastward on great circles around the zodiac. However, totally unlike the sun and moon, the planets sometimes slow down in their eastward motion, come to a full stop, reverse their motion to westward, stop again, and then resume their "normal" eastward motion. This important peculiarity is called *retrograde* motion, in contrast to *direct* motion. Figure 1-7 shows Mars' path through the stars during part of 1971. The Greeks called the sun, moon, Mercury, Venus, Mars, Jupiter, and Saturn "wanderers", which is what the Greek word "planet" meant. The explanation of their wanderings and particularly why five of them, alone among the heavenly bodies, can reverse their motion became the central problem of astronomy from antiquity through the time of Copernicus.

To study the motions of the planets in more detail, it is convenient to divide them into two groups: Mercury and Venus, which are never found far from the sun; and Mars, Jupiter, and Saturn, which can be found at any distance along the zodiac from the sun.

Venus shares the westward diurnal motion of the sun and the stars, as well as the annual eastward motion of the sun around the zodiac. Venus moves with the sun, but is not always the same distance from the sun. It moves back and forth across the sun, sometimes ahead, sometimes behind. Venus never gets more than about 45° from the sun, which means it never sets more than three hours after the sun sets or rises more than three hours before the sun rises.

Let us follow Venus through one cycle, beginning when it is 45° west of the sun. This is its *greatest western elongation*. In this configuration Venus is a "morning star", rising in the morning about three hours before the sun. Venus dominates the morning sky, so bright that on a moonless night it can cast a shadow. Like the sun, Venus moves eastward among the stars but slightly faster than the sun so that it approaches closer and closer to the sun, rising each day nearer and nearer to sunrise. Eventually it overtakes the sun, gets lost in its glare, and reappears east of the sun as an "evening star." (It was a great step forward in ancient astronomy when the morning star, Lucifer, and the evening star, Hesperus, were understood to be one and the same

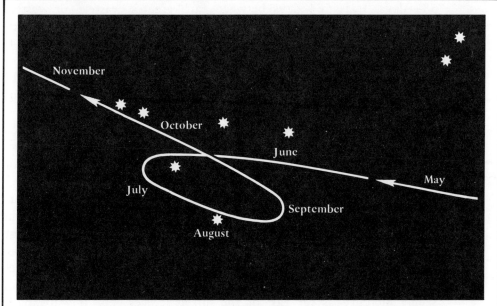

**FIGURE 1-7**
*Mars Retrogressing Through the Stars of the Constellation Capricorn in 1971. Mars Was at Opposition on August 10, and Was Over 100 Times Its Normal Brightness.*

planet, Venus.) Venus continues moving eastward faster than the sun until it reaches its *greatest eastern elongation;* it is then 45° east of the sun, and sets about three hours after the sun. Venus then slows its eastward motion, so the sun starts to catch up. Soon Venus stops moving eastward and starts moving westward among the stars. This is its retrograde motion. While Venus is moving westward, the sun is, of course, still moving eastward along the ecliptic so that they approach each other very rapidly. In a few weeks, Venus becomes a morning star again, stops its westward, retrograde motion, and starts moving eastward again; and the cycle repeats itself. Notice that Venus is in the middle of its retrograde motion when it passes the sun.

The time during which Venus moves westward relative to the sun, and during most of which it is also moving westward relative to the stars, is comparatively short, only 144 days. From its greatest western elongation to its greatest eastern elongation, during which time Venus is always in direct, not retrograde, motion, moving eastward among the stars faster than the sun, takes about 440 days, about three times as long.

Relative to the stars, Venus' motion is complicated. Relative to the sun, Venus' motion is very regular. Its period of oscillation is

584 days. It is 584 days between each greatest eastern or western elongation. Venus passes the sun twice each 584 days. This configuration is called *conjunction,* meaning "together with" the sun. The 584-day period is called the *synodic* period of Venus, the time between every other conjunction. (You may recognize the Greek word "synod", when all the bishops meet together. Venus' synodic period is the period between meetings of the sun and Venus.)

Mercury moves in a way similar to Venus, but it moves faster and stays even closer to the sun. Mercury's greatest elongation is not always exactly the same; while the average is 23°, Mercury can on occasion be as far as 28° from the sun. Nevertheless, either figure is considerably smaller than Venus' greatest elongation, which means Mercury is notoriously difficult to see, since it is usually hidden in the sun's glare, and is visible only near its greatest elongation. If you observe what appears to be a bright star low in the western sky about an hour after sunset and if you are familiar enough with the constellations to know that no bright star should be there, you are probably seeing Mercury as far east of the sun as it ever gets. It will rapidly change position with respect to the stars, disappear in the sun's glare, and reappear six or seven weeks later as a bright morning star. Mercury's synodic period is 116 days.

Unlike Mercury and Venus, Mars, Jupiter, and Saturn do not follow the sun around the sky. Since they do not stay close to the sun, they are often visible throughout the night, rather than for a few hours after sunset or before sunrise. Even though their motions are not uniform, we can make a short listing of the average time each one takes to traverse all twelve constellations of the zodiac:

*Mars* takes 1.9 years,
*Jupiter* takes 11.9 years,
*Saturn* takes 29.5 years.

Like Venus and Mercury, Mars, Jupiter, and Saturn also stop their eastward motion from time to time, retrogress westward, and then resume their eastward motion through the stars. They move westward, or retrogress, only a small part of the time so the net result is an overall eastward motion. Mars, for example, retrogresses for only about two months every two years. Even though Mars, Jupiter, and Saturn do not follow the sun around the sky, the times when they retrogress are still governed by the sun. Unlike Venus and Mercury, which retrogress near the sun, Mars, Jupiter, and Saturn are in the middle of their retrogressions when they are farthest from the sun, a half-circle (or 180°), around the zodiac from the sun. This configuration is called opposition; at opposition Mars, Jupiter, or Saturn are highest in the sky at midnight. The synodic period for these planets is the interval between oppositions, between two successive times when the planet is directly opposite the sun on the celestial sphere. The synodic period for Mars is 780 days; for Jupiter, 399 days; and for Saturn, 378 days.

It should be obvious why the planets are in opposition at the time intervals stated above. The sun moves around the ecliptic once a year. A fixed star, therefore, is in "opposition" once a year, every 365 1/4 days. However, the planets also drift eastward. Thus, when the point opposite the sun comes back a year later to where a planet was the year before, the planet has lagged behind; and it takes a while to catch up. The slowest moving planet, Saturn, does not lag much and the interval between Saturn's oppositions is only slightly more than a year. Mars and Jupiter take longer to catch up. Interestingly, all three are always brightest at opposition when they retrogress, and when they are farthest from the sun.

Such in outline is a summary of the motions of the heavenly bodies available to any careful, systematic observer who uses the naked eye. Differing only by a little detail, such was the knowledge passed on to the ancient Greeks from the Egyptians and Babylonians.

The word "astronomy" comes from a Greek word meaning, roughly, the arrangement of the stars. "Cosmology", literally meaning the study of order or harmony, has come to mean a description of the universe, an explanation of the facts of astronomy that we have just outlined. The ancient Greeks took it upon themselves to make this explanation, to create a cosmology. They asked not only "how?" but "why?" Here are some of the questions they asked; our next few chapters will discuss the answers given these questions through the centuries: Why do all the stars trace out perfectly regular circular diurnal motions throughout the sky? Why are the stars' motions circular? Why do the planets, sun, and moon have irregular motions relative to the stars? Why does the sun move faster on one part of its trip around the ecliptic, in winter, than in summer? Why do Mars, Jupiter, and Saturn retrogress when in opposition; why do Mercury and Venus retrogress every other conjunction? Why are the planets' retrogressions linked to the sun and not to the stars? Why do the stars shine with a constant brightness and the planets do not? Finally, why are Mars, Jupiter, and Saturn brightest during retrogression when they are in opposition, farthest around the ecliptic from the sun?

How the Heavenly Bodies Move

*Summary*

*All the heavenly bodies move as if they were on a giant sphere that rotates westward steadily every 23 hours, 56 minutes. We are at the center of the sphere; at any time we see only that half of the sphere that is above our horizon. The axis of rotation of the sphere passes very close to Polaris, the North Star.*

*The fixed stars are just that—fixed on the celestial sphere. The sun, moon, and planets share in the celestial sphere's diurnal rotation, but also move, generally eastward, on the sphere. They are not fixed.*

*The sun moves along the ecliptic, a great circle that makes an angle of 23 1/2° to the celestial equator. It goes around the ecliptic once a year at an almost steady rate, moving a little faster in winter than in summer. The moon is the fastest moving celestial body. It goes around its path once a month, and its phases (new, first quarter, full, etc.) depend on how far it is from the sun. If the moon and the sun happen to be at the same place on the celestial sphere at the same time, there will be an eclipse of the sun.*

*There are five visible planets, which are also not fixed on the celestial sphere. Mercury and Venus stay close to the sun, moving first on one side of the sun then on the other. Since they stay close to the sun, the* average *time it takes them to circle the celestial sphere is the same as that of the sun, 1 year. Relative to the sun, their motions are quite regular. Venus reaches as far as 45° from the sun, and takes 584 days to complete one cycle. Mercury never gets more than 28° from the sun, and completes its cycle in 116 days. Mars, Jupiter, and Saturn circle the celestial sphere in 1.9 years, 11.9 years, and 29.5 years, respectively, but do not follow the sun.*

*For a few weeks every year—every second year for Mars—the planets retrogress; that is, they move westward relative to the stars. Mercury and Venus retrogress when they cross the sun moving westward; Mars, Jupiter, and Saturn retrogress when they are in opposition, farthest from the sun. Opposition occurs in the middle of their retrogression. Mars, Jupiter, and Saturn are brightest at opposition.*

*The moon and the planets stay near the ecliptic but do not follow it exactly. Each of their paths around the celestial sphere is a great circle that intersects the sun's path, the ecliptic, at small angles of no more than 7°. Thus, the moon and planets stay near but not exactly on the ecliptic.*

# Appendix to Chapter 1

*The Precession of the Equinoxes and the Astrological Calendar*

Our description of the sun's motion has omitted a tiny effect unknown to the most ancient astronomers. The North Star was not always at the North Celestial Pole, although the celestial pole (the center of the circles of the daily rotation of the stars, sun, moon, and planets) has always been at the same place in the sky. The map of the stars has been gradually sliding around on the celestial sphere. The constellations along the equator change too. The motion is easily described. It is as if the celestial sphere rotates constantly, every 23 hours, 56 minutes, while the map of the stars slowly slides around so that the celestial poles draw out circles on the star map. The radius of the circles is 23 ½°—exactly the same as the inclination of the ecliptic to the equator, and the poles take about 26,000 years to go once around the circle. The ecliptic and the paths of the moon and all the planets are carried along with this very slow motion of the star map. Therefore, the constellation in which the celestial equator crosses the ecliptic also changes; it will go all the way around the ecliptic in 26,000 years, so it will change by one constellation in a little more than 2,000 years. When the sun, moving northeast along the ecliptic during late March, crosses the celestial equator, it is the beginning of spring (the vernal equinox). In ancient Greek times, when modern astrology was codified, the sun was in the constellation Aries at the beginning of spring. Now, due to the precession of the equinoxes, the sun is in the constellation Pisces on March 21, but astrologists in the newspaper columns still keep the old names. Thus, on March 21, the sun is in the *sign* of Aries, but an astronomer will tell you that it is in the *constellation* Pisces. The precession of the equinoxes is so slow that a single person would not notice it during his lifetime. It was probably first discovered by the Greek astronomer Hipparchus in the second century B.C., while comparing his observations with even more ancient Babylonian records.

## QUESTIONS

1. Using the model of the celestial sphere as a guide, describe the motions of the stars seen by an observer who is facing due east in the Earth's northern hemisphere.
2. Using the fact that Venus' greatest elongation is 45°, can you show that the earliest Venus can rise in the morning is about three hours before the sun?
3. Do you think the fact that out of several thousand visible heavenly bodies only seven objects, the sun, moon, and five visible planets, fail to move in absolutely perfect diurnal circles should be a threat to the usefulness of the celestial sphere model?
4. Why is the time the celestial sphere takes to make one rotation four minutes less than a full 24 hours?

5. Actually, the sidereal day (the period of rotation of the celestial sphere) is a few seconds shorter [longer] than 23 hours, 56 minutes. Why? Calculate the length of the sidereal day more accurately.
6. Suppose that some year the moon is exactly full on December 21, the first day of winter. When and where will it rise? *E 7AM*
7. In what part of the Earth can the sun sometimes be straight overhead? Why?

**ARISTOTLE**
*(Courtesy of the Spada Gallery, Rome, Italy and by permission of the Houghton Library, Harvard University)*

CHAPTER

# 2

# THE GREEK COSMOS
## The Beginning of Science

Our subject now is the classical example of posing and solving a scientific problem; namely, explaining the apparent motions of the heavenly bodies. The ancient Egyptians, the Babylonians, and the Mayans named and counted the stars and planets, and guessed where they would be; however, only the ancient Greeks attempted to construct an intellectual universe of logically related natural laws that underlie the real universe around us. Our scientific tradition began with the Greeks.

Just why there suddenly appeared in the 6th century B.C. a new philosophy of nature, separate from religious cosmology and based upon the use of reason, has long excited the curiosity of historians. The first school of natural philosophy is traditionally thought to have been founded around 575 B.C. by Thales in the city of Miletus, on the Mediterranean coast of present-day Turkey. However it started, Greek science grew rapidly. The early phases of Greek science (600–400 B.C.) were devoted to formulating all possible questions about nature, probing them for logical consistency and completeness of logical alternatives. These early Greek scientists played around with many different questions, trying them on for size. While the search for ever more refined answers is unceasing, it is fair to say that those earliest originators of scientific speculation asked all the right questions, questions that continued to be asked thereafter, and are still asked today.

The Greeks recognized that various separate scientific problems could be linked together to form a discipline. For example, from the many specific problems associated with the measurement of areas, they abstracted a general discipline, geometry, which is concerned with the properties of figures and their shapes. The study of light, how it propagates and is reflected and refracted, forms a discipline to which they gave the name optics. The fact that so many modern scientific disciplines (dynamics, kinematics, biology, and cosmology) have names of Greek origin attests to their recognition that classes of problems should be studied together.

The Greeks thought about thought itself. They defined rules of thought, logic, by which conclusions could be drawn rigorously from premises and evidence. By the fourth century B.C., the philosophers Socrates, Plato, and Aristotle could draw on a vibrant scientific tradition. During this period scientific problems were discussed in broadly philosophical terms.

The period 300–100 B.C. following the decline of Athens, the "Hellenistic" period, is generally regarded as the high point of technical Greek science. At this time, Greek learning spread to all parts of the Mediterranean world, and even as far east as India. In the Hellenistic period, the scientific emphasis was not so much upon all-encompassing philosophical understanding but upon the accurate solution of specific problems of physics. Euclid, whose great book on geometry we still read today, lived at the beginning of the Hellenistic period; Archimedes, a great physicist and a major influence upon Galileo, was murdered by Roman soldiers in 212 B.C. after the fall of his home city, Syracuse in Sicily, for whose defense he had constructed many engines of war.

The last period, the Graeco-Roman, from 100 B.C. to perhaps 600 A.D., was a period of slow, steady decline in which Greek knowledge penetrated the Roman empire, was assembled and codified, and then translated into Latin. It was eventually transmitted down through the centuries, first by Arabic and then by medieval Christian scholars. The work of the greatest astronomer of this period, Ptolemy of Alexandria, is a good example. Ptolemy summarized and organized the work of his Greek predecessors, refined greatly their calculational techniques, and passed it all on in a giant work on astronomy, the *Almagest,* which remained the ultimate authority for 1500 years.

Over most of its long life, Greek science developed cumulatively and sequentially. The great Greek thinkers were not isolated men forced to rediscover what had gone before, they built on the work of their predecessors. Much of the credit for the preservation and amplification of this coherent intellectual tradition belongs to Plato's Academy founded in 388 B.C., and Aristotle's Lyceum, founded in 335 B.C. Plato's Academy survived for 917 years, about as long as the oldest present-day European universities. The scientific activity at the Lyceum is thought to have stimulated the foundation of the Great Museum at Alexandria in Egypt, where Ptolemy worked. At its peak it may have had a library of half a million books and a research staff of one hundred.

### PYTHAGORAS

The 6th century B.C. produced a handful of men under whose influence the human race still lives: Confucius in China, Buddha in India, Zoroaster in Persia, and Pythagoras from Samos, an island near the coast of Greece. To Pythagoras, we owe the ideas that numbers and nature are related, and that mathematics is the proper tool to describe natural phenomena. He had an answer to the question, "what is everything made of?" Everything is made of number; or, as Galileo was to echo over two thousand years later, "The book of nature is written in the language of mathematics." Pythagoras was fascinated by how numbers appeared in nature, for example, by the fact that musical notes are harmonious if they are played from strings the ratio of whose lengths were small whole numbers. Strings whose lengths were in any other ratio produced dissonance. Eventually, a Pythagorean Brotherhood, a cult in which mathematics and science served the functions religion did for other groups, emerged to preserve and protect the secrets of nature. One entered the brotherhood only following rigorous purification of mind and body. Its members were sworn to secrecy—one was put to death for revealing that the square root of two is *not* a ratio of whole numbers. Science was not for the uninitiated.

Pythagoras' greatest achievement is a theorem about right triangles, such as that shown in Fig. 2-1. In a right triangle, the angle

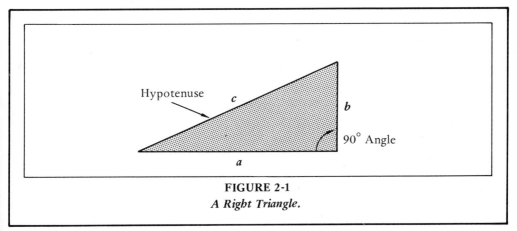

FIGURE 2-1
*A Right Triangle.*

between the sides $a$ and $b$ is 90°. It amazed Pythagoras that if one knows the lengths of sides $a$ and $b$, one can calculate the length of side $c$, which is called the hypotenuse, opposite the right angle of 90°. The ancient Egyptians, by trial and error, learned some of the rules. For example, if the length of sides $a$ and $b$ are 3 and 4 feet, respectively; then, the length of the hypotenuse, $c$, is 5 feet. In short, it is always true that if $a = 3$ and $b = 4$ then $c = 5$. It is also true that if $a = 5$ and $b = 12$ then $c = 13$. (Here we are using the letters $a$, $b$, and $c$ to stand for the lengths of the sides named $a$, $b$, and $c$.) Pythagoras abstracted this collection of rules into a theorem that is always true for any $a$, $b$, $c$, even if $a$, $b$, and $c$ are fractions or decimals instead of whole numbers. Pythagoras' famous theorem is:

$$a^2 + b^2 = c^2$$

where $a^2$ is modern shorthand for "$a$ times $a$". How could Pythagoras know this was always true? Instead of a collection of facts about the lengths of the sides of specific right triangles, Pythagoras created a logical proof about *all* triangles, a theorem that could be used over and over again. A proof of Pythagoras' theorem is given in the chapter Appendix.

### THE SHAPE OF THE EARTH

In astronomy one can ask two questions: "What is the nature of the heavens and the objects therein?" and "What is the relation between terrestrial and celestial phenomena?" To the Greek mind, deeply imbued with the power and rigor of geometry, the first question was rephrased and restricted to: "If you could stand outside the universe and look at it, what would it look like? What would its geometrical shape be?" In other words, they sought a simple geometrical model that would enable them to remember astronomical data without recourse to Babylonian tables for each and every event.

The second question concerning the relationships between celestial and terrestrial phenomena can be broken down into a number of subsidiary questions: What is the shape of the Earth? Where is the Earth located, relative to the heavens? Should terrestrial and celestial phenomena follow the same rules of behavior? Are the objects in the sky the same as those on Earth? Are the motions observed in the sky consistent with motions observed on Earth? These last three questions will be discussed somewhat later, when we come to the physics of

Aristotle. The science of the universe's structure we call cosmology; the science of the rules of behavior of objects in the universe we call physics.

Soon after the time of Pythagoras, the Greeks established something by no means obvious to the average man: the earth is a sphere floating in space. Even the Babylonians had thought the earth was flat. We do not know precisely who invented this idea of a spherical earth, but it seems never to have been seriously disputed in classical times. (It is nonsense that Columbus had difficulty financing his voyages because everybody believed the earth to be flat. Columbus underestimated the radius of the earth so he believed he could sail around it in a reasonable time. His proposal was unpopular with those who knew the real size of the earth.)

The way the Greeks concluded the earth is round is typical of their imaginative construction of new facts based upon deductions from observed facts using mathematics. Three phenomena led them to a round earth: the disappearance of ships below the horizon, the change in altitude of the north celestial pole with the observer's latitude, and the shape of the earth's shadow upon the moon during a lunar eclipse.

If you stand on a mountain top and watch a boat sail away, it will disappear, hull first and then finally the mast. Similarly, if you watch a mountain near the shore from a departing ship, the mountain eventually disappears, base first and then the peak. The explanation is obvious: the earth's surface is curved. In Fig. 2–2, the circle is the earth, and $M$ is the top of a mountain whose height is greatly exaggerated. When the ship is at $A$, the mountain top is visible; when the ship is at $B$, the mountain top is invisible because it is blocked by the curved earth. When the ship is at $S$, the mountain top disappears below the ship's horizon.

To believe this proof of the curvature of the earth's surface at all, we must assume, as the Greeks did, that the light travels in straight lines from the mountain top to the eyes of the observer on the ship. Furthermore, strictly speaking, this only proves that the earth's surface is curved in some way. To establish that the earth is a sphere, it is necessary to show that the curvature is the same everywhere, which means that this observation would have to be repeated at many points on the earth's surface.

The second argument for the earth's being round is astronomical. In our description of the stars' apparent motions, we were deliberately vague about the altitude of the north celestial pole above the horizon, since it is different in different places. The further north one goes, the higher the pole star is above the horizon. At the earth's north

# The Greek Cosmos

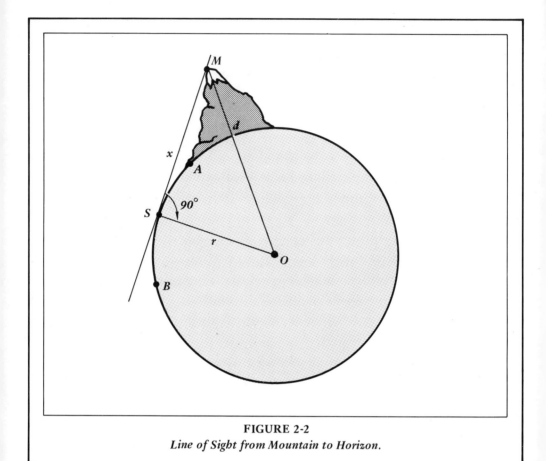

**FIGURE 2-2**
*Line of Sight from Mountain to Horizon.*

pole, the pole star is at the zenith (directly overhead), and all stars move in horizontal circles. Half-way between the pole and the equator, the celestial pole is 45° above the horizon. At the equator, the celestial pole lies on the horizon. South of the earth's equator, the north celestial pole is not visible, but the altitude of the south celestial pole increases as one moves further and farther south.

When we discussed the stars' motions, we introduced the celestial sphere for convenience. Let us analyze in terms of the celestial sphere how the altitude of the celestial poles changes with position on the earth. We put the earth at the center of the celestial sphere, as in Fig. 2-3. From any point on the earth, one can see all the stars above one's horizon.

Now the Greeks knew that, as far as anyone could tell, exactly half the celestial sphere was up at any given time. This meant that the

# The Shape of the Earth

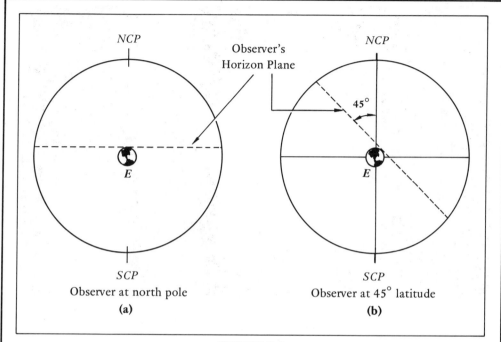

**FIGURE 2-3**
*E Denotes the Earth, at the Center of the Celestial Sphere; NCP and SCP, the North and South Celestial Poles.*

earth must be very small compared to the celestial sphere. The alternative is illustrated in Fig. 2-4. Figure 2-3a shows why the pole star is directly overhead for an observer at the north pole, while Fig. 2-3b shows that the elevation of the pole star about the local horizon is 45° when the observer is at a latitude of 45°, half-way between the equator and north pole. It is easy to see that the pole star will appear farther above the horizon the farther north one goes.

Once again, we have not proven that the earth is a sphere, only that its surface is curved. To prove the earth a sphere, we would have to show that the elevation of the pole star increases by the same amount each time the same distance northward is traveled. The difficulty of this measurement limited the accuracy obtainable by the Greeks. Today, we know the north celestial pole moves one degree further above the horizon for every 69 miles the observer moves north.

The third argument for a spherical earth comes from the shape of the earth's shadow on the moon during a lunar eclipse. Of course, to use this argument one already has to know the explanation for

31

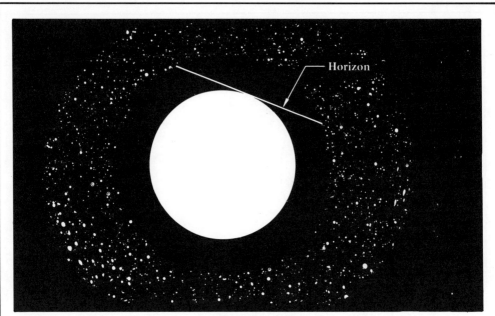

**FIGURE 2-4**
*If the Stars Were Close to the Earth, Fewer Than Half Would Be Visible. The Inner Circle Is the Earth, The Outer One the Stars.*

eclipses of the moon. Other civilations had little understanding of eclipses as natural phenomena; they were supernatural portents, to be feared, and with luck, to be predicted. Even the Babylonians, for all their careful observations, apparently never came near the correct— and to us obvious—explanation of the moon's phases. (If this surprises you, remember that they did not even believe that the world was round.) The key bit of information is the way the moon's phases correlate with the relative position in the sky of sun and moon, which we described in the first chapter. Figure 2.5 shows the earth and the moon at various positions in its orbit around the earth; the sun is so far away that it is off the page. In any position, the half of the moon that is toward the sun is lit, the other half dark. We cannot see the whole lit side except when the moon is directly opposite the sun, at full moon. At new moon, when the moon is between the earth and sun, it is invisible; somewhat later, after the moon moves a little, we can see a small portion of the moon's lit side, which looks like a crescent. Halfway between new and full moon, at first and third quarters, we can see half the lit side. Now for a very important question. Suppose the moon's orbit were always in the plane of the earth and sun; i.e., the

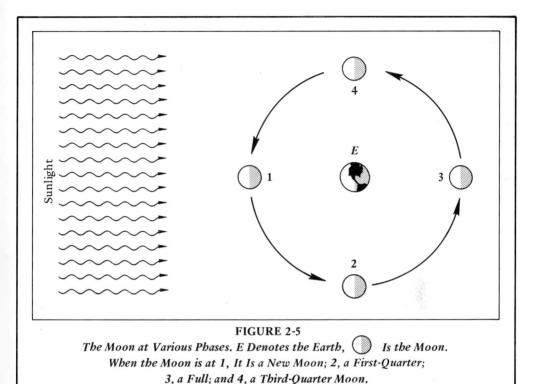

**FIGURE 2-5**
*The Moon at Various Phases. E Denotes the Earth,* ◯ *Is the Moon. When the Moon is at 1, It Is a New Moon; 2, a First-Quarter; 3, a Full; and 4, a Third-Quarter Moon.*

moon moved along the ecliptic. Then we should never see a full moon, for the moon would always be in the earth's shadow at full moon. Why do we see a full moon at all? It must be that the moon's orbit is not in the ecliptic but rather makes a slight angle to it. Therefore, in Fig. 2-5, the sun should really be slightly above or below the page, so that the earth's shadow usually misses the moon. However, twice a month the moon crosses the ecliptic. If it happens to be full moon then, the moon will enter the earth's shadow and an eclipse of the moon will occur. Similarly, if a new moon falls on the ecliptic, the moon is between the earth and sun, and those in the moon's shadow see an eclipse of the sun. The ecliptic is where there can be eclipses. This fact was known even before the reasons for it were understood.

When as a full moon crosses the ecliptic, the eclipse is at first only partial. Here, the outline of the earth's shadow on the moon is clear. It is circular. This was the most direct proof available to the Greeks that the earth is really a sphere, rather than, say, egg-shaped. A much larger portion of the earth's surface than was easily accessible to ancient travellers forms the earth's shadow on the moon.

## ERATOSTHENES AND THE SIZE OF THE EARTH

If we know how far a ship can be seen from a mountain before it disappears below the horizon, we can calculate the radius of the earth. Suppose the mountain is two miles high. Let us designate the unknown radius of the earth by $r$. Now, refer to Fig. 2-2, and consider the triangle $MSO$. The length $OS$ is the unknown radius, $r$. From a two-mile high mountain, a ship can be seen as far as 125 miles away. Therefore, the length $MS$ is 125 miles. Since the mountain is two miles high, the length $MO$ is $r + 2$ miles. Now, since $S$ is on the horizon, the line $MS$ just grazes the surface of the sphere; that is, it is tangent to the sphere. A tangent to a sphere is always perpendicular to its radius. Thus, $MS$ is perpendicular to $OS$. Therefore, the angle at $S$ is a right angle, $MSO$ is a right triangle, and Pythagoras' theorem must apply to $MSO$, and:

$$r^2 + 125^2 = (r + 2)^2$$

This algebraic equation for $r$ is easy to solve. The answer is $r = 3905$ miles.

There is nothing special about the particular numbers we used in this calculation, which is important enough to repeat in general. Let the height of the mountain be $d$, and the distance to the horizon, $x$. In our previous example, $d$ was 2 and $x$, 125 miles. If $x$ and $d$ are known, $r$ can be calculated by Pythagoras' Theorem:

$$r^2 + x^2 = (r + d)^2$$

whose solution is:

$$r = \frac{x^2 - d^2}{2d}$$

Now, we make an important approximation. If the height of the mountain, $d$, is much smaller than $x$, the distance to the horizon; then, $d^2$ is much smaller than $x^2$ (that means we can almost forget about the $d^2$ in the above formula). The equation:

$$r = x^2 / 2d$$

is almost right and much simpler to use. For example, in our calculations above, $x^2$ was $(125)^2 = 15{,}625$, while $d^2$ was $(2)^2 = 4$; thus, for all practical purposes we could have ignored the 4. We will encounter this method again; for Galileo used it to calculate the height of the mountains on the moon. It also will appear in our discussion of Newton's law of gravity. Details are included in the chapter Appendix.

A second, more accurate way to measure the size of the earth comes from the fact that the north celestial pole is the same number of degrees above the north horizon as the observer's latitude on the earth. Therefore, since the elevation of the pole increases by 1° for each 69 miles you go north, 69 miles must be the size of one degree of a circle around the earth. There are 360° in a circle, so the circumference of the earth is 69 × 360 = 24,840 miles. Now, the circumference is $2\pi$ times the radius $r$, where $\pi$ is approximately 3.14. Thus, the radius of the earth by this method is:

$$\frac{24,840}{2\pi} = \frac{24,840}{2(3.14)} = 3960 \text{ miles}$$

in approximate agreement with the answer obtained from our first method. The ancient Greeks could not have done as well with either method, since the measurements required were difficult to make accurately.

In contrast to the other arguments for a spherical earth, the shape of the earth's shadow on the moon cannot be used to measure the size of the earth until you already know how far away the moon is. Aristarchus of Samos (310 – 230 B.C.), perhaps the most imaginative astronomer of classical Greece, turned the problem around. Already knowing the size of the earth, he computed the distance to the moon! Without accurate instruments, without most of modern mathematics, Aristarchus answered the question every child asks, "How far away is the moon?" One Appendix section to this chapter explains Aristarchus' calculation.

The most famous ancient measurement of the earth was made by Eratosthenes of Alexandria in the third century B.C. He placed a vertical stick in the ground in Alexandria and measured the length of its shadow at high noon on the first day of summer. He knew that the Egyptian city of Syene, the modern Aswan, was due south of Alexandria. Furthermore, on the first day of summer the sun was directly overhead at high noon at Syene. At that time, a vertical stick would cast no shadow there. The rest of the year, the sun is always south of the zenith at noon, even at Syene. Furthermore, Eratosthenes knew very accurately the distance between the two cities. This was enough to calculate the size of the earth. The geometry is shown in Fig. 2-6.

Since the sunlight falling on the two cities can be thought of as traveling in parallel straight lines (because the sun is so far away), the angle at the top of the stick in Alexandria, marked $B$ in Fig. 2-6, and the angle $SOA$ at the center of the earth, are equal. This is an example of an elementary theorem of geometry. By measuring the length of the sun's shadow at noon in Alexandria on the day that the sun was

## The Greek Cosmos

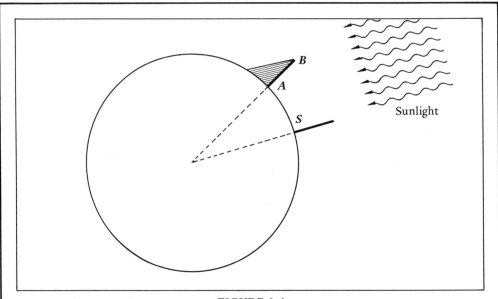

**FIGURE 2-6**
*Eratosthenes Measurement. O is the Center of the Earth; A, Alexandria; and S, Syene. Shown Grossly Exaggerated Are Two Vertical Sticks at S and A. The Stick at S Casts No Shadow, but That at A Does.*

directly overhead at Syene, Eratosthenes measured the angle at $B$ to be 7 1/5 degrees, or 1/50 of a circle. Therefore, angle $SOA$ is also 7 1/5 degrees, and the arc $SA$ is 1/50 of a circle. The circumference of the earth, therefore, is fifty times the distance from Alexandria to Syene. Eratosthenes's estimate of the size of the earth was probably accurate to within a very few miles.

These calculations of the size of the earth illustrate the power of mathematical reasoning to make statements about the physical world precise and logically consistent, a power first used by the ancient Greeks. The Greeks were very good at geometry; Eratosthenes' arguments, as well as the previous ones, are deeply entwined with geometrical reasoning. Wherever geometry could be used in physics, the Greeks made striking progress; however, wherever the wisdom of hindsight now indicates that other forms of mathematics unknown to the Greeks were required, Greek physics faltered.

### THE TWO-SPHERE UNIVERSE

In this section, we describe the two-sphere universe, a conceptual simplification of the Greek model of the universe suggested by the historian of science, T. S. Kuhn. Its purpose is to explain the relationship of the earth and stars. The stars are on the surface of a large celestial sphere, while the earth is a second very much smaller sphere whose center coincides with that of the celestial sphere. The celestial sphere rotates at a completely uniform rate, completing a revolution once every 23 hours, 56 minutes, carrying the fixed stars with it, while the earth remains motionless at the center. Someone standing upon the surface of the spherical earth can see that half of the celestial sphere above his/her horizon; as the person moves, their horizon changes and so does the visible portion of the celestial sphere. The elevation of the pole changes as one moves north or south. The pole star does not move during the night, and all other stars rotate in circles at the same rate about the pole star.

The Two-Sphere Model is a natural way to summarize the observations. Furthermore, the evidence available to the Greeks indicated to them that the universe was actually like the model. For example, the fact that the stars never appear to change their positions relative to one another strongly suggests they are firmly attached to a rotating solid object. What else could be responsible for the fact that they all rotate about the pole star at the same rate, night after night, year after year?

The fact that the stars never change brightness suggests they do not move nearer or further from us. If they are attached to the surface of a sphere, they never change their distance from the center, the earth, as the sphere rotates. If the earth were not at the center of the celestial sphere, a star could brighten and then fade during the passage of one night, as the sphere carries it first nearer and then further from the earth. This does not happen. The fact that the stars do not change their apparent brightness over the course of a year suggests that the earth remains at the center of the universe during the year. The fact that the stellar images are all the same size also suggests that all the stars are attached to the surface of the sphere.

Above are what in today's terms would be called strictly scientific arguments for the validity of the two-sphere cosmology. Probably the most compelling arguments stem from philosophy, at least in accounting for the long historical life of this conception of the universe and its variant, the Ptolemaic System (to be discussed shortly). Greek philosophy attempted to create a structure of rational thought that encompassed all aspects of human existence. To the extent that scientific thought was integrated into a wider philosophical structure, it gained authority and meaning. Obviously, the Greeks did not make the distinction between science and philosophy we do today.

The school of Plato (427-347 B.C.) emphasized the mathematical beauty of the two-sphere universe. Though not a mathematician of the first rank himself, Plato was impressed by the power and rigor of geometry. In particular, he was interested in the relation between abstract geometric conceptions and our materialization of them in the real world. For example, no one can draw perfectly parallel lines; however, one can certainly conceive of them. From their definition, that they never meet, we can derive many properties. Thus, in geometry the important thing is the one abstract ideal of parallel lines, not the imperfect physical approximations. Plato carried this reasoning into a more general philosophical context. To any imperfect example of an object, say a tree, or even an idea like duty or beauty, there corresponded the perfect ideal, the idea, or form, of "tree" or "beauty". In Plato's philosophy, abstractions were superior to the examples from which they were abstracted. The idea, the abstraction of perfection, was Plato's reality. Plato was a secular philosopher; but in his mind, and those of his followers down through the centuries, the beauty and perfection of ideals gave to them a special holiness. Today we do not call abstractions holy, however austere and beautiful.

Plato's comments upon the celestial spheres, while not central to his philosophy, had a powerful effect. The sphere is the only geometrical object that can be in perpetual uniform motion without changing its location. Its motions are eternal and self-sufficient. Plato, therefore, defined uniform circular motion as the most perfect, and the only one appropriate to the heavens. The majesty and simplicity of celestial motions stood in stark contrast to the chaotic, complex motions we see around us in our imperfect terrestrial world. A universe of two spheres, in themselves perfect geometric objects, with the centers of the earth and the celestial sphere in exactly the same place, has maximum symmetry and simplicity. Heaven and earth are in harmony when they have the same center. Man's abode, a sphere, is a small copy of the heavens, also a sphere. Man is fittingly placed at the center of the universe. The beauty and correspondence with philosophical conceptions of perfection of the two-sphere universe seemed to add to its correctness. This form of reasoning is unfamiliar today, but it persisted in one form or another in science all the way down to the time of Galileo.

### THE UNIVERSE OF EUDOXUS

This two-sphere model accounted nicely for nearly all the motions of the heavenly bodies. But, there was trouble in identifying the real sky with Plato's philosophical paradise, for seven exceptions were known to the ideal of uniform, circular motion; namely, the sun, the moon, and the five planets. Many times in the history of physics, exceptions to a general theory have stimulated great ingenuity in accounting for the anomalies within the overall framework of the general theory. Eudoxus' work, which we are about to describe, is one of the earliest examples of this process. The Greeks called it "saving the appearances".

Eudoxus (408-355 B.C.) was an ingenious mathematician. Plato, an ardent and discriminating admirer of mathematics, called him one of the two greatest mathematicians of the age. Biographical information on Eudoxus is scanty. It is said he went to Athens at age 23, where he studied in Plato's academy. There he probably was influenced by Plato, who pointed out that the sun and planets (to be discussed shortly) violated the rule of the perfect circular motion. This motivated Eudoxus to take up the problem of the non-uniform planetary motions.

We know what Eudoxus' research program was. The bodies in the heavens should be more nearly ideal (perfect) than those on earth. The stars have ideal circular motions. The sun and planets certainly do not have a single uniform circular motion. However, being celestial bodies, their motions should certainly partake of some aspects of perfection. One way out of the dilemma would be if some combination of uniform circular motions would account for the observed irregular motions.

Here is how it worked out for the sun. Since the sun has a westward diurnal motion identical to that of the stars, it must clearly be attached in some way to the stellar sphere. On the other hand, the sun also moves slowly eastward along the ecliptic (a path inclined 23 1/2° to the celestial equator), completing a full circuit in a year. Eudoxus' solution was to attach the sun to the equator of a second solar sphere that rotates eastward on its axis once a year; the axis of the second solar sphere is firmly attached to the celestial sphere 23 1/2° from the north celestial pole, so that the solar sphere's equator is the ecliptic, as illustrated in Fig. 2-7. Let us see how the system works. Since the solar sphere is attached at its pivot points to the celestial sphere and since the sun is firmly attached to the solar sphere, the sun is carried about by the 23 hr 56 min westward diurnal rotation of the outer, celestial sphere. This accounts for the sun's daily motion. However, the inner solar sphere is slowly rotating eastward. This accounts for the slow annual motion of the sun relative to the background of the fixed stars.

In this way Eudoxus accounted for the apparent motion of the sun. He adhered to Plato's dictum that celestial motion should be the perfect motion of uniformly rotating spheres. Unfortunately his solution

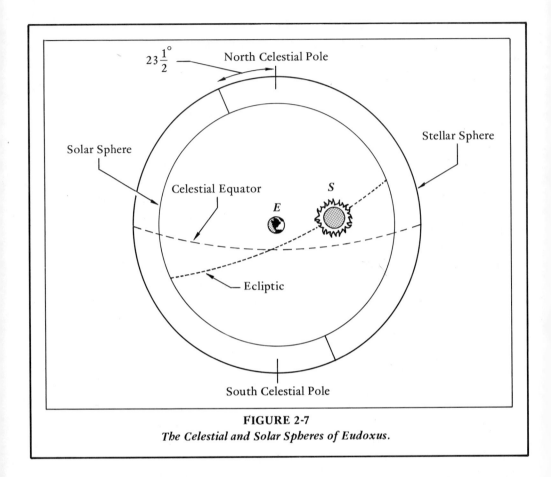

**FIGURE 2-7**
*The Celestial and Solar Spheres of Eudoxus.*

disagrees slightly with a well-known observation, the sun's apparent motion is not exactly uniform; in Chapter 1 we learned that the sun moves along the ecliptic slightly more rapidly in winter than in summer. The period from September 21 to March 21 is three days shorter than the period from March 21 to September 21. Winter is shorter than summer by three days.

This small deviation from uniform motion meant that Eudoxus' simple scheme was not exactly right. Eudoxus and his student, Callipas, rectified the situation by adding yet another sphere rotating with respect to the solar sphere. By placing the sun on this new sphere and choosing this sphere's axis of rotation and speed of rotation appropriately, they could account for the observed motion of the sun. Thus, Plato's rule of uniform circular motion was bent by Eudoxus' devices, but not broken. But, from this time on the astronomer's task took a slightly different direction: to concoct schemes, however complicated, that would explain and predict accurately the observed motions, to "save the appearances", while retaining uniform circular motion as an ideal.

In a similar fashion, the moon was fixed to the equator of its own sphere and made to revolve around the earth. The moon's sphere was much smaller than that of the sun, since the moon was known to be much closer. The moon's sphere was attached to the stellar sphere, to give it the westward diurnal rotation, but it also rotated eastward once a month so that the observed motion of the moon was a compound of the daily and monthly rotations. This model was very successful for it accounted for the phases of the moon.

A conceptual problem remained with the moon and all the planets. How could the moon's sphere be attached directly to the stellar sphere without the sun's sphere getting in the way? One argument was that Eudoxus' construction was just elegant mathematics and one really should not worry how the spheres were attached. On the other hand, Aristotle, whom we shall come to next, thought of Eudoxus' spheres as real material objects, and devised the necessary system of "gears" to make it all work.

How could one possibly say that the planets obey the rule of perfect circular motion when they are observed to stop and reverse their direction of motion through the heavens during retrogression? Even today, we can imagine Eudoxus' feeling of triumph when he discovered that he could construct for each planet a system of linked spheres, all of which rotate uniformly about a suitably chosen axis, which reproduces retrograde motion. The axis of the outermost sphere of each planet's system is firmly attached to the celestial sphere so that the entire system of spheres, with its planet attached to the innermost, shares in the diurnal rotation of the stars. The outermost sphere rotates eastward to give the planets average eastward motion through the constellations; other spheres can be added to account for small variations in the general eastward motion. Most importantly, Eudoxus found that he could add a westward rotating sphere that would reproduce retrogression, at approximately the proper places and times. In effect, the eastward motions of the outer sphere would occasionally be overcome by the westward motion of the retrogression sphere. How this worked in detail must have been very complicated.

Of course, there were difficulties with Eudoxus' model of the universe. His method was arbitrary and incomplete. It was arbitrary because there was no systematic way of choosing in advance the orientations of the rotational axes and speeds of rotation of the various spheres in a planet's sphere system. Eudoxus merely showed that if you used enough spheres by trial and error you could build a model that reproduced the complicated motions of the sun, moon, and planets, including planetary retrogression. Eudoxus could not even show that if you found a system of spheres capable of reproducing a planet's motion that it was the only such solution of the problem. Eudoxus' method was incomplete because it left several questions unresolved. We have already mentioned one difficulty. Each planet's sphere system must be connected to the outermost celestial sphere. Must the rotation axis of an inner planetary sphere penetrate all the planetary spheres beyond it to reach the celestial sphere? Perhaps Eudoxus was constructing only a mathematical model, not a physical model of the heavens.

This question leads naturally to the next question: What is the ordering of the planets' spheres? Eudoxus' method simply did not say how far away the planets are, which ones are nearer, and which are farther from the earth. The method never could say how far away they were, but a little common sense might help resolve the ordering of their distances. Clearly, the moon was the closest planet—it has a big visible disc. And the sun must be further away than the moon, to explain the moon's phases. Beyond this, it seems to have been decided that the fastest moving heavenly bodies were closest to the earth. Thus, the moon and its system of spheres were closest. Similarly, the slowest moving, Saturn, was furthest out next to the celestial sphere followed, going inward, by Jupiter and then Mars. Since Venus and Mercury never get very far from the sun, they and the sun all traverse the zodiac in one year on the average. Here, it was not clear how to proceed. Different people made different choices. When all we can measure is angles and not distances, is it really important for a theory to specify how far away the planets are when it does a good job predicting their angular positions?

Eudoxus' method completely fails to account for the observed fact that the planets change their brightness. While Eudoxus could produce a planetary retrogression, he could not say why the outer planets, for example, are brightest during retrogression. Since his planets were attached to earth-centered spheres, their distance from earth never varied, and he could not even in principle—no matter how many spheres he introduced—ever produce a planetary brightness variation.

All these criticisms should not obscure our admiration of Eudoxus' mentality. It required immense ingenuity and geometrical imagination to construct a model of the universe that actually agreed fairly well with observation and yet retained the Platonic imperative of uniform circular motions in the heavens.

ARISTOTLE'S SYNTHESIS OF PHYSICS AND COSMOLOGY

The survival of Greek geocentric cosmology for over two thousand years—five times as long as modern physics—was due above all to the philosopher Aristotle (384–322 B.C.) and his followers through the centuries. Aristotle found in the Eudoxan universe a framework for his laws of physics. They only made sense if the earth were at the center of the universe. On the other hand, the universe made sense using Aristotle's physics. With the structure of the universe and the laws of physics entwined, the combined system was all the more convincing.

Aristotle's physics was appealing. His method was to apply logic to commonsense observations about nature. He trusted only the evidence of the human senses, and was naturally suspicious of observations gathered by anything but direct human perception. With Aristotle's powerful logic as guide, the least sophisticated student could deduce from his own experience sweeping generalities about the universe, and the earth's place in it. Aristotle's physics was a small part of a vast philosophical program linking all realms of human thought such as ethics, metaphysics, logic, physics, politics, and more. This was certainly the first and possibly the last time that all branches of human knowledge were formulated and linked by one man. With physics linked to cosmology, and both to the rest of human thought, the whole structure took on a coherence and strength that its individual parts sometimes lacked. Therefore, while people later criticized specific details of Aristotle's physics, for many centuries they were quite unwilling to see his key physical statements undermined, fearing to damage the edifice of his thought.

Aristotle's father had been personal physician to King Philip of Macedonia; Aristotle himself was tutor to Philip's son, Alexander the Great. Aristotle was the greatest of Plato's many great students. He enrolled in Plato's Academy at seventeen and remained there twenty years until Plato died. He spent the next twelve years tutoring, and working for, Alexander the Great. While Alexander led his conquering armies over a great part of the known civilized world, Aristotle studied the people, customs, thought, and plants and animals in the conquered provinces. In 335 B.C., Aristotle formed his own school, the Lyceum, where he remained until his death. Many of Aristotle's works are thought to stem from students' notes upon his lectures at the Lyceum.

Much of Aristotle's philosophy is concerned with correcting what he considered to be imbalances in his master's (Plato's) thought. For example, Aristotle was too keen an observer of the natural world to believe that Plato's severe idealizations and abstractions could ever be entirely sufficient, or rich enough in possibility, to describe nature. Irregularity and complexity was the rule in the world around him. And so, Aristotle's scientific work is based upon attentive commonsense observation of nature. He would derive his physics solely from the evidence of his senses. Where such observation revealed a persistent pattern, then and only then would he try to abstract from experience a general rule of behavior. If Plato represented deductive thought (logical inference from axioms) to future generations, then Aristotle represented inductive thought (inference from observed facts).

Aristotle acquired from his father a love of biology and medicine; and through his father's instruction he learned to be a systematic and perceptive observer of nature. Certainly, biology was Aristotle's first and deepest scientific love, so much so that biological analogies penetrated all his thinking. In his *Politics,* he views the state as an organism. Aristotle transferred two aspects of biological thought into his physics of motion. First, just as plants and animals can be classified, so also can types of motion. Second, just as seeds develop naturally into trees, so also must the natural motions of bodies in the universe be towards a definite goal.

Aristotle founded the science of the classification of biological species. It was important to him to give things that were different, different names. Similarly, he made a distinction between types of motions; for example, the motions of light and heavy bodies on earth and the motions of celestial bodies in the heavens. Moreover, each biological species seemed to have a natural mode of development. Seeds grow into plants, eggs into chickens, and caterpillars into butterflies. Is not the purpose of a seed to become a plant; the egg, a chicken? For Aristotle, all things strive towards perfect correspondence with their ideal nature. While Plato was interested in this final achievable state, Aristotle concerned himself with the more tangible problem of how things tried to reach the ideal. Thus, for Aristotle, motion is the striving to achieve a natural end, a conception far more general than ours today, involving all sorts of changes that we do not associate with motion—growth or decay, change of properties, change in quantity, as well as change in location. Furthermore, Aristotle made a distinction between "natural" and "unnatural" motions. While a seed will naturally grow into a tree if left undisturbed, a blight or any other external agent may force it to grow unnaturally into a stunted tree. Similarly, those motions—changes of position—which do not carry the body nearer its natural place are *un*natural and *require* some external agent to carry them out.

In biology just precisely what natural development is depends upon the plant or animal. In physics, therefore, the type of motion depends upon the types of objects that are moving. Thus, we must first ask, "what is everything made of?" And, only then, "how do the different things move?"

Long before Aristotle, Greeks had speculated that there might be a few fundamental building blocks of matter from whose combinations all the different materials in our environment could be fabricated. Evidently, matter could be transformed from one state to another, especially under the influence of heat. Water turns to steam, and ice, water; eggs boil and wood burns. Is there some underlying simplicity? Pythagoras of Samos taught that everything was made of number. In the sixth century B.C., Thales of Miletus had suggested that everything is made of water. Before Plato, the philosopher Democritus taught that matter was made up of atoms; microscopic, invisible, and indivisible pieces of matter that combined in different ways to form different materials.

Aristotle rejected Democritus' atoms but he believed that there were a few basic substances—elements out of which everything is made. They were earth, water, air, and fire. Earth is cold and dry, whereas, water is cold and moist. Air is warm and moist. Fire is hot and dry. Earth and water are heavy. Air and fire are light. Everything around us possesses these properties in varying degrees. We perceive them with and they are defined by our senses: sight, smell, touch, taste, and hearing. If we were to break down an object into its elements, we would find according to Aristotle that it contained so much earth, so much water, so much air, and so much fire. In what proportion? This is determined by its properties. For example, a heavy, dry object must be primarily made up of earth.

We may now define natural motion. Let us begin with heavy bodies: What is it they always do when left alone? Fall to earth. Since there is no exception to this rule, we conclude from observation alone that the natural motion of heavy bodies is to fall. Further observation tells us that heavy bodies always fall in straight lines. If they are dropped and not thrown, and if the wind is not blowing, heavy bodies fall straight down—towards the center of the earth. They keep falling until their natural motion is interrupted by collision with the ground; if they were not so stopped their natural motion would continue towards the center of the earth. But, the center of the earth is the center of the universe. Therefore, from observation we conclude that the natural motion of heavy bodies is in straight lines towards the center of the universe. However, this is the natural motion only for heavy bodies made primarily of earth. Clearly, flames and smoke always rise through air, and air always rises through water; whereas, a stone falls through water. An object's natural motion depends upon its composition. Earth falls, fire rises, water falls, but not below earth. The natural motions of compound materials depend upon their composition, to what extent they are made of earth, water, fire, or air.

How fast do things fall? Here Aristotle proposed two general rules. First of all, the heavier a body is, the more it is made of earth; and the result is that the more earth's natural tendency to fall overcomes the tendence to rise of the body's light elements, the faster the heavy body falls. Aristotle became very precise on this point. A body twice as heavy will fall to earth twice as fast. Secondly, a stone falls faster in air than in water. Thus, the speed with which a heavy object falls depends upon the density of the resisting medium. The denser the medium, the slower the fall. Could a body ever fall in a vacuum? In other words, could it fall when there is no resisting medium? With nothing to resist its motion, a body would fall infinitely rapidly. But, if it were moving infinitely fast, it would be in two places at the same instant of time. This is logically impossible. We must always take into account the resistance of the surrounding medium in discussing the motion of any and all objects.

For Aristotle, there could be no such things as a vacuum—"Nature abhors a vacuum." Not only would the speed of bodies moving in a vacuum be infinite, which is ridiculous, but what we mean by space is defined by the relationships between the things in it. If there is nothing in it, it has no properties; it does not exist. A vacuum is *logically* impossible! Aristotle believed that the absolute impossibility of empty space refuted the idea that matter is made up of atoms. If there were atoms, what would be between the atoms? Empty space, a vacuum? But, empty space can not exist; therefore, atoms can not either. We will see that disputes over the reality of empty space became more and more central to the development of physics as the centuries passed.

Now, let us turn to unnatural motions. There is nothing to explain in natural motion, it always happens. What needs explaining is any motion that is not up and down, but horizontal. These unnatural motions must be caused by external agents, which, using a modern term loosely, we will call "forces". (The notion of forces was not rigorously defined until Newton's time.) For example, a heavy object rests on the ground. It is in its natural state, since it is as close to the center of the universe as it can get. No one has ever observed a heavy stone move unless it was pushed. As soon as you stop pushing, the heavy stone stops moving. Therefore, Aristotle reasoned, a direct application of force is required to cause unnatural or "violent" motion.

How much force is required? If you double the force, you double the speed. Two horses can pull a cart twice as fast as one horse. When the horses stop pulling, the cart stops. Of course, a real cart takes a while to come to its natural state of rest, but Aristotle did not consider this to be an essential feature of the motion. He did not pay much attention to the way motion changed from natural to violent, other than to state a force was required. Should one really take the fact that there *are* a few instants when the cart is changing its motion, slowing down when the horses are doing nothing, as a serious criticism of an otherwise logical presentation of the facts?

It is difficult to find in our everyday experience a motion caused by a force that is not applied directly. To move a stone, you must push it with your hands. Aristotle elevated this into a general rule: "Action at a distance is impossible". He rejected "occult"—which simply means hidden—causes of phenomena, such as the idea that the moon causes the tides. It is not logical to talk about a cause of motion unless you can observe how that cause acts directly on the body being moved.

Now we can discuss Aristotle's cosmology. First of all, are the objects in the heavens the same as on the earth? Obviously not, since their natural motions are different. Celestial bodies move in circles and do not fall toward or away from the earth. Since the natural motion depends on the kind of material, the celestial spheres must be constructed of a new fifth element not found on earth, aether. In the Latin of medieval Aristotelians, aether was also called "quintessence", which simply means "fifth element". It must be weightless; it does not fall towards the center of the universe. It is invisible, but it carries the stars around the heavens. The outermost celestial sphere is in perfect uniform circular motion, which, because of its perfection, is its natural motion. The "Prime Mover" of the universe resides at the outermost celestial sphere.

The planets' motions are more complicated, but being celestial objects their natural motion is still circular, and they must also reside on spheres made of the celestial material. In fact, since there can be no vacuum, the entire universe is filled with celestial material, from the moon's sphere to the stars' sphere.

Since all motions require a direct application of force, the planets' motions must be caused by contact of the various spheres with one another. In fact, Aristotle inserted extra celestial spheres to Eudoxus' 27, spheres that carried no planets and whose function it was to fill up the universe and to act as gears linking the motions of the various planetary spheres. All this continued down to the sphere of the moon. How wonderful that all the heavenly motions could be derived from the perfect motion of the outermost celestial sphere, which was natural unto itself.

Aristotle was convinced by his predecessors' arguments that the earth is round:[1]

> If the earth were not spherical, eclipses of the moon would not exhibit segments of the shape which they do . . . If the eclipses are due to the interposition of the earth, the shape must be caused by its circumference, and the earth must be spherical. Observation of the stars also shows not only that the earth is spherical, but that it is of no great size, since a small change of position on our part southward or northward visibly alters the circle of the horizon, so that the stars above our heads change their positions considerably, and we do not see the same stars as we move north or south. Certain stars are seen in Egypt and in the neighborhood of Cyprus which are invisible in more northerly lands, and stars which are continuously visible in the northern countries are observed to set in the others. This proves both that the earth is spherical and that its periphery is not large, for otherwise such a small change of position could not have had such an immediate effect. For this reason those who imagine that the region around the Pillars of Heracles joins on to the regions of India, and that in this way the ocean is one, are not, it would seem, suggesting anything utterly incredible. They produce also in support of their contentions the fact that elephants are a species found at the extremities of both lands, arguing that this phenomenon at the extremes is due to communication between the two. Mathematicians who try to calculate the circumference put it at 400,000 stades.

Moreover, Aristotle gave good reasons why the earth must be round. All the materials on earth would, if left to their own natural motions, arrange themselves in a natural order. Suppose all the heavy matter were originally distributed throughout the universe. It would all move to the center, collide, and accumulate near the center. If the earth were not at first a sphere, some heavy matter would be farther from the center of the universe than other matter, and would move downhill until there were no lower portions of the surface available (until the earth became a sphere). Thus, the earth must be a sphere. The other elements would arrange themselves in like fashion: first earth, then layers of water and air, and finally one of fire just below the lunar sphere. However, it is not exactly like this, because the rotating lunar sphere is connected to the terrestrial elements, jostling them and mixing them up. Man's sphere has not yet arrived at a state of perfect order.

### HERAKLIDES, ARISTARCHUS, AND THE SUN-ORDERED UNIVERSE

Since the Greeks at one time or another asked most important scientific questions, it is not surprising that serious consideration was given to the question, "Is the sun, rather than the earth, at the center of the universe?" Heraklides of Pontus (388–310 B.C.), another of Plato's students, argued that it was the earth and not the heavens that rotated. It was much simpler to imagine one body, the earth, rotating than to assume all the stars whirled about the earth at immense speed. Heraklides took a radical further step. Instead of requiring all the objects in the universe to move about the earth, Heraklides proposed that Mercury and Venus orbit the sun, since they never appear far from the sun in the sky. The sun and other planets—Mars, Jupiter, and Saturn—then orbited the earth. Almost two thousand years later, Tycho Brahe would propose a similar plan of the universe.

The same Aristarchus of Samos, who calculated the distance to the moon, made the most radical proposal of all. If the earth could rotate on its axis, why could it not also move through the heavens? Could not the earth also be a planet like the others, and did not *all* the planets orbit the sun? We know of Aristarchus' hypothesis from a report of it in the *Sandreckoner*, in which Archimedes tried to enumerate the grains of sand in the universe.[2]

> Aristarchus of Samos brought out a book consisting of some hypotheses, in which the premises led to the result that the universe is many times greater than now so-called. His hypotheses are that the fixed stars and the sun remain unmoved, that the earth revolves about the sun in the circumference of a circle, the sun lying in the middle of the orbit . . .

To understand why Archimedes thought that a sun-centered universe had to be many times greater than an earth-centered one, we must understand how the Greeks attempted to measure the distance to the stars. (If one could establish that all the stars were the same distance from the earth, then, surely, the earth is at the center of the universe.) Their method was one of triangulation. Suppose for the sake of argument that one star were closer to the earth than all the others on the celestial sphere. If you could measure its angular position, relative to the celestial sphere, from two vantage points a known distance apart, then, you could calculate the distance to the star. Figure 2–8 illustrates the situation. Observers at $A$ and $B$ will see the star in front of different constellations on the celestial sphere. Knowing the distance, $AB$, and the change in apparent angular position, the distance to the star could be calculated in principle. The Greeks knew the stars were very far away so that the angular shifts, or parallax, would be very small and hard to measure (see chapter Appendix). The measurement gets easier the larger the baseline, $AB$. If the earth were at the center of the universe, the largest such distance that could be used as a baseline would be the diameter of the earth, some 7900 miles. Although they tried many times, the Greeks never succeeded in measuring

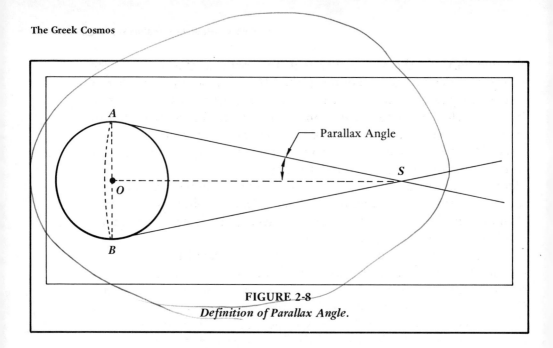

**FIGURE 2-8**
*Definition of Parallax Angle.*

a stellar parallax. This meant that the stars were sufficiently far away that the parallax angle was less than the few minutes of arc resolvable by the human eye. Now, if Aristarchus were correct, it would be possible to find a much larger baseline for the parallax measurement, since one could look for apparent shifts in the star's position at six-month intervals. Then, the baseline, *AB,* would be the diameter of the earth's orbit around the sun, which we know today is 186 million miles. Aristarchus himself believed the sun was at least twenty times as far away as the moon. He had calculated that the moon is 60 earth radii away from the earth. Thus, Aristarchus' model required a baseline at least 1200 times ar large as an earth-centered model. If Aristarchus' theory were correct, the inability to measure parallax implied that the stars would have to be at least 1200 times farther away than they would have to be if the earth were at the center of the universe. This was hundreds of times greater than any other known distances. Surely, if the earth orbited the sun, a parallax would be observable. It was of course possible the stars were fantastically far away so that the largest parallax shifts were less than four minutes of arc, and, thus, unobservable. But, even though the bold imaginations of the Greeks conceived the idea, no good evidence existed for a belief that we live on a minor planet of an unimportant star with trillions of miles of emptiness between it and its nearest neighbor.

The Greeks never did learn how far away the stars are, only that they are very far away. Since they did not know the stars' distances, they could not tell whether any of them were at different distances from the earth. Nothing contradicted the idea—and it was by far simplest to assume it—that the stars were at the same distance, implying the earth is at the center of the universe. The same argument applies to stellar brightness variations. The naked eye can not perceive the tiny variations in brightness that would indicate the earth and stars are moving closer or farther apart. Furthermore, Aristotelian physics provides good arguments why the earth *must* be stationary at the center of the universe. If the earth were rotating or moving, a ball thrown upward would land at a place different from where it was thrown. Once the object loses contact with the earth, the earth cannot exert a force on it—no action at a distance, and it can no longer share the earth's motion. Since this does not happen, the earth does not move. Similarly, one might expect to feel strong winds if the earth moved through the air surrounding it. Again, this does not happen. All in all, the sun-centered universe was at a competitive disadvantage. Aristarchus' universe was not so well integrated with physics and philosophy as Eudoxus', nor was anyone motivated enough to calculate accurate planetary positions using Aristarchus' scheme to show they reproduced the observations as well as Ptolemy's earth-centered system, which we discuss next.

So, ultimately, this was the Greek universe: a tight little ball. The universe with the earth at its center and with the moon, sun, and planets all solemnly revolving on transparent spheres; it was orderly and intimate. For those whose minds turned to astrology, Aristotle's system of linked spheres made it possible to visualize how motions in the heavens could directly affect human fate. One's relation to the heavens was precisely defined.

The heavens were regions of unchanging perfection, while all was corruption and decay on earth. The celestial material had attained a state of unchanging perfect circular motion, to be contrasted with the vulgar up and down, and even worse, sideways, motions characteristic of materials on earth. This Greek distinction between the heavens and earth found its way into medieval Moslem and Christian theology, where it was associated with the distinction between sin and salvation. It was God who moved the perfect celestial sphere, and His angels the planet's spheres. Hell was at the center of the earth. Man lived on the earth's surface, between heaven and hell, between sin and salvation. Dante's *Divine Comedy* is a medieval Christian's grand tour through Aristotle's universe. With theological fervor and scientific logic mutually supporting each other, Aristotle's vision of the universe became a Christian vision. All this occurred because Aristotle was the first to fuse physics and cosmology into one unified world view.

# The Greek Cosmos

*Summary*

To the ancient Greeks we owe our scientific tradition, the idea that we can understand the world through rational thought and logical, mathematical, reasoning rather than through myth and tradition. Pythagoras and his followers discovered much of geometry and arithmetic that became the principal technical tool for making precise, numerical statements about the physical world.

The early Greeks explained the phases of the moon, and discovered that the earth and the moon were both spheres. There were three principle arguments: the way ships disappeared over the horizon, the fact that the north celestial pole rose one degree in altitude every 69 miles one went north, and the shape of the earth's shadow on the moon during an eclipse. From the first two phenomena, the early Greeks could compute the size of the earth. From the third, Aristarchus, somewhat later, computed the distance to the moon. The key formula that will recur frequently relates $r$, $x$, and $d$ in Fig. 2-2. It is:

$$x^2 = 2rd$$

The most accurate ancient measurement of the size of the earth was made by Erathosthenes of Alexandria, by measuring how far the sun was south of the zenith on the first day of summer at Alexandria.

The philosopher Plato lent his authority to the idea that the stars lie on a sphere that has the same center as the earth but is much larger. This celestial sphere rotates every 23 hours 56 minutes about the stationary earth. This was only fitting, since the sphere has a perfect shape, the only object that can rotate without changing how much space it occupies. Plato thought that all heavenly motion should be spherical rotation. His student, Eudoxus, elaborated on the scheme, and constructed a system of heavenly spheres that reproduced the observed motions of the sun, moon, and planets, as well as the stars.

With this background, Aristotle answered the two questions of physics. He taught that everything is made of four elements. In order of heaviness, they were earth, water, air, and fire. The natural motion of these elements is to fall straight down toward the center of the universe, or to rise straight up from it. The heavier an object is, the faster it falls. The denser the resisting medium, the slower an object falls. A vacuum cannot exist. The heavenly bodies are made of a fifth element, aether or quintessence, which, partaking of heavenly perfection, has circular motion as its natural motion. In the later middle ages, Aristotle's ideas acquired the support of Christian religion.

There was another tradition, but it was ultimately unsuccessful.

Democritos taught that matter was made up of atoms. Heraklides believed that the earth rotated, instead of the celestial sphere; and Aristarchus advocated a picture of the sun, moon, and planets more or less like the one we believe today. But, thoughtful people rejected these ideas. Aristotle's physics made sense in an earth-centered universe, but not in another. And finally, the absence of parallax proved that the earth was not moving around the sun.

## Appendix to Chapter 2

*Pythagoras' Theorem*

There are many proofs of Pythagoras' Theorem. Here is one that is simple to understand. Although we use modern algebraic notation, the proof is similar in principle to the original.

In Fig. 2-9, we have drawn a square whose sides have length $a + b$, and another one, inside it, whose sides have length $c$. What is the area of each of the four little triangles around the edges? The area of a right-angled triangle whose short sides are $a$ and $b$, as in the diagram, is, $(1/2)ab$. [To prove this, draw a rectangle whose sides have lengths $a$ and $b$. Its area is the product, $ab$; that is what we *mean* by area. Now draw a diagonal; it divides the rectangle into two right triangles with sides of lengths $a$ and $b$; so the area of each triangle is $(1/2)ab$].

Since four of these triangles are in the big square, their total area is $4 \times (1/2)ab$, which is $2ab$. The area of the inside square is $c^2$. Therefore, the total area is $2ab + c^2$.

But, the area of the big square is just the square of the length of its side, which is $a + b$; namely, $(a + b)^2$. The area calculated both ways must be the same; therefore:

$$(a + b)^2 = 2ab + c^2$$

A general rule of algebra is that for any numbers $a$ and $b$:

$$(a + b)^2 = a^2 + 2ab + b^2$$

For example, $(3 + 5)^2 = 9 + 30 + 25$. Therefore, making this substitution for $(a + b)^2$, we get:

$$a^2 + 2ab + b^2 = 2ab + c^2$$

Finally, subtract $2ab$ from both sides of this equation to get:

$$a^2 + b^2 = c^2$$

This is Pythagoras' Theorem.

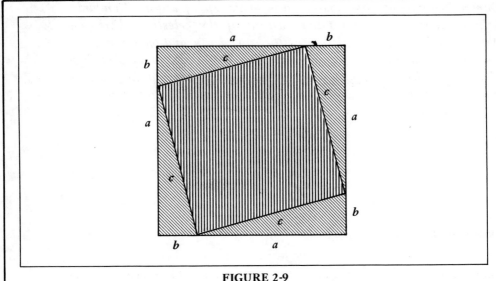

**FIGURE 2-9**
*Diagram for Proof of Pythagoras' Theorem.*

*Calculation of the Earth's Size: Algebraic Details*

In the discussion of the first method for computing the earth's size, some mathematical details were omitted for ease of reading, but the student should be able to follow the argument. For the triangle in Fig. 2-2, Pythagoras' theorem reads:

$$r^2 + 125^2 = (r + 2)^2$$

Remember that for any two numbers, $a$ and $b$, the square of their sum is:

$$(a + b)^2 = a^2 + 2ab + b^2$$

and apply this rule to the case where $a = r$, $b = 2$:

$$(r + 2)^2 = r^2 + 4r + 4$$

Now, the Pythagoras Theorem for our triangle can be written:

$$r^2 + 125^2 = r^2 + 4r + 4$$

Next, we do a little algebra. Subract $r^2$ from both sides:

$$125^2 = 4r + 4$$

and then solve for $r$ by subtracting 4 from both sides and then dividing the equation by 4:

$$r = \frac{125^2 - 4}{4}$$

This is the desired formula for $r$. Since $125 \times 125$ is $15{,}625$:

$$r = \frac{15625 - 4}{4} = \frac{15621}{4} = 3905$$

In general, we start with:

$$r^2 + x^2 = (r + d)^2$$

but,

$$(r + d)^2 = r^2 + 2rd + d^2$$

so,

$$r^2 + x^2 = r^2 + 2rd + d^2$$

Subtract $r^2$ from both sides:

$$x^2 = 2rd + d^2$$

The exact solution is:

$$r = \frac{x^2 - d^2}{2d}$$

Since $2rd + d^2$ can be written $d(2r + d)$, we see that provided $d$ is much smaller than $2r$, $d^2$ will be much smaller than $2rd$, and:

$$x^2 = 2rd$$

is a useful approximation.

## Aristarchus' Method for Measuring the Distance to the Moon

Aristarchus attempted to measure the distance to the sun and the moon. His measurement of the sun's distance was very sensitive to the details of his method, and he got it quite wrong. Aristarchus reported that the sun was about 20 times as far away as the moon; the correct answer is closer to 400.

Nevertheless, his method works very well for calculating the distance to the moon and the moon's size. The argument employs only elementary algebra and geometry, and a few simple naked-eye astronomical observations.

Let us introduce some notation. Let $D_s$ and $D_m$ be the diameters of the sun and the moon, respectively; and $R_s$ and $R_m$ are their distances respectively, from the earth. Figure 2-10 shows the sun, the earth, and the moon when it is on the far side of the earth; in our notation:

$$D_s = AB$$

$$R_s = HI$$

$$R_m = HG$$

## The Greek Cosmos

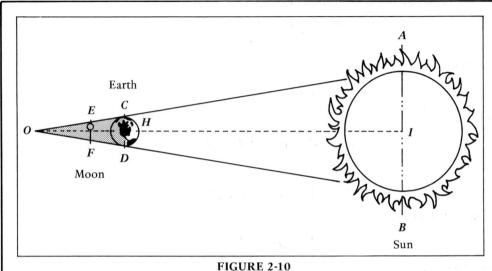

**FIGURE 2-10**
*The Sun, Moon, Earth, and the Earth's Shadow.*

But Fig. 2–10 is greatly exaggerated; both the moon and the sun are much further away from the earth, relative to its diameter, $CD$, than shown. Figure 2–11 shows more realistically the small ratio, $D_m / R_m$. The dashed line is part of the moon's orbit. We know from observation that the angle subtended by the moon (i.e., the angle made by the two solid lines in Fig. 2–11 at the earth), is almost exactly $1/2°$.

The moon's orbit is a circle around the earth, containing 360 degrees, and, therefore, 720 half-degrees. Thus, the circumference of this circle is $720 \times D_m$. The distance to the moon, $R_m$, is the radius of this circle so the circumference is $\pi \times$ *diameter of circle*, or $2\pi R_m$; thus:

$$2\pi R_m = 720 D_m$$

or,

$$\frac{R_m}{D_m} = \frac{720}{2\pi} = 115$$

In other words, the distance to the moon is about 115 times the moon's diameter.

We know that during a solar eclipse the moon just barely covers the sun. This is just an accident of nature, but allows us to repeat the above calculation, replacing the moon by the sun. The sun, too, subtends an angle of $1/2°$, as seen from the earth; and, therefore, the same argument shows that its distance is 115 times its diameter:

$$R_s = 115 D_s$$

Now, we make an important approximation: The sun is very far away, compared both to the moon and the size of the earth. This means that the angle subtended by

Appendix to Chapter 2

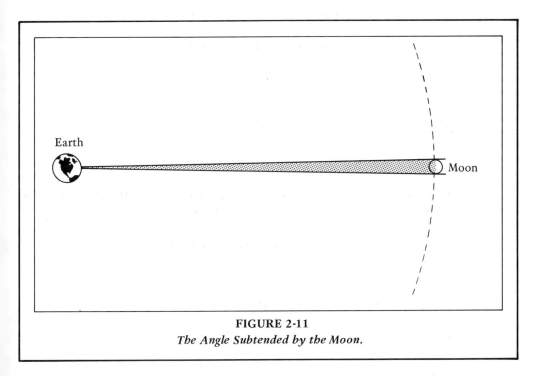

**FIGURE 2-11**
*The Angle Subtended by the Moon.*

the sun from any place near the earth or the moon will still be 1/2°. In Fig. 2-10, the region *COD* contains the earth's shadow. The point *O* is at the shadow's tip. From *O*, the angle subtended by the earth is the same as the sun's, so the angle the earth makes at *O* is also 1/2°. Repeating the above argument, we find that the distance to the tip of the shadow, *OH*, is related to the diameter of the earth, *CD*, by the same ratio 115. But, this time we know the earth's diameter. Take it to be 4,000 miles. Then, *OH* = 115 × 8000 = 920,000 miles. The moon, of course, must be closer than this, since it sometimes enters the earth's shadow. (After you have understood this method, you might want to go back and use Aristarchus' value for the distance of the sun, $R_s = 20\, R_m$, or the modern value, $R_s = 400\, R_m$, exactly.)

In Fig. 2-10, the moon is shown just as the total part of a lunar eclipse is beginning. During a central eclipse, one in which the moon passes through the center of the earth's shadow, the moon takes about one hour to enter the shadow and remains eclipsed for one hour 40 minutes, or $1\,^2/_3$ hours. From the time the moon begins to enter the shadow to the time it begins to emerge is $2\,^2/_3$ hours. Since it moves its own diameter in one hour, we conclude that the distance across the earth's shadow at the moon, *EF* in Fig. 2-10, is $2\,^2/_3$ the moon's radius:

$$EF = 2\frac{2}{3} D_m$$

or

$$EF = \frac{8}{3} D_m$$

Now, the angle subtended by $EF$ at $O$ is also $1/2°$. Therefore, as above:

$$GO = 115 EF = 115 \times \frac{8}{3} \times D_m$$

but,

$$115 D_m = R_m$$

therefore,

$$GO = \frac{8}{3} R_m$$

Now, $GH$ is the distance to the moon, $R_m$, and $GO + GH$ is the distance $OH$ that we have already calculated to be 920,000 miles; therefore:

$$\frac{8}{3} R_m + R_m = 920,000$$

or,

$$\frac{11}{3} R_m = 920,000$$

The solution is:

$$R_m = 250,000 \text{ miles}$$

This is the distance to the moon! Its diameter is:

$$D_m = \frac{R_m}{115} = 2,182 \text{ miles}$$

We have used some approximate figures, which, nevertheless, are as good as those the ancient Greeks had. Modern accurate values are 239,000 miles and 2,130 miles. The important point in all this is that a few observations, knowledge of the earth's diameter, and some elementary mathematics were enough for Aristarchus to compute the moon's distance and size.

### Stellar Parallax and the Distances of the Stars

In the sun-centered universe that Aristarchus advocated, the nearer stars, as viewed from the moving earth, might be expected to show a shift of their apparent positions against the background of the more distant stars as the earth moved around the sun. No parallax can be seen with the naked eye; to the ancients, and later to opponents of Copernicus' heliocentric theory, the apparent absence of this effect was a strong, and quite scientific, argument against the idea that the earth orbits the sun, since it proved that if the earth did revolve around the sun, as we now believe, the stars must be terrifyingly distant.

The important angle to measure is the angular shift in a star's position during a half-year. This is the same as the angle as $S$, in Fig. 2-8. (Conventionally, this "parallax angle" is not the angle $ASB$ but the angle $OSB$, where $O$ is the sun. This "parallax angle" is half the angle $ASB$ in Fig. 2-6.)

In Fig. 2-8, the large circle is the earth's orbit. The sun's position is labelled $O$; $A$ and $B$ are the positions of the earth, say, on March 21 and September 21. The distant stars are far off the page.

The parallax angle, as defined above, is the angle $ASO$, or half the angle $ASB$. Since it is very, very small, we can make another one of our physicist's approximations. Draw the curved line between $A$ and $B$, which is the portion of a circle that passes through $A$ and $B$ and whose center is the star $S$. The length of the curved line is the same fraction of the circumference of the complete circle, as the angle at $S$ (angle $ASB$) is a fraction of $360°$. For example, if the angle at $S$ is one degree, the length of the arc is 1/360 of the circumference of a full circle. Our approximation is that the length of this segment of arc is almost the same as the distance, 186,000,000 miles, from $A$ to $B$. The difference is much exaggerated in the diagram. In reality, this error in this approximation is fantastically small. (We made an identical approximation in the Appendix section on the moon's distance.)

Now, the best the unaided human eye can distinguish is two stars separated by about four minutes of arc, but not closer separations. Therefore, it was clear to the ancient Greeks that if the earth goes around the sun, the angle at every star, $S$, must be less than four minutes, since no parallax was observed. In this way one can establish a minimum distance from the earth to the star ($AS$, $BS$, or $OS$ will do equally well). All that one has to do is suppose that angle $ASB$ is just barely invisible, i.e., just four minutes. According to our approximation, the arc $AB$ is almost exactly the same as the straight line $AB$, which is 186,000,000 miles.

With this assumption one can calculate the distance to the star. The distance $AB$, 186,000,000 miles, divided by the circumference of the large circle, $2\pi d_s$ ($d_s$ is the distance to the star, therefore, the radius of the large circle), is the same ratio as 4 minutes divided by 60 $\times$ 360 minutes, the number of minutes in a circle:

$$\frac{186,000,000}{2\pi d_s} = \frac{4}{60. \times 360}$$

Therefore (solving for $d_s$):

$$d_s = \frac{186,000,000 \times 60 \times 360}{8\pi} \text{ miles}$$

Word out this distance in miles. It is a "lower limit" to the stars' distance. They may be farther but not closer.

## QUESTIONS

1. How far can you see from an airplane five miles above the ground? (This is a typical cruising altitude for a jet liner. Ignore the effect of the air, which is not taken into account in our formula.)

2. Two people are standing on a perfectly smooth earth. To simplify the arithmetic, suppose their eyes are exactly 5.28 feet above the ground (this is exactly 1/1000 of a mile). According to the formula, how far apart can they be and still see each other? (Do not forget, in this problem, the distance between the two men is $2x$.)

3. The same two people (eyes 5.28 feet high) go to the moon. The moon has no air so the formula corresponds more closely to physical reality. They measure that they can see each other's eye until they are 2.94 miles apart; then they calculate the radius of the moon. What is their answer?

4. Now remember that our formula, $x^2 = 2rd$, is an approximation that should be good when $x$ and $d$ are very small compared to the large distance $r$. We shall see that approximations of this sort play a very fundamental role in the application of mathematics to physics; more examples will occur later. Let us see how big the "error" is. The exact geometrical formula is:

$$x^2 = 2rd + d^2$$

In the above applications, we have neglected $d^2$, because it is small compared to $x^2$ and $2rd$.

Consider the problem of computing the maximum distance, $x$, from which one can see a two-mile-high mountain. Calculate or estimate the error made in ignoring the term $d^2$.

5. From how far away can one see an artificial satellite whose height is 100 miles above the earth's surface? What error is made by ignoring the term $d^2$?

6. To explain the phases of the moon in an Eudoxan universe, a minimum of four spheres are needed; the earth, the moon's sphere, the sun's sphere, and finally the outer stellar sphere. Explain in this context why the full moon appears in different constellations of the zodiac in different months and why the full moon returns to the same constellation of the zodiac after approximately one year.

7. Having read the Appendix section on stellar parallax, work out the distance to the nearest star assuming that its parallax is two minutes of arc.

8. The actual parallaxes of a few nearby stars are much smaller than four minutes of arc. They were first observed in powerful telescopes invented in the 1830s, proving that the earth does move, if not relative to the center of the universe, at least relative to a few nearby stars.

The star with the largest parallax is called Alpha Centauri. It is near the south celestial pole, and cannot be seen in the United States. Its parallax is 3/4 second of arc! How far away is Alpha Centauri?

### REFERENCES

1. Aristotle, *On the Heavens*, translated by W. K. C. Guthrie, in the Loeb Classical Library (Harvard University Press, Cambridge, 1939). Copyright 1939 by the Harvard University Press. Reprinted by permission.

# References

2. Archimedes, *History of the Planetary Systems from Thales to Kepler,* quoted by J. L. E. Dreyer, (Dover, New York, 1953).

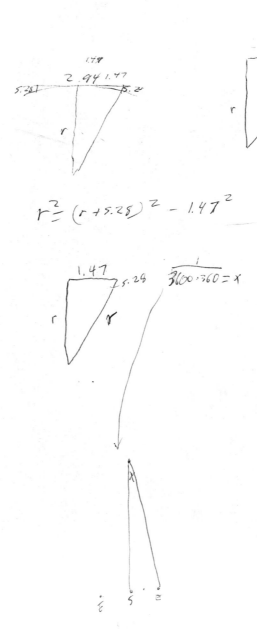

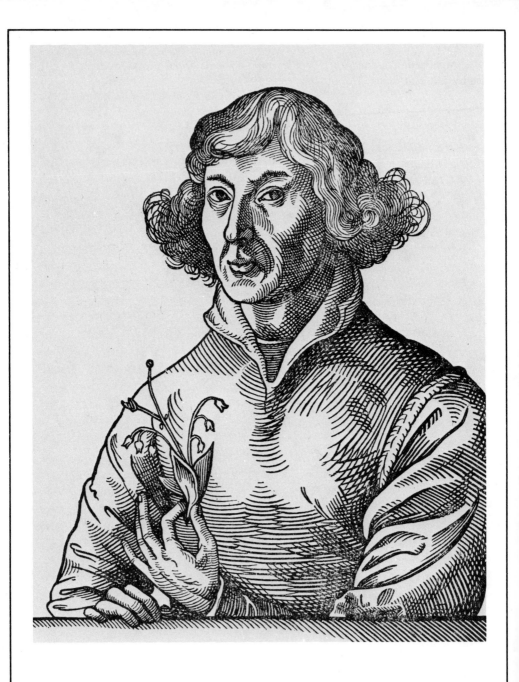

**COPERNICUS**
*(Courtesy of Owen Gingerich, Smithsonian Astrophysical Observatory, and by permission of the Houghton Library, Harvard University)*

CHAPTER

# 3

# PTOLEMY AND COPERNICUS
## Technical Astronomy in the Greek Tradition

Eudoxus and Aristotle created a consistent and spiritually appealing view of the universe. Yet to preserve consistency, Aristotle had to be satisfied with a qualitative physics, one which only gave plausible reasons for the motions in the sky. Even Eudoxus' marvelous geometrical proof that the planetary motions could be thought of as a combination of perfect circular motions did not lead to practical techniques for computing accurate planetary positions.

As time passed, Greek astronomers began to feel that the philosophers would never produce a physics that could explain planetary motions to the accuracy of their observations. Consequently, after Aristotle, a second Greek astronomical tradition developed, which limited its interest to the practical problem of tabulating and predicting the planets' motions. This second tradition reached its highest expression in the work of Ptolemy, whose greatest work, the *Almagest,* became the handbook of mathematical astronomy for thirteen centuries. Both Greek astronomical traditions were bequeathed to posterity, but they led separate lives. Aristotelian cosmology appealed to philosophers and theologians while Ptolemaic astronomy prevailed unchallenged over that very much smaller group of people who calculated planetary positions. So accurate were Ptolemy's calculational techniques that mathematical astronomy remained unchanged until Copernicus.

## PTOLEMAIC ASTRONOMY

Ptolemy worked at the Museum in Alexandria around 150 A.D. In all his work, Ptolemy combined a certain practical sense of the true complexity of the real world with typically Greek mathematical rigor. He wrote books on geography, the mathematical techniques of mapping, on astrology, and on optics. His astrology was fully as influential as his astronomy. Ptolemy's optics was the starting point for Kepler's research on optics some fourteen centuries later. In the *Almagest,* he refined, extended, and summarized a technique for computing accurate planetary positions originated by Hipparchus some four hundred years earlier.

Ptolemy's was a system of wheels rather than spheres. In other words, Eudoxus' three-dimensional spheres were not really needed; two-dimensional circles, which can be drawn on a flat piece of paper, would do. Hipparchus had applied this basic idea to the sun's motion several centuries before Ptolemy. We discussed Eudoxus' theory for the sun in Chapter 2. Eudoxus had placed the sun on the equator of its own rotating sphere. The sun's sphere was attached to the celestial sphere so that the sun's orbit, the ecliptic, makes an angle of 23 1/2° to the equator. Hipparchus noticed that the sun never wanders off the ecliptic; it always remains in the same plane. He could, therefore, replace Eudoxus' solar sphere with a wheel centered on the earth and tilted at 23 1/2° to the celestial equator. If the wheel completes one rotation every 356 1/4 days, the sun's average annual motion around the ecliptic would be described correctly.

One circle could not reproduce the different lengths of the seasons. Eudoxus' follower, Callipas, had added another sphere to make winter shorter than summer. Hipparchus needed a similar device. He proposed that we attach the sun not to the rim of the big annual-motion wheel centered at the earth, but to the rim of a smaller wheel whose center was attached to the annual-motion wheel, as in Fig. 3-1. The big and small wheels were called the *deferent* and the *epicycle,* respectively. If the sun moves around its small epicycle twice a year, twice while the deferent carrying it rotates once, we can reproduce the seasons' inequality. For six months, the sun's motion around the epicycle and the epicycle's motion around the deferent are in the same direction. They therefore add. The other six months, the motion of the epicycle must be subtracted from the motion of the deferent. When the motions add, the sun, viewed from the earth, appears to move quickly through the background stars, and it is the short winter; when the

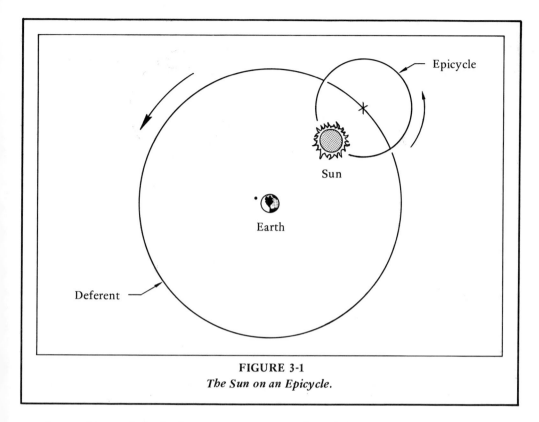

**FIGURE 3-1**
*The Sun on an Epicycle.*

motions subtract, it is the longer summer. By choosing properly the size of the epicycle relative to the size of the deferent, the exact difference in the seasons' lengths could be reproduced. Hipparchus also achieved similarly striking successes using epicycles for the moon.

Ptolemy's greatest contribution was to realize that epicycles naturally account for retrograde motion, the feature most difficult to reconcile with Plato's requirement of perfect circular celestial motions.

All heavenly bodies, including the planets, share the daily rotation of the stars. Since a rotating celestial sphere accounts perfectly well for diurnal motion, the interesting part of a planet's motion is how it moves relative to the background of fixed stars. This is the motion we shall be referring to from now on. Let us see how Ptolemy's system worked for Saturn, the slowest moving planet. Figure 3-2 shows the Ptolemaic model for Saturn's motion. Saturn's deferent rotated slowly about the earth to account for its average motion. Since Saturn takes 29.5 years to circle the zodiac, Ptolemy chose the big wheel—the deferent—to rotate once every 29.5 years. If there were no retrograde motion, Saturn could have been attached to the deferent. Then, it

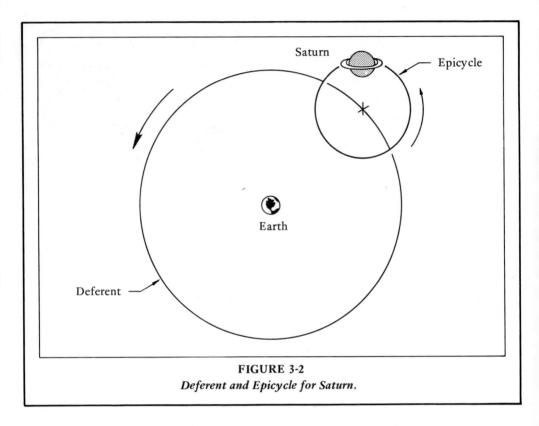

**FIGURE 3-2**
*Deferent and Epicycle for Saturn.*

would travel eastward through the stars at a constant speed, spending about 2 1/2 years in each constellation of the zodiac. But Saturn does retrogress. To understand how epicycles reproduced retrograde motion, remember that without the epicycle, Saturn would move eastward with a constant speed. Next, consider the other extreme. Suppose for a moment that Saturn had no average eastward motion, that the big wheel, the deferent, were stationary and the planet always remained in the same part of the sky. As viewed from the earth, it would appear to move first eastward, then westward, then eastward again; it would oscillate about its center on the epicycle. Now imagine starting the big wheel again. Superimposed on the east-west-east-west oscillation is a steady eastward drift. Each time Saturn starts to move westward, it is a little farther east than at the previous retrogression. Viewed from the earth, Saturn's motion would look something like Fig. 3-3.

As described thus far, Saturn's motions would be all in a line, whereas small loops as in Fig. 3-3 are actually observed. Ptolemy easily reproduced this fact by slightly tilting the epicycle to the deferent.

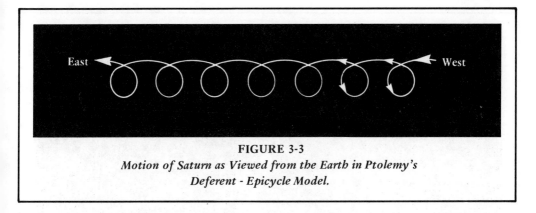

**FIGURE 3-3**
*Motion of Saturn as Viewed from the Earth in Ptolemy's Deferent - Epicycle Model.*

Viewed from above, Saturn's path looked like the curve in Fig. 3-4.

How did Ptolemy choose the period of rotation of Saturn around its epicycle? Each time Saturn was on the part of its epicycle nearest the earth, it was in the middle of its retrograde motion, which happens when Saturn is in opposition. Therefore, the time for Saturn to go around its epicycle once (relative to the line joining the earth and the center of the epicycle) had to be the time between oppositions—Saturn's synodic period of 378 days. The speed Saturn rotated around its epicycle was fixed by the requirement that each retrogression coincide with an opposition.

How big were the wheels? There was no way to tell how far away the deferent was. But, once that was decided, the size of the epicycle was determined.

Figure 3-5 shows why. A line $EC$ from the earth ($E$) to the center of the epicycle ($C$) follows Saturn's average motion. A line from the earth to Saturn, $ES$, makes an angle with $EC$ that is continually changing as Saturn moves around its epicycle. At any time, this angle is the distance in the sky (measured in degrees of arc) that Saturn deviates from its average position, which it would have had if it were moving eastward uniformly. Figure 3-5 shows Saturn when it has the maximum deviation* from its average position. The angle $SEC$ at maximum deviation was known from observation. This allowed Ptolemy to fix the size of the epicycle relative to the deferent.

Ptolemy constructed a similar system of deferent and epicycle for Jupiter and Mars. The deferent rotated with the average period of rotation—12 years for Jupiter and 1.9 years for Mars, and the planet rotated around the epicycle whose center is attached to the deferent with precisely the right speed so that the planet was in the middle of

---

*This angle of maximum deviation for an outer planet like Saturn does not have a specific name, but is obviously analogous to an inner planet's maximum elongation.

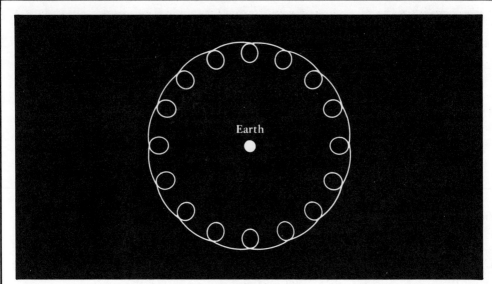

**FIGURE 3-4**
*Saturn's Path in Ptolemy's Model, Resulting from the Combined Motions of the Deferent and the Epicycle.*

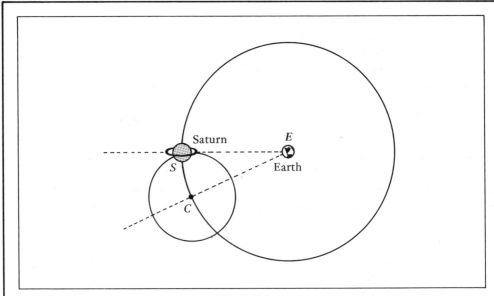

**FIGURE 3-5**
*Saturn at Its Greatest Distance from Its Average Position.*

retrograde motion every 399 days for Jupiter and every 780 days for Mars, the periods between oppositions.

Epicycles and deferents could also explain the motions of Venus and Mercury. Since these planets are never far from the sun in the sky, they average one year to travel around the zodiac. Thus, Ptolemy made the centers of Venus' and Mercury's epicycles rotate around their deferents once a year. Furthermore, since Venus' greatest eastward and westward elongations are equal, 45°, the center of Venus' epicycle could be lined up with the sun at all times. The geometry is sketched in Fig. 3-6. Venus and its system of wheels is entirely within the sun's system. Venus rotates counterclockwise about the epicycle, and the epicycle counterclockwise about the deferent. This direction is eastward. The center of Venus' epicycle, marked $C$, is on the line between the earth and sun; and it always remains there, following the sun about in its annual motion. To keep up its alignment with the sun, the center of the epicycle rotates about the earth with exactly the sun's speed. Therefore, the points $C$, $T$, and $S$ move around the sky with the sun, but their positions relative to one another remain unchanged. Venus rotates about the epicycle once every synodic period, once every 584 days or 83 1/2 weeks. When Venus arrives at $T$ in its motion around the epicycle, its angular distance from the sun reaches its maximum. Venus then is at its greatest elongation, 45°.

When Venus is at $S$, it is a morning star west of the sun. At $T$, it is an evening star. Roughly speaking, the epicyclic motion from $S$ to $T$ is "direct motion" since it is in the same direction as the eastward motion of the whole epicycle around the deferent. When Venus moves from $T$ back to $S$, its epicycle motion is westward relative to the sun. For a good portion of the time Venus spends between T and S, its westward epicycle motion is faster than its eastward deferent motion, and so Venus appears to retrogress.

We have said that the angle $CET$ is 45°. The line $ET$ in Fig. 3-6 is tangent to Venus' epicycle, and we know, therefore, that the radius $CT$ is perpendicular to $ET$. Thus, the triangle is a right triangle and the angle $ETC$ is 90°. Since the three angles of a triangle add up to 180°, and $CET$ plus $ETC$ makes 135°, the remaining angle, $TCE$, is 45°. By entirely similar reasoning, the angle $ECS$ is 45°. Adding $TCE$ and $ECS$, we conclude that Venus traverses 90° of its full epicycle in moving westward from $T$ to $S$, and 270° moving eastward from $S$ to $T$. Thus, Venus should take three quarters of its 83 1/2 week synodic period (more than 60 weeks) to go from east to west relative to the sun. From greatest eastern elongation when Venus is an evening star to greatest western elongation when it is a morning star takes only about

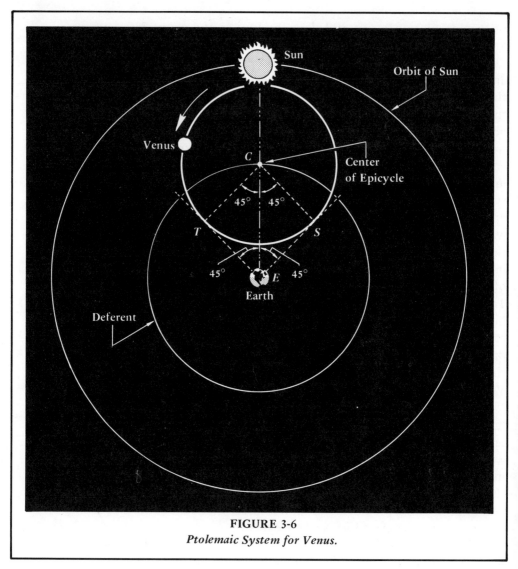

**FIGURE 3-6**
*Ptolemaic System for Venus.*

twenty weeks, or one quarter of its synodic period. Thus, not only do epicycles describe retrograde motion, but when the size of the epicycle is fixed by the requirement that the greatest elongation be 45°, the striking irregularity of Venus' motion—60 weeks eastward, 20 weeks westward—can also be explained. *No new device is needed.**

A powerful attraction of Ptolemy's basic system was its simplicity—that retrogression was a natural consequence of a simple combination of perfect circular motions. Furthermore, the fact that the outer planets are brightest during retrogression seemed to fall

---

*Similar considerations lead to a good theory for Mercury, but since the mathematical details are somewhat more complicated, this is discussed in a chapter Appendix.

right out of the theory. Recall that by the very design of the deferent–epicycle system, a planet is in the middle of its retrograde motion when it is nearest the earth. Now, if the Ptolemaic orbits are physically real and not just a mathematical device, the planets should appear brightest to an observer on earth during retrogression.

The simplicity of the Ptolemaic system vanished in practice, for while the combination of deferent plus epicycle reproduced the basic features of retrograde motion, the planets' motions still deviated in fine details from the simple calculations. Most of Ptolemy's efforts were spent in constructing ingenious mathematical schemes to make the theory account more exactly for the observations. One device was to set the planet rotating around another small or minor epicycle, whose center was in turn rotating about the basic major epicycle, which, of course, rotated around the deferent. Adding more and more epicycles tended to increase the accuracy of the theory and the amount of fine detail it accounted for.

Another problem Ptolemy struggled with was the fact that even with the minor epicycles the planets do not really move with constant speed around their deferents. Rather, they seem to speed up and slow down slightly in a continuous fashion. Now, everyone until the time of Sir Isaac Newton had insurmountable difficulties with the problem of continuously varying speed. Ptolemy developed several ingenious schemes whereby a planet's speed would appear to vary continuously to an observer on earth, but whose computation still involved only easy-to-compute, celestially perfect, uniform circular motions. One way to do this was to put the earth slightly off the center of the deferent circle, as in Fig. 3-7. The planet and all its epicycles have uniform motion about the deferent's empty center; to an observer on the earth the planet would appear to be moving faster when it is closest to the earth (ignoring the epicyclic motions). Another device was to renounce the idea of uniform speed around the circle, but to say that there is one point, the "equant point", around which the planet has a uniform speed. In Fig. 3-8, imagine a clock hand rotating uniformly about the equant point, $Q$. The center of the planet's major epicycle follows the point where the clock hand intersects the circle. This keeps the earth at the center, but produces varying speeds as viewed from the earth. The planet appears to move fastest when it is farthest from the equant, slowest when it is closest. The equant was a device particularly repugnant to Copernicus.

Slowly but surely the beauty of the Ptolemaic system ebbed away. With epicycle piled upon epicycle, the spirit if not the letter of Plato's laws that heavenly motions be circular was violated. With the earth slightly off the center of a planetary deferent, the celestial material

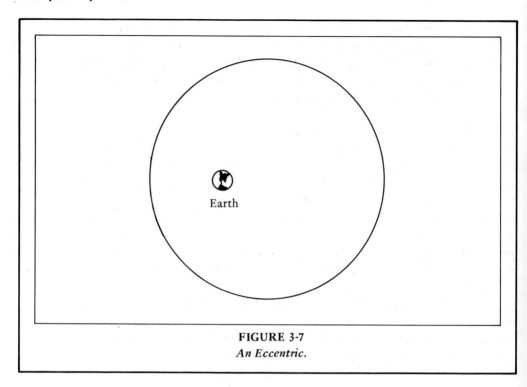

**FIGURE 3-7**
*An Eccentric.*

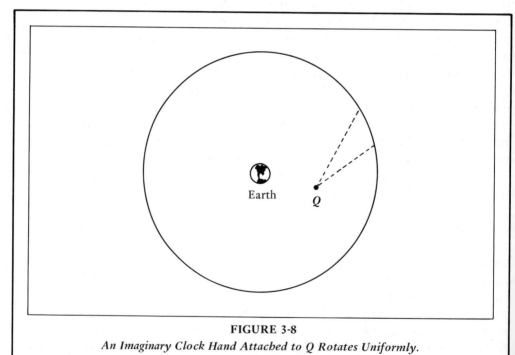

**FIGURE 3-8**
*An Imaginary Clock Hand Attached to Q Rotates Uniformly.*

no longer rotated in circles about the center of the universe, as Aristotle had maintained was natural for celestial material. With equants, the earth could still be at the center, but then the motion around the deferent was non-uniform and, therefore, imperfect. Ptolemy's devices were resoundingly successful in accounting for small deviations from the basic simple theory. They saved the appearances. The question was, should we revise our philosophy and physics because of a few small irregularities in the planets' motions?

Ptolemy's theory had its arbitrary aspects. For example, Ptolemy, just as Eudoxus, constructed the theory for each planet's motion separately. No unified prescription existed to delineate how to build each planet's theory. Furthermore, as the centuries passed and Ptolemy's mathematical devices continued to be used, different theories of the same planet were constructed. Two different astronomers could use two different combinations of epicycles, equants, and whatnot to arrive at models of a planet's position that agreed with observation. What then? Which one was really real?

By the thirteenth century, the complexities of the Ptolemaic system outweighed whatever fundamental simplicity had tantalized Hipparchus and Ptolemy. In 1252, Alphonso the Wise, King of Castile, sponsored a new set of astronomical tables based on a late version of Ptolemaic theory. These Alphonsine Tables were for several hundred years thereafter the best available. From the struggle to produce the Tables, the king himself learned enough astronomy to remark:

> If the Lord had consulted me before embarking upon the Creation, I should have recommended something simpler.

By the seventeenth century it was clear that the Ptolemaic system was a construction of astronomers, not a reasonable model of the universe. In Book VIII of John Milton's *Paradise Lost,* Adam asks the angel Raphael to explain to him how the universe is laid out. Here is the angel's reply:

> From man or angel the great Architect
> Did wisely to conceal, and not divulge,
> His secret to be scanned by them who ought
> Rather admire; or, if they list to try
> Conjecture, he his fabric of the Heavens
> Hath left to their disputes, perhaps to move
> His laughter at their quaint opinions wide
> Hereafter, when they come to model Heaven
> And calculate the stars, how they will wield
> The mighty frame, how build, unbuild, contrive
> To save appearances, how gird the sphere
> With centric and eccentric scribbled o'er,
> Cycle and epicycle, orb in orb.

## COPERNICUS

Ptolemy was the last great astronomer of classical antiquity; Copernicus, the first in modern times. It is astounding that we can so easily pass over the thirteen centuries separating them. While these centuries had many stargazers, none could add significantly to Ptolemy's accomplishments, none could challenge Ptolemy's basic assumptions; none, that is, until Copernicus. And yet today, some still argue that Copernicus was really the last Ptolemaic astronomer rather than the first modern astronomer, since the full force of Copernicus' ideas was not felt until a century after his death. Copernicus' aim was to use Ptolemaic techniques in constructing planetary orbits with the sun, rather than the earth, at the center. Two considerations were uppermost in his mind: would sun-centered calculations either be easier, or would they agree with observation better, than earth-centered calculations?

Nicolas Copernicus was born in Torun, Poland, in 1473. His studies at the University of Cracow were without distinction. Upon graduation, his uncle Lucas, who was a Bishop, appointed him Canon of Frauenberg Cathedral, a job with good income and virtually no duties. Copernicus immediately left for Italy where he studied for ten years at the Universities of Bologna and Padua, again without special distinction. In the intellectual ferment of Renaissance Italy, he encountered the speculation that perhaps the sun, and not the earth, was at the center of the universe. But, no calculations existed like Ptolemy's to back it up. Apparently, Copernicus decided during his stay in Italy to make the hypothesis of a sun-centered universe the basis of a new calculation of planetary positions. He was to pursue this goal, in solitude and isolation, for 36 years.

In 1506, he returned to Poland. Lucas freed him from his onerous duties at Frauenberg to make him his private physician. His obligations—medicine and seeing to some of Lucas' diplomatic correspondence—left him plenty of leisure. With his leisure he prepared an outline of his system of the universe, which never was published in his lifetime. In 1512, Bishop Lucas died, and Copernicus was finally obliged to take up his official duties at Frauenberg Cathedral. In 1530, or thereabouts, he wrote his book, *On the Revolutions of the Heavenly Spheres,* only to lock it away. He would take it out now and again to fuss with it, add a point here and there. But he would not publish. Nevertheless, people knew that Copernicus was working on radically new planetary calculations. Despite himself, he had a reputation. Rheticus, a young German firebrand, became his disciple. Having learned of Copernicus' work, Rheticus had come to Poland to find out more about it. In 1539, he published his own sketchy account of the Copernican system. Even with this prod, Copernicus was reluctant to let his complete work out. Finally, in 1541, Rheticus edited the manuscript, and took it back to Germany to be set in type. The two most influential books in the history of modern science, Copernicus' *Revolutions* and Isaac Newton's *Principia* would never have seen the light of day unless others had seen to the editing and publication of the manuscripts. Copernicus saw a printed copy of his life-work only on his deathbed.

The historian of science, T. S. Kuhn, has called the chain of events initiated by Copernicus *The Copernican Revolution;* yet, by all accounts, Copernicus—sour, solitary, hesitant, and conservative—was no revolutionary in spirit. Unlike other great scientists, his interests were exceptionally narrow, limited to computing planetary orbits from a sun-centered point of view. Why then, after thirteen centuries of Ptolemy, would someone as unlikely as Copernicus produce a revolutionary treatise? At least two kinds of answers are possible to this question. First, one can argue that Copernicus' followers were in fact more revolutionary than he. We will see some truth in this view. On the other hand, perhaps the spirit of the times was radical enough to force even Copernicus to think in a new way. He was a contemporary of Michelangelo, of Columbus, and of Martin Luther. The art of painting had changed; the new interest was in perspective, the accurate depiction of space. Did the artistic interest in three-dimensional space motivate Copernicus to create a three-dimensional picture of the sun, moon, and planets in their orbits? New continents were being explored; even the geography of the earth was not as Ptolemy had written. Before Copernicus' death, Magellan had added a convincing proof to the traditional abstract arguments that the world was round: his ship had sailed around it. It was a time for new ideas. Even the once unchallenged theology of the Catholic Church was attacked by Martin Luther. And new ideas spread rapidly, for now books were printed.

## ON THE REVOLUTIONS OF THE HEAVENLY SPHERES

In the preface to his work, *De Revolutionibus Orbium Caelestium*, Copernicus tells us why he devoted thirty-six years to reformulating planetary astronomy. First of all, the calculations were becoming inaccurate. Small errors in either observation or calculation that were unnoticed by Ptolemy led to increasingly noticeable deviations as time passed, just as a watch that misses only a second a day will still be off by several days after a thousand years have passed. Moreover, over the centuries copying errors had crept into the tables of observations.

His interest seems to have been stimulated by a request from Pope Leo X to join a commission studying the reform of the calendar. The calendar, established by Julius Caesar, was off by a few days; the beginning of spring no longer occurred on March 21. Copernicus declined, saying that astronomy was in such a scandalous state that the accurate calculations needed for calendar reform were unavailable. Copernicus also thought that the precession of the equinoxes should first be understood much better, with fifteen more centuries of observations to add to the knowledge of the ancients.*

Copernicus then remarked that the attempts of his contemporaries to modernize the calculations of planetary positions were not successful. The Eudoxan system was unable to produce accurate planetary positions, however satisfying philosophically it might be. Those who used the Ptolemaic system produced results that were more accurate but still not good enough. Copernicus reserved his bitterest criticism for the devices and artifices of Ptolemy's followers. Not one but many Ptolemaic theories had sprung up, distinguished by different arbitrary choices of epicycles, equants, and eccentrics, but not by greater calculational ease or accuracy. There was no way, no reasonable criterion, for choosing among the various systems. Of the many different calculational schemes, which, if any, was physically real? Copernicus' final criticism was esthetic: Ptolemaic systems had no beauty or symmetry. Ptolemy's artifices repulsed him.

In book I of *De Revolutionibus,* Copernicus outlines the philosophical and physical reasoning underlying his gigantic computational effort. We recount this briefly, not because his reasoning would be considered correct today, but because it places him in an Aristotelian tradition in all but planetary orbit calculations. First of all, the stars are fixed on a sphere. His reason for this is the usual one: celestial objects partake of perfection, spheres are the most symmetrical, perfect geometric objects. The earth is spherical; this was known, experimentally, from the time of the Greeks. The motions of the planets ought to involve uniform, circular motions. While the planets' motions are irregular, they *are* periodic and recurrent. To Copernicus only a combination of circular motions seemed able to explain recurrent planetary motions. Finally, Copernicus states an Aristotle-like argument: the natural motion of spheres is rotation about an axis.

Since the earth is a sphere, we must consider the possibility that it rotates on its axis. Let us disregard for the moment *physical* arguments

---
*The details are explained in the chapter Appendix.

such as Ptolemy's—if the earth rotated all the objects would be flung into space—and ask whether any purely *astronomical* evidence argues against the earth's rotation. Here the answer is no: the diurnal motion of the stars could be explained either by a rotation of the celestial sphere or one by the earth in the opposite direction. The two schemes are compared in Fig. 3-9. In the Greek System, the earth was fixed at the center. Consequently, the horizon of any observer was also fixed. The celestial sphere rotated westward, carrying the stars with it, with a twenty-four hour period. On the other hand, if the earth rotated eastward, carrying the horizon around with it, once every $23^h\ 56^{min}$, the results would be entirely equivalent. On the basis of the observations of diurnal motions, there was *no* way to choose between the two hypotheses. Copernicus' system reversed the roles of the celestial and terrestrial poles. In the Greek earth-centered system, the celestial sphere rotates about an axis that passes through the north and south celestial poles. In Copernicus' system, it is the other way around. The celestial poles just happen to be above the earth's poles. Similarly, the celestial equator is the circle in the sky directly over the earth's equator. Nevertheless, despite this reversal, the two models made the same predictions. However, one feature of an esthetic nature pleased Copernicus. A diurnal rotation is found in all the stars, planets, and, indeed, every object in the universe; it seemed simpler to assign it once to the earth. Physically, it was ridiculous to have the distant stars whizzing around the earth every 23 hours and 56 minutes, all at the same enormous speed.

If the earth rotates on its axis, why not also consider the possibility that it has other motions as well? If the earth moves through the heavens, it is a heavenly body; therefore, its motion is likely to be circular. Most likely, the center of its motion is the center of the universe. One of the ancient arguments against the earth's motion had been that the fixed stars had no parallax. If the earth moves in a large circular orbit about the center of the universe, the argument went, you could observe the same star from two different points in the earth's orbit to get a parallax measurement. The new baseline would be the diameter of the earth's orbit, much larger than the diameter of the earth. Their inability to measure stellar parallax convinced the ancients that the earth could not move. However, argued Copernicus, suppose the stars were *very* far away, much farther away than anyone thought? The absence of parallax implied *either* that the earth was at the celestial sphere's center *or* that the accuracy of the measurements was not great enough to detect a very distant celestial sphere. Consider for a moment this second possibility. Then we can ask once again: "Is there any astronomical evidence against the earth's motion?"

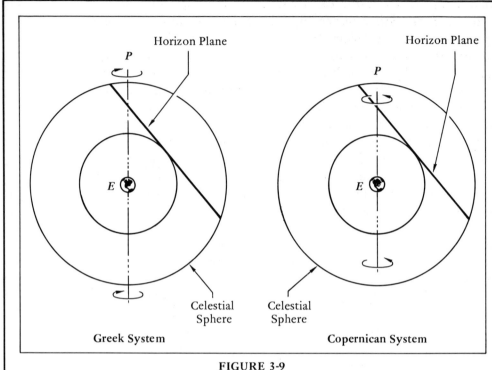

**FIGURE 3-9**
*Motion of the Stars in the Greek and Copernican Systems.*

In Ptolemy's earth-centered system, the sun was special, even though it was just one of the seven planets circling the earth. Venus and Mercury circle the sky with the same average period as the sun—one year. Ptolemy forced the centers of their epicycles always to line up with the sun, and, therefore, to rotate around the deferent once a year. The epicycles of Mars, Jupiter, and Saturn were also synchronized with the sun's motion so that they would be in the middle of retrograde motion when they are in opposition. Perhaps this is no accident; perhaps the sun not only organizes the planets' but also the earth's motion; perhaps it is the sun that is at the center of the universe. But if the sun is at the center of the universe, why do we see it move relative to the stars? Copernicus answered that the earth is a moving observation platform, so that we see the sun against different stars in the background at different times, as shown below in Fig. 3-10.

Is there any difference in the astronomical observations of the solar position with respect to the stars between the two systems? The answer is no. The annual motion of the sun, relative to the background stars, is entirely equivalent in the two systems. No observational

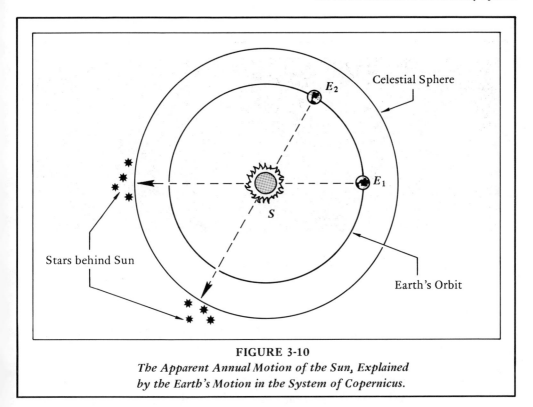

**FIGURE 3-10**
*The Apparent Annual Motion of the Sun, Explained by the Earth's Motion in the System of Copernicus.*

way was available to choose between them. However, choosing the earth to move meant you had to include a 365 day periodicity only once—for its motion—rather than for *all* planets.

Finally, it was necessary to explain why the ecliptic is tilted at an angle of 23 1/2° to the celestial equator. Copernicus' earth is tilted, as in Fig. 3-11. Its axis of rotation is not perpendicular to the plane of its orbit around the sun—the ecliptic plane, but it is inclined at an angle of 23 1/2°. As the earth moves, its axis always points to the celestial poles. That is why the sun appears north of the celestial equator in northern summer, and south in winter. Copernicus gave the earth a third motion to keep its axis in the same direction.

Another feature of Copernicus' system was a simple picture of the precession of the equinoxes. We have already noted that it is likely that this tiny phenomenon, and the related difficulties with the calendar, were Copernicus' principal motivation to suggest a heliocentric universe. In order not to lose the main thread of our argument, the precession of the equinoxes and the problems of the calendar are deferred to the Appendix.

Copernicus' greatest triumph was a simple account of retrograde motion. The central idea was that all the planets, including earth,

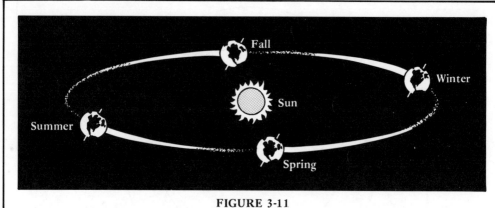

**FIGURE 3-11**
*The Earth's Axis Always Points in the Same Direction As It Moves Around the Sun.*

orbit the sun. To understand observations of the other planets, one has to realize that not only are they moving but that our observation platform, earth, is moving also. Earth's motion is mixed into the planetary observations. Here was Copernicus' great opportunity to simplify astronomy. The major epicycles in Ptolemy's calculations were really due to the earth's motion. Copernicus could reduce the number of circles needed, using one for the earth alone, rather than one for each of the other five known planets. Moreover, Copernicus' system *required* a certain ordering of the planets' orbits around the sun. The order of the planets is as follows: Mercury, Venus, Earth, Mars, Jupiter, and Saturn (see Fig. 3-12). The inner planets really move inside the earth's orbit, the outer planets outside; and Copernicus' system required that planets closer to the sun move faster.

Here are a few paragraphs of Copernicus' own description, from the introduction to his book:[1]

> We therefore assert that the center of the Earth, carrying the Moon's path, passes in a great orbit among the other planets in an annual revolution round the sun; and that near the sun is the center of the universe; and that whereas the sun is at rest, any apparent motion of the sun can be better explained by the motion of the earth. Yet so great is the universe that though the distance of the earth from the sun is not insignificant compared with the size of any other planetary path, in accordance with the ratios of their sizes, it is insignificant compared with the distance of the sphere of the fixed stars.
>
> I think it is easier to believe this than to confuse the issue by assuming a

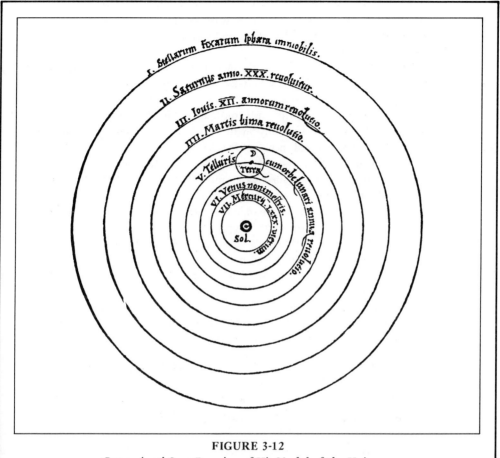

**FIGURE 3-12**
*Copernicus' Own Drawing of His Model of the Universe.*

vast number of spheres, which those who keep the earth at the center must do. We thus rather follow Nature, who producing nothing vain or superfluous often prefers to endow one cause with many effects. Though these views are difficult, contrary to expectation, and certainly unusual, yet in the sequel we shall, God willing, make them abundantly clear, at least to mathematicians.

Given the above view, . . . then the order of the spheres, beginning from the most distant, is as follows. Most distant of all is the sphere of the fixed stars, containing all things, and being therefore itself immovable. It represents that to which the motion and position of all the other bodies must be referred. Some hold that it too changes in some way, but we shall assign another reason for this apparent change, as will appear in the account of the earth's motion. Next is the planet Saturn, revolving in 30 years; next comes Jupiter, moving in a 12 year circuit; then Mars, who goes around in 2 years. The fourth place is held by the annual revolution in which the earth is contained, together with the orbit of the Moon as on an epicycle. Venus, whose period is nine months, is in the fifth place, and sixth is Mercury, who goes round in the space of 80 days.

In the middle of all sits Sun enthroned. In this most beautiful temple could we place this luminary in any better position from which he can illuminate the whole at once? . . . .

So we find underlying this ordination an admirable symmetry in the universe, a clear bond of harmony in the motion and magnitude of the orbits such as can be discovered in no other wise, for here we may observe why the progression and retrogression appear greater for Jupiter than Saturn, and less than for Mars, but again greater for Venus than for Mercury; and why such oscillation appears more frequently in Saturn than in Jupiter, but less frequently in Mars and Venus than in Mercury; moreover why Saturn, Jupiter and Mars are nearer to the earth at opposition to the sun than when they are lost in or emerge from the sun's rays. Particularly Mars, when he shines all night, appears to rival Jupiter in magnitude, being only distinguishable by his ruddy color . . . . All these phenomena proceed from the same cause, the earth's motion.

That there are no such phenomena for the fixed stars proves their immeasurable distance, compared to which even the size of the earth's orbit is negligible, and the parallactic effect unnoticeable . . . .

The important general rule is that the farther out a planet is the slower it goes. Thus, consider the outermost planet known to Copernicus, Saturn. If the earth were not moving, we would see Saturn slowly go around the sky, against the background of the stars. The time it would take to go around would be the time it takes Saturn to complete one orbit. Therefore, the period of Saturn is the apparent average time it takes Saturn to go all the way around the zodiac, namely, 29.5 years. In fact, since the earth's orbit is very much smaller than Saturn's, Saturn's motion does appear fairly steady; the changing position of the observation platform—the earth—is not very important.

What *is* the effect of the earth's motion? Saturn moves very slowly indeed, so for a moment consider it to be stationary. It is still much closer than the distant "fixed" stars so as the earth goes around the sun the parallax effect is very large. Saturn would appear to oscillate back and forth, once each year, relative to the star background. The earth and Saturn both go around the sun in the same direction. Seen from the sun, they would both appear to be moving "eastward," Saturn slowly, the earth much more rapidly. In Ptolemy's system, retrograde motion was produced by combining two motions, the eastward motion around the deferent, and circular motion around the epicycle. This epicyclic motion usually added to the eastward motion. In Copernicus' system, retrograde motion is also a result of compounding two motions; but now these are the motions of Saturn and the motion of the earth. Saturn's gives a steady eastward drift. The earth's, like the epicycle, gives a shift in Saturn's apparent position (against the background of the distant stars) that sometimes adds to Saturn's eastward motion, and sometimes cancels it to produce apparent westward (retrograde) motion.

The observations of Jupiter and Mars are explained in the same way. Figure 3-13 shows the earth's and Mars' orbits drawn roughly to scale. Figure 3-13, plus the fact that Mars moves more slowly than the earth, not only explains Mars' retrograde motion but also explains why retrogression happens at opposition and why Mars is brightest then, all without major epicycles. Figure 3-14 shows part of the orbits in more detail.

In Fig. 3-14, both planets move counterclockwise, or eastward as viewed from the sun. When the earth and Mars are on opposite sides of the sun, the earth is moving in a direction opposite to Mars' direction, so the apparent speed of Mars is then faster than it would be as observed from a stationary platform like the sun. But, when the earth is between the sun and Mars, as in Fig. 3-14, the earth's motion causes Mars to appear to move backward, i.e., westward—against the distant stellar background. Why does Mars not appear just to slow down rather than actually stop and move westward? Because inner planets move faster than outer ones! In Fig. 3-14, the earth's positions are spaced farther apart than Mars' positions, since the earth is moving faster. Normal, direct, eastward motion is observed from 1 to 2. As the earth overtakes Mars (2 to 3), Mars appears to move westward, and retrogression occurs. Finally, (3 to 4) normal eastward motion is resumed.

So far Copernicus had merely reproduced the motions that Ptolemy had accounted for with his epicycles. The elimination of the epicycles might have been convincing enough. But, now something marvelous has occurred. Retrogression happens automatically when the earth is between Mars and the sun, i.e., when Mars is in opposition! During the middle of retrogression (between 2 and 3) in Fig. 3-14, Mars is opposite the sun; it is, therefore, highest in the sky at midnight. Ptolemy had to impose this rule arbitrarily; it is not arbitrary in the heliocentric system of Copernicus. Furthermore, we can understand why Mars—or for that matter Jupiter or Saturn—is brightest near opposition. At opposition, the earth is closer to Mars than at any other time and so it appears brighter.

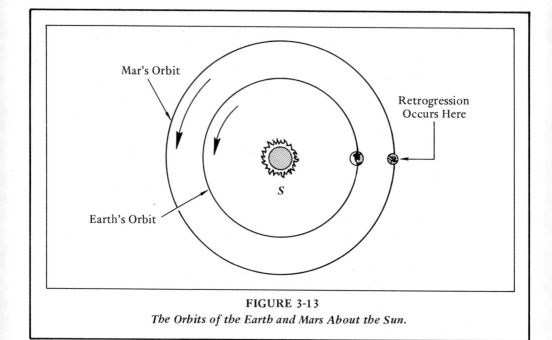

**FIGURE 3-13**
*The Orbits of the Earth and Mars About the Sun.*

The apparent retrograde motions of the inner planets, Mercury and Venus, can also be explained as a combination of their own and the earth's orbital motion.

Figure 3-15 shows the apparent positions of Venus in the heliocentric model. Venus moves faster than the earth, so the spacings between its positions are larger than the earth's at the same time. Venus appears to be in direct motion from positions 1 to 2 and from 3 to 4, but retrogresses between positions 2 and 3, when Venus overtakes the earth. In this way Copernicus' system explains why retrogression occurs when it does.

Now we come to the quantitative, numerical description of Copernicus' system. Two features are important: the sizes of the planets' orbits, and their periods of revolution about the sun. The fact that the radii of the planets' orbits could be determined was a novel feature of Copernicus' system, and one which he held to be very important, as is evident from the passage quoted from his introduction. Recall that in Ptolemy's universe there was no very good reason for the order of the planets and the sun and moon, as one goes out from the earth. In Copernicus' heliocentric universe, not only is the order fixed, but one can easily calculate the distance of each planet from the sun.

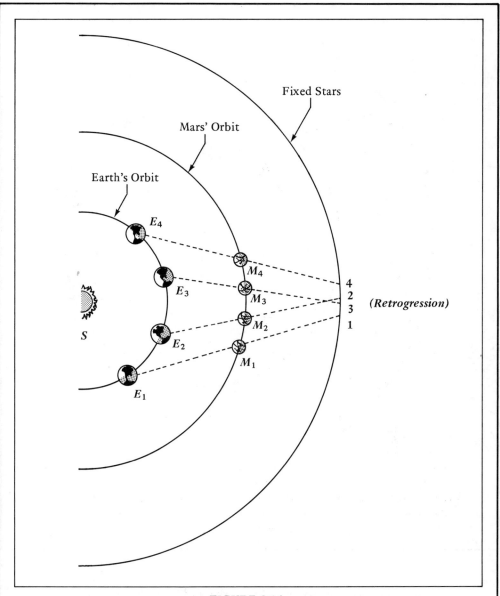

**FIGURE 3-14**
$E_1$, $E_2$, $E_3$, and $E_4$ Are the Earth's Position at Four Points on Its Orbit, Equally Spaced in Time. $M_1$, $M_2$, $M_3$, and $M_4$ Are Mars' Positions at the Same Four Times. The Numbers, 1, 2, 3, and 4, Are the Apparent Positions of Mars among the Stars, as Viewed from the Earth, at Those Times.

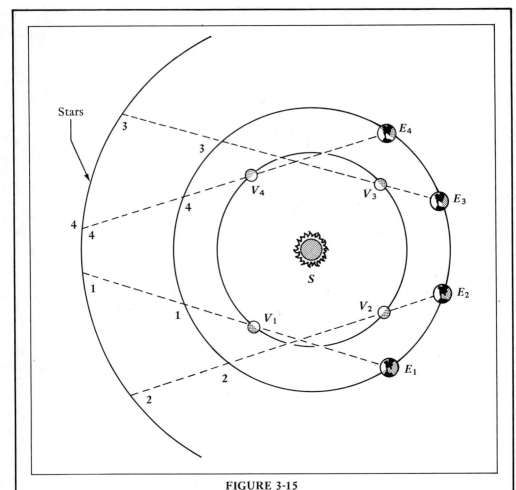

**FIGURE 3-15**
$V_1$, $V_2$, $V_3$, and $V_4$ Are Venus' Positions at Four Different Times, and $E_1$, $E_2$, $E_3$, and $E_4$ Are the Earth's at Those Times. The Numbers, 1, 2, 3, and 4, Indicate Venus' Apparent Positions against the Stellar Background.

This collection of six numbers, the radii of the planets' orbits, was a new and important mathematical fact known about nature. Copernicus could not calculate planetary distances in miles, since he did not know how far the earth is from the sun. But, he was able to calculate the radii relative to the earth's distance from the sun. This distance is still a convenient unit for measuring distances in the solar system, and it is called the astronomical unit (A.U.). We will measure all distances in A.U. We now know that 1 A.U. is 93 million miles, but Copernicus thought it was considerably smaller.

We illustrate the calculation for Venus. Venus' maximum elongation is about 45°; the configuration of maximum elongation is illustrated in Fig. 3-16. Since Venus (V) is at greatest elongation, the line EV must be tangent to the circle of Venus' orbit. The angle at E is observed to be 45°. Since VS is a radius, VS is perpendicular to the tangent VE, and the triangle SVE is a right triangle, with the right angle at V. The sum of the angles of a triangle is always 180°. If one angle is 90°, and one is 45°, the third must be 180° − 90° − 45° = 45°. Therefore, the angle at S is also 45°. Since the angles at S and E are the same, SVE is an isosceles right triangle: VS = VE. But Pythagoras' theorem tells us:

$$(VS)^2 + (VE)^2 = (SE)^2$$

And, since for Venus, VS = VE:

$$2(VS)^2 = (SE)^2$$

Now, let $R_v$ denote the radius of Venus' orbit, and $R_e$ the radius of the earth's orbit. In the diagram, $R_v$ = VS, $R_e$ = SE, and we rewrite the above equation with this new notation:

$$2R_v^2 = R_e^2$$

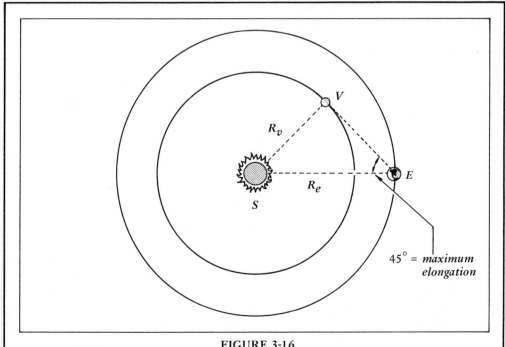

**FIGURE 3-16**
*Venus at Greatest Elongation.*

Dividing both sides by $2R_e^2$, we get:

$$\frac{R_v^2}{R_e^2} = \frac{1}{2}$$

And, taking the square root:

$$\frac{R_v}{R_e} = \frac{1}{\sqrt{2}}$$

This calculation determines only the ratio $R_v / R_e$. Since $1/\sqrt{2}$ is about 0.71 and $R_e = 1$ astronomical unit by definition, $R_v = 0.71$ A.U.

A similar argument can be made for each of the other planets. In general, some elementary trigonometry is needed. The case of Mercury is worked out in a chapter Appendix. The distances of the six planets from the sun are given in Table 3.1. The unit of length is the astronomical unit.

The other important collection of numbers is the list of the periods of revolution of each planet about the sun, i.e., the planets' "years." The easiest period for earth-bound astronomers to measure is the synodic period—between oppositions for the outer planets, and between greatest eastern, or western, elongations for the inner planets.

TABLE 3.1. *The Distances of the Planets from the Sun (in A.U.).*

| Planet | Distance from Sun |
| --- | --- |
| Mercury | 0.39 |
| Venus | 0.71 |
| Earth | 1.00 |
| Mars | 1.52 |
| Jupiter | 5.20 |
| Saturn | 9.54 |

But, the real period of revolution of a planet is its own year, its period relative to the stars, called its "sidereal" period. Copernicus demonstrated that from the knowledge of a planet's observed synodic period, one could *deduce* its sidereal period. Let us consider the outer planets first. The period between oppositions for Mars, Jupiter, and Saturn are 780 days, 399 days, and 378 days, respectively. Saturn's synodic period is just 13 days longer than one earth year. We understand this as follows: If Saturn were not moving, but were at rest like a fixed star, it would be in opposition every 365 days. The extra 13 days must be the time it takes the earth to catch up with Saturn because of Saturn's slow motion around the sun. In the extra 13 days, the earth moves 13/365 of the way around its orbit, while Saturn moves hardly at all. Therefore, in one earth year, Saturn moves 13/365 of the way around its orbit also. How long does it take Saturn to go all the way around? Since it completes 13/365 of its orbit each earth year, one orbit must take 365/13 earth years, or about 29 earth years. It is no surprise that this is the average time it takes Saturn to make one trip around the zodiac. This calculation is not exactly correct, since we ignored the fact that during the 13 days it takes the earth to catch up with Saturn, Saturn is slowly moving; if this fact is taken into account and if the calculations are performed accurately, the sidereal period of Saturn comes out almost exactly 29.5 years. The method is explained in the Appendix.

Jupiter's period is calculated to be 12 years; Mars' period is very close to two years. If it were exactly two years, the interval between oppositions would also be two years: One year after opposition, the earth would have gone all the way around the sun, and Mars half-way, so that they would be on opposite sides of the sun; one year later, they would be together again. Actually the period between oppositions is 780 days, a bit more than two years; the sidereal period of Mars come out to be 1.9 years, slightly less than two years. The Appendix shows how to do the computation exactly.

How to calculate the sidereal period of the inner planets is less obvious, since it is not their average time to circle the sky. For example, we know that Mercury, the sun, and the earth repeat the same configuration every 116 days. Now, Mercury's period of revolution must be shorter than 116 days, since when Mercury has gone round the sun once the earth has also moved, so it takes a while for Mercury to catch up. We can make a rough estimate of Mercury's real period by noticing that 116 days is slightly less than one-third of a year, so Mercury has to go around the sun an extra one-third of its journey to catch up with the earth. Therefore, 116 days is 4/3 of Mercury's period of revolution, $116 \times 3/4$ or 87 days.

For the inner planets, just as for the outer planets, this approximate argument can be replaced by an exact rule that is quite simple. In this way the periods of revolution of all the planets can be calculated. (Mercury, Saturn, and Mars were particularly simple examples.) Table 3.2 is an extension of Table 3.1. The first column, labeled $R$ (for radius), repeats the distance of each planet from the sun. The next column gives the period of revolution (labeled $T$ for time). A new

TABLE 3.2. *Radii of Orbits, Periods, and Speeds for the Six Planets, according to the Copernican Theory.*

| Planet  | Radius R   | Sidereal Period T | Speed (A.U./yr) V |
|---------|------------|-------------------|-------------------|
| Mercury | 0.39 A.U.  | 88 days           | 10.15             |
| Venus   | 0.72 A.U.  | 225 days          | 7.22              |
| Earth   | 1.00 A.U.  | 1.0 years         | 6.28              |
| Mars    | 1.52 A.U.  | 1.9 years         | 5.03              |
| Jupiter | 5.20 A.U.  | 12.0 years        | 2.72              |
| Saturn  | 9.54 A.U.  | 29.5 years        | 2.03              |

regularity has appeared: the farther from the sun, the longer it takes a planet to orbit. We will use the letter, $V$ (from "velocity"), for speed. Remember the definition of speed: for an object moving with a steady speed, its speed is the distance it goes in a certain time, divided by that time. The most convenient time interval to use is the planet's period of revolution, $T$. How far does it go in time $T$? The distance is the circumference of its orbit, which, according to the rule relating the circumference to the radius of a circle, is $2\pi R$. For example, Saturn's $R$ is 9.54 A.U., so the circumference of its orbit is 2 × 3.14 × 9.54 A.U. = 60 A.U. Saturn's speed is, therefore, 60 / 29.5 or 2.03 / A.U. per year. In general, a planet's speed is given by the formula:

$$V = 2\pi \frac{R}{T}$$

The speed comes out in astronomical units per year.

Table 3.2 contains an important new fact. The calculated speed of a planet is smaller the farther it is from the sun—a striking check of Copernicus' heliocentric model. First, he showed that retrograde motion could be explained if the planets' speeds decreased with their distance from the sun. Then, using his model, he found that the observations could be explained only if each planet had a special orbital radius, $R$, and a sidereal period, $T$. From $R$ and $T$, he could calculate their speed. If the calculated speeds did not decrease with distance from the sun, his model would have been wrong.

### COMPARISON OF PTOLEMAIC AND COPERNICAN ASTRONOMY

How was one to decide between the two great world systems? Copernicus believed that the greater harmony of his system was in its favor. What can he have meant? Both models explained the observed apparent motions of the sun and stars. In fact, for explaining an individual planet's motion, about the same difficulty is encountered in each system. Two circles were required to explain retrogression—an epicycle and deferent for Ptolemy, the planet's orbit and the earth's orbit, for Copernicus.

Copernicus cannot have meant that his calculations were simpler than Ptolemy's, for they were not. While Copernicus had hoped that assigning the function of Ptolemy's epicycles to the earth's motion around the sun would reduce the number of circles needed, trouble came when he tried to increase the accuracy of his computations. He was forced to use many of Ptolemy's mathematical devices after all—a bitter irony. His computations were neither simpler nor much more accurate than Ptolemy's, yet as complicated. The main difference was that his were sun-centered, Ptolemy's earth-centered.

For Copernicus, the true harmony was that his theory could systematically derive more facts from fewer assumptions. For example, Ptolemaic astronomy constructed a separate theory for each planet, whereas, for Copernicus, each planet including the earth was on the same footing. He started the theory for each planet the same way. Ptolemy had to introduce arbitrary linkages between the planets and the sun in order to produce retrogressions at the proper times; for Copernicus, the occurrence of retrogressions at alternate conjunctions for inner planets (and oppositions for outer planets) was built into the structure of the theory. Nothing extra had to be assumed. The planets make irregular trips around the sky, their apparent periods vary from trip to trip; Copernicus recognized these variations to be due to the earth's motion. A planet's own year, its *sidereal* period, is the same, trip after trip.

For the first time, the conception of physical size and distance was introduced into planetary astronomy. Ptolemy could describe only the relative sizes of epicycle and deferent, but could not say anything about their absolute sizes. Ptolemy could only guess at the order of the planets. For Copernicus' method to work at all, each planet's orbit about the sun had to have one and only one radius. With the size of the planet's orbits fixed, and their years constant, there seemed to be more order than previously suspected. It was possible to talk of a solar *system*.

On the other hand, many deep and serious objections were raised against the Copernican hypothesis. For example, how harmonious is it when Copernicus requires the moon to circle the earth, and all other planets, the sun? How can there be two centers of circular motion in the universe? The Copernican system absolutely required that Venus have phases, just like the Moon's. The geometry is sketched in Fig. 3-17.

Shown is Venus at four points in its Copernican orbit. We have kept the earth fixed for the sake of illustration. Shading indicates the

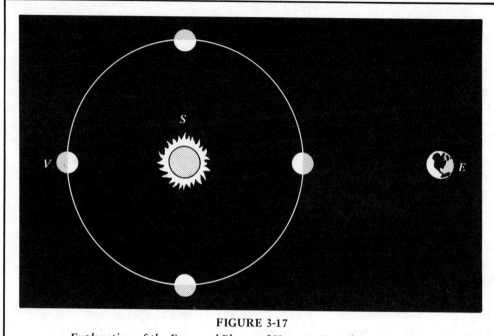

**FIGURE 3-17**
*Explanation of the Expected Phases of Venus as Seen from the Earth.*

parts of Venus in shadow. Venus can have a full phase at the conjunction when it is on the opposite side of the sun from the earth, and a sequence of partial phases leading up to and following its full phase. A similar diagram would show that a full phase is impossible in the Ptolemaic system. In Copernicus' time, when only naked-eye observations were possible, phases of Venus were not observed. However, seeing a full phase of Venus would mean that Copernicus' system was "really real," and not just a convenient mathematical scheme.

Copernicus' system directly contradicted deeply held conceptions about physics. For example, the fact that the earth has to be a planet contradicted the Aristotelian principle that the objects in the heavens are absolutely different from those on earth. It directly threatened the Greek definition of the five elements out of which all matter should be constructed. If the earth rotates on its axis so rapidly, what holds it together? What keeps objects on its surface from flying off into space? Why are there not fearsome winds?

One deep question concerned the possibility of the earth's motion. If you drop a ball and the earth is moving, why does the ball still fall straight down? After all, when you stand on the earth, there is a contact whereby the earth's motion is communicated to you and all other objects attached to it. On the other hand, when the ball falls, nothing communicates the earth's motion to it; thus, it should appear to an observer moving with the earth that the ball falls sideways. Since all bodies are observed to fall straight down, the earth must not move. In fact, in a Copernican universe, why should bodies fall at all, if the earth is no longer at the center of the universe? If the sun is the center of the universe and the earth is heavy, why does the earth not fall into the sun? Copernicus' narrow intellectual vision served him well, for he disregarded these questions. Was it a sign of the times that he could so easily suppress these questions? We shall see that of all these questions the one "Why do bodies fall as they do if the earth is not the center of the universe?" proved to be the most fruitful. If Copernicus were right, a new physics would be needed.

The most unsettling objection was that Copernicus required his universe to be frighteningly large; the same objection that had been made when Aristarchus first posed the idea. For the stars to show no parallax in the heliocentric system, they had to be very far away—much further than people had thought. When people thought the earth stationary, the largest possible baseline for taking a parallax measurement was the diameter of the earth. But if the earth orbited the sun, then, a baseline equal to the diameter of the earth's orbit could be achieved simply by taking measurements of the stars six months apart. Copernicus' universe required an expansion of the distance scale by at least the ratio of the diameter of the earth's orbit to the earth's diameter, a huge number in anyone's cosmology. Still, no one observed parallax.

If the earth rather than the celestial sphere rotated, the stars did not have to move. Since the stars did not move, they did not have to be embedded in a sphere. They could equally well be evenly distributed throughout space. The universe could, in fact, be infinite. The conservative Copernicus never gave up the celestial sphere. But Giordano Bruno, an Italian philosophical commentator, made this step:

> There is a single general space, a single vast immensity which we may freely call Void: in it are innumerable Globes like this on which we live and grow; this space we declare to be infinite, since neither reason, convenience, sense-perception, nor nature assign to it a limit; for there is no reason, nor defect of nature's gifts, either of active or passive power, to hinder the existence of other worlds throughout space, which is identical in natural character with our own space, that is everywhere filled with matter or at least aether.

For this heresy, among many others, Bruno was burned at the stake in 1600.

Since a sun-centered cosmology had existed since Aristarchus, why, then, do we consider Copernicus great? First of all, we must admire his deep geometrical intuition. To see, by an inversion of point of view, that a sun-centered system contained the possibility of accounting for *all* the known astronomical motions was an extremely important perception.

Perhaps even more important, Copernicus actually undertook the immense effort of constructing accurate sun-centered calculations. His results were certainly no worse that Ptolemy's. But now two systems were proposed, each with equivalent conceptual content and accuracy. Could one say anymore that an earth-centered universe, so tied to philosophy and theology, was definitely correct?

> ... *The new philosophy calls all in doubt,*
> *The element of fire is quite put out;*
> *The Sun is lost, and th'earth, and no man's wit*
> *Can well direct him where to look for it.*
> *And freely men confess that this world's spent*
> *When in the Planets, and the Firmament*
> *They seek so many new; then see that this*
> *Is crumbled out again to his Atomies,*
> *'Tis all in pieces, all coherence gone;*
> *All just supply, and all Relation: ...*
>                                        John Donne (1611)

*The fool will upset the whole science of astronomy, but as the Holy Scripture shows, it was the sun and not the earth which Joshua ordered to stand still.*
                                        Martin Luther (On Copernicus)

### Summary

*About 150* A.D., *Ptolemy of Alexandria worked out the great technical codification of Greek geocentric cosmology that dominated Roman, Moslem, and Christian astronomical thinking until the Renaissance.*

*These are the basic features of Ptolemy's system: the sun and the moon are attached to big wheels that rotate, relative to the celestial sphere, with periods of a year and a month, respectively. The planets also have wheels, centered on the earth. For the outer planets these big wheels, called deferents, rotate with the average period these planets take to circle the zodiac. The inner planets, like the sun, have deferents that rotate once a year.*

*The planets are not attached to the deferents, but to epicycles, whose centers are on the deferents. The epicycles rotate relative to the* deferents with the planets' synodic period. For the inner planets, the center of the epicycles are lined up with the sun. The rotations of the outer planets' epicycles are timed so that they are in the middle of retrogression when in opposition. The whole system is enclosed by the celestial sphere, and rotates westward once a day. This system explained the major features of the motions of the heavenly bodies. In particular, it predicted retrograde motion to occur when, in fact, it does. Also, it provided a plausible explanation of the fact that the outer planets brighten around opposition. In detail, however, Ptolemy needed to add other devices such as minor epicycles, equants, and eccentrics to his system to account exactly for the observed motions of the heavenly bodies.

What Ptolemy did for the cosmology of Eudoxus and Aristotle, Copernicus did for the heliocentric model of Aristarchus. Copernicus lived in Poland in the fifteenth century A.D. In his system, the sun was the center of the universe; and the six planets, Mercury, Venus, Earth, Mars, Jupiter, and Saturn, circled it. The farther out a planet is, the slower it moves. Since the other planets are observed from a moving platform, the earth, their motions, including retrograde motion, can be understood in this scheme also.

Copernicus' heliocentric picture of the universe accounted for the observed motions of sun, moon, and planets, as well as Ptolemy's. Its most obvious advantage was that it no longer needed major epicycles. It also required not only definite, calculable, periods for the planets to circle the sun but that they have definite, calculable, distances from the sun. To account for the observations exactly, however, Copernicus had to add many minor epicycles; and his system was not clearly simpler than Ptolemy's.

## Appendix to Chapter 3

### The Construction of Venus' Epicycle

As an example of the mathematical calculations needed to construct Ptolemy's epicycle, consider the problem for Venus. The geometry is sketched in Fig. 3-6.

Venus and its system of wheels lies entirely inside the sun's circle. The center of Venus' epicycle, marked $C$, is always on the line from the earth to the sun. When Venus is at the point marked $T$, its apparent distance from the sun is as great as it ever gets. Then, Venus is at a greatest elongation, which is observed to be about 45°. To make the problem simple, let us assume that it is exactly 45°. The angle $TEC$ is, therefore, 45° ($E$ is the earth).

Now, $EC$ is the radius of the deferent, and $TC$ is the radius of the epicycle. Their ratio is not arbitrary. Observe that the line, $ET$, can not cross into the circle, since then it would make an angle smaller than the largest one possible. Therefore, $ET$ is tangent to the small circle. $CT$ is a radius of this circle and so it is at right angles to $ET$. Angle $ETC$ is, therefore, a right angle.

Now, we can "solve the triangle." Since the angle at $T$ is $90°$ and the angle at $E$ is $45°$, the angle at $C$ must also be $45°$. (Remember that the sum of the three angles in a triangle is always $180°$.) Therefore, the triangle $CTE$ is an isosceles right triangle, and $CT$ and $ET$ have the same length. (This feature is special to Venus.) By Pythagoras' theorem:

$$(CT)^2 + (ET)^2 = (EC)^2$$

But, we have just proved that:

$$CT = ET$$

therefore,

$$2(CT)^2 = (EC)^2$$

Divide both sides by $2(EC)^2$:

$$\frac{(CT)^2}{(EC)^2} = \frac{1}{2}$$

Finally, take the square root:

$$\frac{CT}{EC} = \sqrt{\frac{1}{2}} = \frac{1}{\sqrt{2}}$$

The ratio of the radius of the epicycle to the radius of the deferent is $1/\sqrt{2}$, which is about 0.7.

Now imagine that the deferent and the epicycle are both rotating counterclockwise. (This direction is, therefore, eastward.) The deferent rotates at exactly the speed of the sun to keep up the alignment. The epicycle rotates once each 584 days (83 1/2 weeks). Therefore, the points, $C$, $T$, and $S$, are moving around, but the configuration remains unchanged. When Venus is at $S$, it is a morning star west of the sun. When Venus is at $T$, it is an evening star east of the sun. Roughly, the motion from $S$ to $T$ is "direct motion," while the motion from $T$ to $S$ is "retrograde motion." (This is not exactly right; Venus is not in retrograde motion the entire time it takes to get from $T$ to $S$, since at $T$ and $S$ it is moving eastward just as fast as the sun. The mathematics is a bit complicated. The point is that while Venus moves from $T$ back to $S$, it is moving westward with respect to the sun. But true retrograde motion is motion westward with respect to the stars.)

From $T$ to $S$ is $90°$ of the way around the epicycle. From $S$ to $T$ is the remaining $270°$. That is why it takes Venus three times as long to get from $S$ to $T$ as it does to get from $T$ to $S$.

The scheme for Mercury is similar, but it takes elementary trigonometry to solve the triangle analogous to $ETC$ in Fig. 3-6. Since Mercury's greatest elongation is only $23°$, its epicycle must be smaller than Venus', as in Fig. 3-18.

# Appendix to Chapter 3

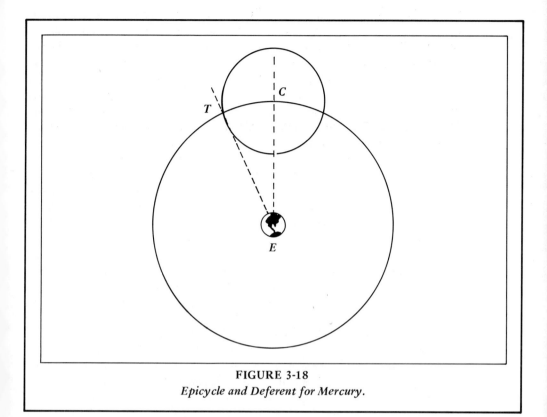

**FIGURE 3-18**
*Epicycle and Deferent for Mercury.*

The angle at $E$ is $23°$. The line $ET$ is again tangent to the epicycle, so the angle at $T$ is $90°$. Therefore, the angle at $C$ is $67°$ ($90 + 23 + 67 = 180$). Thus, it is easy to see that Mercury moves westward, relative to the sun, a larger fraction of the time than Venus does; namely, $67/180$, compared to $45/180$, or about $0.35$ compared to $0.25$. The ratio of the radius of the epicycle to the radius of the deferent is $CT/CE = \sin 23°$.

The important technical point here is that in Ptolemy's model if you know the maximum distance an inner planet gets from the sun (greatest elongation) and if you assume that motion on the epicycle is uniform (the epicycle does not speed up or slow down), then you can calculate how long the planet takes to go from greatest western elongation to greatest eastern elongation. The answer agrees rather closely with astronomical observation.

## The Precession of the Equinoxes and the Calendar

In Copernicus' system, the precession of the equinoxes has a simple description. Recall that the celestial poles do not really stay fixed forever, but move very slowly among the stars. By Copernicus' time, the nature of this motion was understood in detail. The celestial pole traces out a small circle in the sky, and the center of that circle is the point that is exactly 90° around from the ecliptic; in other words, if we had drawn our celestial latitudes and longitudes so that the ecliptic were the equator the center of the circle traced out by the North Celestial Pole would have been called the north pole. Furthermore, the radius of that circle is exactly 23 1/2°, the same as the angle between the ecliptic and the equator. Therefore, the axis of the ecliptic always points to a place in the sky 23 1/2° away from the celestial pole. The angle that this axis makes with the axis of the equator is, therefore, 23 1/2°, and so the equator and the ecliptic always make this angle with each other, even though the point where they cross is slowly moving. The precession of the equinoxes is very slow, taking about 26,000 years to make one full circuit.

To account for these observations, all Copernicus had to do was to make the direction of the earth's axis change very slowly, with a 26,000-year period, conical movement, so that the axis, extended to the stars, traces out the circle described above. If you draw a line from the earth's center perpendicular to the ecliptic, the earth's axis always makes an angle of 23 1/2° with that imaginary line.

This description of the precession of the equinoxes in a heliocentric system was a totally original discovery. It was a small, but important, phenomenon, which the new physics to come would have to explain along with the rest.

Copernicus' description of the precession of the equinoxes seemed simpler and easier to visualize than the description in a geocentric system, although of course they are equivalent as far as observations are concerned. Instead of having the celestial poles meander around the sphere of stars, we have the axis of the earth *precessing*—that is the technical words for this type of motion, which occurs in any type of gyroscope—about a direction perpendicular to the earth's orbit. This motion, too, is governed by the sun. In fact, Copernicus says in his introduction that it was the study of the precession of the equinoxes that led him to consider a sun-centered universe.

A related problem was that of the calendar. The trouble with the old system started because the calendar lagged behind the seasons. By the early sixteenth century, the vernal equinox did not happen on March 21, as it was supposed to, but eleven or twelve days earlier. Now, it was fairly easy, with a few thousand years' observations, to determine exactly how long it takes the sun to go around the ecliptic. This is the period of the earth's revolution about the sun; namely, the time it takes for an imaginary line drawn from the earth to the sun (and extended) to return to the same place among the stars. This period of time, called the sidereal year, or sidereal period of the earth, is 365.256 days. But this is not exactly the interval between two successive beginnings of spring, due to the precession of the equinoxes. Spring begins when the sun appears to cross the celestial equator (this is when a line drawn from the sun to the center of the earth passes through the equator and makes a right angle with the earth's axis). But, the place where the celestial equator crosses the ecliptic is slowly moving around the ecliptic, once around in 26,000 years, or 360 × 1/26,000 degrees each year. The upshot is that the calendar year—the time from the beginning of one spring to the beginning of the next—is only 365.242 days, about 20 minutes shorter than the sidereal year.

Measurements of the calendar year, which in effect are measurements of the rate of the precession of the equinoxes, are difficult, and were in very bad shape in the early sixteenth century. Copernicus was one of the astronomers asked by the Pope to join a project that would improve the calendar. Copernicus declined, stating that it was necessary to have better astronomical theories and observations.

The precession of the equinoxes was not the only reason the calendar had gotten behind, but it added to the confusion. The Julian calendar, established by Julius Caesar, had been in effect for fifteen centuries. The Julian calendar had assumed that the correct calendar year should be exactly 356 1/4, or 365.25 days. Since each year had to have a whole number of days, most years had 365 days, but a leap year was added every fourth year so that the average was exactly 365.25. But, the correct value is 365.242 days, which is 0.008 (= 8/1000) days shorter. Now, each day has $24 \times 60 = 1440$ minutes so that the calendar was off on the average by $8/1000 \times 1440 = 11\ 1/2$ minutes each year. This is too small to be noticed in one man's lifetime, but in 400 years amounts to $400 \times 11\ 1/2 = 4600$ minutes, or 77 hours, a little over three days. Our modern calendar is called the Gregorian, after Pope Gregory XIII, who established it in Catholic countries in 1582. It eliminates three leap years every four-hundred years by declaring the years like 1700, 1800, and 1900, years that are multiples of 100 *not* to be leap years, even though these years are multiples of 4. The years that are multiples of 400, like 1600 and 2000, are nevertheless leap years. Another way of stating these facts is to note that while the year according to the Julian calendar is exactly 365.25 days the "tropical", or calendar year, the period between successives equinoxes, is 11 minutes and 14 seconds shorter. Because of the precession of the equinoxes, this is less than the "sidereal" year, the time it takes the sun to go around the ecliptic, which is 9 minutes and 10 seconds longer than the Julian year.

The Greek and Russian Orthodox Churches still use the Julian calendar, which is why they celebrate Christmas and Easter on different dates from the western Christians.

## *The Sizes of the Planets' Orbits*

The calculation of the size of Venus' orbit was especially simple, since accidentally Venus' greatest elongation is a simple angle, 45°. (Of course, even this is only an approximation.) Let us see how to calculate it when we are not so lucky. The configuration for Mercury looks like Fig. 3–19.

The two circles are Mercury's orbit and the earth's. When the line, *EM*, is tangent to the circle of Mercury's orbit, Mercury is at greatest elongation. Since *EM* is a tangent, and *SM* is a radius, *EMS* is a right-angled triangle, with the right angle at *M*. The angle at *E* is Mercury's greatest distance (in degrees) from the sun. Now, the distance Mercury gets from the sun each time around is slightly different. Although the maximum possible distance is 28°, the average greatest elongation is 23°. We will use 23° to compute Mercury's average distance from the sun. (It is the average distance, not the maximum distance, that we have been listing in the tables.)

Thus, the angle at *E* is 23°. Therefore, from the definition of the trigonometric sine:

$$(MS) = (SE) \sin 23°$$

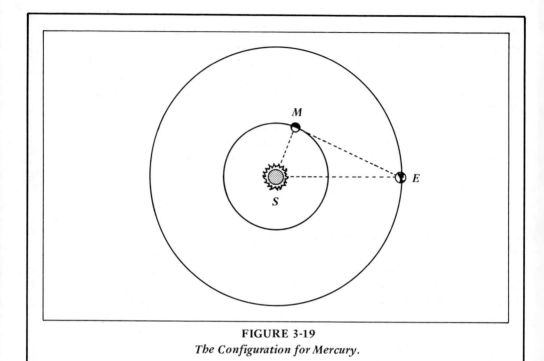

**FIGURE 3-19**
*The Configuration for Mercury.*

Let $R_m$ and $R_e$ be the distances of Mercury and the earth from the sun:

$$R_m = SM$$
$$R_e = SE$$

and so;

$$R_m = R_e \sin 23°$$

Since $\sin 23° = 0.39$, $R_m = 0.39 R_e$. Measured in astronomical units:

$$R_m = 0.39 \text{ A.U.}$$

The radii of the orbits of the three outer planets proceeds in a similar way. We will not go into the details here. The important point is that they can be computed from observations of simple features of those planets' motions in the sky.

### Calculation of the Sidereal Periods of the Planets

Consider an outer planet like Saturn. The time between its oppositions is 378 days. Figure 3-20 shows the earth and Saturn at two successive oppositions. At the first one, the earth is at $E_1$, and Saturn is at $S_1$; 378 days later, Saturn has moved to $S_2$, while the earth has moved once all the way around and up to $E_2$.

Now, we must get a formula that will tell us how far the earth has gone

# Appendix to Chapter 3

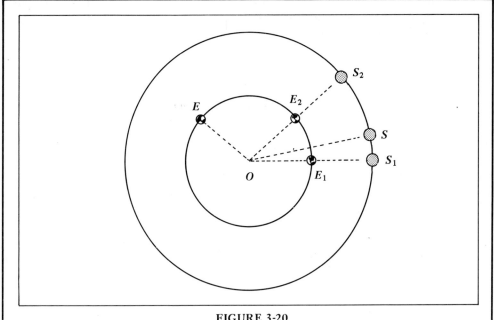

**FIGURE 3-20**
*The Orbits of the Earth and Saturn.*

around after any time since the first opposition. Let $t$ be the time since the first opposition, and $E$ the earth's position. At $t = 0$, the earth is at $E_1$. At a later time, $t$, the earth is at some place, $E$, and makes an angle, $E_1OE$, with line $OE_1$. Call this angle, $A_e$. $A_e$ grows with the time. Therefore, $A_e$ is some number multiplied by $t$. But we know that $A_e$ is $360°$ when $t = 0$ year. So the formula is:

$$A_e = \frac{360}{1} t$$

where $t$ is measured in years.

A similar formula works for the angle that Saturn makes with the original direction at a time, $t$, after opposition. Call this angle $A_s$; in the diagram it is $SOS_1$. We do not know Saturn's period, so let us give it an algebraic symbol, $T_s$. The correct formula must be:

$$A_s = \frac{360}{T_s} t$$

This is right, since it says that when $t = 0$, $A_s = 0$; and when $t = T_s$, $A_s = 360°$, which is correct. In one "Saturn Year," Saturn goes around $360°$.

When $t = 378$ days (1.036 years), the earth and Saturn are aligned again. At first sight it looks as if the angles, $A_e$ and $A_s$, should be equal at that time. But, remember that $A_e$ is the number of degrees a line from the sun to the earth

101

has swept out. While Saturn has gone from $S_1$ to $S_2$, the earth has gone from $E_1$ all the way around once and then up to $E_2$. So the earth has gone exactly 360° farther than Saturn. We can put this into a formula, when $t = 1.036$ years:

$$A_e = A_s + 360$$

At that time:

$$A_e = 360 \times 1.036$$

$$A_s = \frac{360 \times 1.036}{T_s}$$

Therefore, substituting these values for $A_e$ and $A_s$ into our formula, we get:

$$360 \times 1.036 = \frac{360 \times 1.036}{T_s} + 360$$

Now, it is a simple algebra problem to find the unknown period of Saturn, $T_s$. First notice that the number 360 cancels out of each term; divide the equation by 360 to get:

$$1.036 = \frac{1.036}{T_s} + 1$$

Subtract 1 from both sides:

$$0.036 = 1.036/T_s$$

Solve for $T_s$:

$$T_s = \frac{1.036}{0.036} = 29$$

The correct answer is closer to 29.5. (The error is due to rounding errors; the method is in principle exact but we have to know Saturn's period and the earth's period exactly, rather than just to a whole number of days.)

We can get a formula that will work for any planet. Let $T_e$ be the earth's period (one year). Let $T_{sid}$ be an outer planet's (unknown) sidereal period, and $T_{syn}$ its known synodic period. At any time, $t$, since the last opposition, the outer planet has gone around an angle we can call $A_p$. Then, just as for Saturn:

$$A_p = 360 \frac{t}{T_{sid}}$$

while the earth has gone around an angle:

$$A_e = 360 \frac{t}{T_e}$$

The next opposition occurs when the earth has made exactly one revolution more than the outer planet. At that time, $A_e = A_p + 360$, and $t = T_{syn}$, so that:

$$360\frac{T_{syn}}{T_e} = 360\frac{T_{syn}}{T_{sid}} + 360$$

Divide each term by $360\, T_{syn}$ to get a form that is easy to remember:

$$\frac{1}{T_e} = \frac{1}{T_{sid}} + \frac{1}{T_{syn}}$$

This equation can be solved for the unknown, $T_{sid}$.

Finally, consider an inner planet (Mercury or Venus) whose sidereal period is $T_{sid}$ and whose synodic period is $T_{syn}$. The geometry is the same as in the example we just worked out for an outer planet, except that the roles of the earth and the planet are interchanged. Therefore, for an inner planet, we can write the answer immediately:

$$\frac{1}{T_{sid}} = \frac{1}{T_e} + \frac{1}{T_{syn}}$$

## QUESTIONS

1. If you were a scientist in Copernicus' time and faced with resolving the conflict between the Ptolemaic and Copernican systems, what further research would you undertake or recommend to help clarify the situation?

2. Suppose Ptolemy had discovered a new inner planet whose maximum elongation was always exactly 60°. What would he have determined the ratio of the radius of its epicycle to the radius of its deferent to be? HINT: If the angles of a right triangle are 30°, 60°, and 90°, the short side is exactly one-half as long as the hypotenuse.

3. Suppose this new planet took thirty weeks to go from greatest eastern elongation to greatest western elongation. How long would it take to return to greatest eastern elongation?

4. Suppose Copernicus, not Ptolemy, had discovered this new inner planet, whose greatest elongation was always 60°. What radius (in A.U.) would he have assigned to its orbit?

5. Read the last Appendix section on the exact formula relating a planet's sidereal period and synodic period. What would Copernicus have determined the new planet's sidereal period, i.e., its true period of revolution about the sun, to be?

6. What would be the speed of the new planet in A.U. per year? Would the general rule that planets move slower the farther they are from the sun be violated?

7. Precise values of the synodic periods of Mercury, Venus, Mars, Jupiter, and Saturn are 115.88 days, 224.70 days, 779.95 days, 298.89 days, and 378.10 days, respectively. Calculate accurately the true periods of revolution (the sidereal periods) of the five planets. (You might think for a minute just how accurately you can calculate them with the numbers as given here.)

## REFERENCE

1. *Occasional Notes of the Royal Astronomical Society*, Vol. 2, No. 10 (London, Royal Astronomical Society, 1947). Reprinted by permission.

**JOHANN KEPLER**
*(Courtesy of the Boston Public Library and by permission of the Houghton Library, Harvard University)*

CHAPTER

4

# TYCHO BRAHE AND JOHANNES KEPLER
## Reform of Planetary Astronomy

The Copernican system did not take the world by storm. Copernicus' *De Revolutionibus* was too detailed and technical to reach a wide audience. Nonetheless, it took its place alongside Ptolemy's *Almagest* as a handbook of mathematical astronomy; and for a period professional astronomers used it interchangeably with the *Almagest* as convenience dictated. For example, the astronomer, Erasmus Reinhold, prepared a new set of astronomical tables that replaced the obsolete *Alphonsine Tables*. While Reinhold's *Prutenic Tables* used the Copernican style of calculation, rather than the Ptolemaic, Reinhold never did say whether he thought the Copernican system was physically true. A few speculative people did commit themselves to the reality of the Copernican system, among them William Gilbert, whom we shall encounter again when we discuss magnetism, and Michael Mästlin, Johannes Kepler's teacher. By and large, however, the general attitude regarding the conflict between Ptolemaic and Copernican astronomy was cautious. The European world was locked in a bitter conflict between Catholic and Protestant, which absorbed most of its energy. With an issue so deeply loaded with emotion as the scientific description of God's creation, it was prudent to maintain a position of safe neutrality. Following Copernicus' death in 1543, the tradition of the separation of mathematical and physical astronomy continued. Most astronomers accepted Copernican calculational techniques and planetary tables, just as they had previously accepted and still used Ptolemaic techniques and theory, without committing themselves to the reality of either one.

## TYCHO BRAHE

Neither ease nor accuracy of calculation nor precision of observation permitted a clear-cut choice between the Ptolemaic and Copernican models. In addition, the physical objections to a sun-centered universe remained unanswered. The authority of the ancients remained so great that people did not feel the need to make many new observations. Even Copernicus added only twenty-seven new observations to the tables passed to him from antiquity during the entire thirty-six years he spent on *De Revolutionibus*. Not only were some data inaccurately measured in antiquity, but also some were inaccurately copied as they were transmitted through the centuries. With the wisdom of hindsight, we can see that much of the mathematical complexity of Copernicus' book stems from the fact that he was explaining errors. Reinhold's *Prutenic Tables* were at best a marginal improvement, since they incorporated a minimum of new observations and retained the old inaccurate measurements.

One man reversed the observational tradition of astronomy and founded another, Tycho Brahe (1546-1601). Born into an aristocratic Danish family[*] only three years after the publication of *De Revolutionibus* and brought up by an uncle, Jorgen de Brahe—a vice-admiral in the Royal Danish Navy, Tycho never was intended to be an astronomer, an occupation unusual for one of noble blood. When he was thirteen, Tycho went to the University of Copenhagen to study law so that he could become a diplomat. Near the end of his first year there, an event in the heavens, a partial eclipse of the sun, decided the future course of his life. Most impressive of all to him was the fact that astronomers had predicted the eclipse in advance; later he wrote that it was[1] "something divine that men could know the motions of the stars so accurately that they were able to predict a long time beforehand their places and relative positions." Brahe bought the *Almagest* and started reading. If the solar eclipse stirred the boy's imagination, a conjunction of Jupiter and Saturn in 1563 alerted his mind to difficulties. The *Alphonsine Tables* missed predicting this event by a whole month, and even the Copernican tables missed by several days. This inaccuracy shocked him. So it was that, at age 17, Brahe dedicated his energy and passion to achieving more accurate observations. Tycho

---
[*]Hamlet's friends, Rosenkrantz and Guildenstern, were among his ancestors.

was the first in the modern era for whom sheer accuracy of observation became an all-consuming compulsion; in A. A. Michelson we will meet another.

Brahe's uncle Jorgen had died while rescuing the young king of Denmark, Fredrick II, from drowning. Partly to repay this debt, partly from an interest in learning, and partly from a nationalistic desire to induce the increasingly famous Brahe to do his work in Denmark, Fredrick offered to support Brahe's research. Fredrick gave him the island of Hveen, between Elsinore and Copenhagen, and an enormous income. On Hveen, Brahe constructed a whole castle devoted to astronomy—Uraniborg castle, the castle of the heavens. Figure 4-1 shows Brahe in his observatory in Uraniborg. Each tower and turret was equipped with the most advanced instruments ever made for naked eye observations of the heavens. They were nearly all of Brahe's design. At Uraniborg, Brahe was equipped to satisfy his lust for accuracy.

Tycho remained on his island for twenty-one years, observing the

**FIGURE 4-1**
*Tycho Brahe in His Observatory at Uraniborg Castle.*

heavens, entertaining lavishly, and oppressing his peasants. Eventually, in 1597, an irreconcilable argument with Fredrick's successor, Christian IV, set Tycho wandering about Europe, looking for a more generous patron. In 1600, he settled in Prague, becoming the official imperial mathematician to the Holy Roman Emperor, Rudolph II. Less than two years later, Brahe died. In Prague he had hoped to study the vast quantity of observations he had amassed, and to understand the true motions of the heavenly bodies. But this task was beyond his abilities, and beyond those of his sons to whom he intended to entrust it.

The nova of 1572 made Tycho's reputation. Nova is Latin for "new;" a nova is a bright new star. Tycho's nova suddenly appeared in the constellation Cassiopeia in November, 1572; it gradually grew dimmer until it finally disappeared from view some eighteen months later. For those eighteen months, the entire astronomical community was fascinated with the new star, and the larger intellectual community was equally fascinated with the observational results the eighteen months permitted the astronomers to gather. Why was there so much interest? According to Aristotelian doctrine, the heavenly spheres were perfect and, therefore, unchanging; all change could only take place in the terrestrial spheres below the moon. A new star just could not appear on the celestial sphere. However, perhaps something looking like a star might suddenly brighten in the upper atmosphere just below the moon's sphere where change was permitted. The correctness of Aristotelian cosmology rested on whether the nova was closer or farther than the moon.

It was not difficult for Tycho to check that the nova was farther away than the moon. There is a parallax effect due to the daily rotation of the heavenly bodies, or, in the Copernican system, due to the daily rotation of the earth. (To Tycho it mattered not which explanation was correct.) In Fig. 4–2, the moon is observed twelve hours apart from two places, $A$ and $B$ (assuming that the earth rotates). The line $AB$, the diameter of the earth, is 8000 miles long. Therefore, the moon appears to shift its position relative to the background stars, making a complete cycle once each day. The daily shift of the moon is about $2°$. Brahe's instruments were far superior to those of anyone else. Since he could measure the relative positions of the two heavenly bodies with the error of only a few minutes of arc, he showed conclusively that the nova was not only farther away than the moon, but also that it was much farther away. The new star was definitely in the heavens. Brahe himself speculated that the nova may have been formed out of celestial material, but not so perfectly as other stars and so it soon dissolved. Hipparchus had reportedly observed a new star around

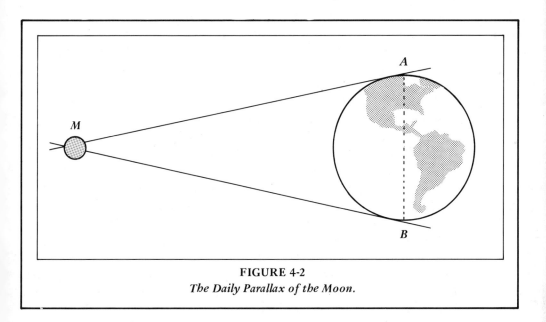

**FIGURE 4-2**
*The Daily Parallax of the Moon.*

125 B.C.; however, since that time the heavens had remained unchanged, so far as anyone in Europe knew. In fact, Chinese and Korean astronomers had reported the outburst of an exceptionally bright new star—a "supernova"—in the year 1054 A.D. The 1054 A.D. supernova was visible *during the day* for over three weeks and at night for many months. Yet, in Europe, people were not excited about it. How things had changed by 1572!*

Brahe's observations of a comet in 1577 confirmed the idea that change was possible beyond the moon. According to Aristotle's

---

*Today we know that novae are not uncommon phenomena. Our modern telescopes can see stars far too faint to be seen with the naked eye; using them, astronomers discover a nova every few years. Certain types of stars are apparently unstable. They suddenly explode, increasing their brightness hundreds of thousands of times, and then gradually simmer down in the course of a few months. Astrophysicists—people who study how stars evolve and why they shine—claim that the sun is not this type of star.

Most novae are so far away that they are too faint to be seen without a telescope. But every so often, a star literally blows itself apart, emitting huge clouds of gas and leaving behind only a small part of the material that was in the original star. Such explosions are much rarer than novae; in our galaxy—the collection of stars that includes those you can see—they apparently happen only once every few hundred years. Such an exploding star is always fantastically bright; brighter than Venus, and even visible in the daytime. Today we call such a rare, bright exploding star a *supernova*. It was presumably a supernova that Tycho Brahe saw in 1572.

The remnants of the 1054 explosion are still a beautiful sight in a telescope. They are the Crab Nebula, a gaseous ring still expanding at a rate fast enough to be observed over the years. In the center there has been discovered recently a very faint and peculiar star. Presumably the remnant of the supernova, it pulsates in brightness about thirty times a second.

doctrine of heavenly perfection, comets must have a meteorological origin. Perhaps they were balls of fire trapped in the upper atmosphere. If so, they had to be just below the moon's sphere, so they could share in the diurnal rotation of the heavens. Brahe measured the comet's parallax; again he found it further away than the moon. He calculated that its orbit was beyond that of Venus, and suggested it might be, "oblong . . . like an oval"—the first time anyone ever said that orbits were not necessarily circular or compounded circles. How could a comet move so freely in the space between the planets, space that was filled with celestial material?

Brahe's greatest influence did not stem from the nova of 1572 or the comet of 1577, but from his recognition of the necessity to base astronomy on systematic observations. Brahe constructed very many measuring instruments, and he studied how his instruments worked. He had a modern point of view concerning instrumental errors. He was careful to tabulate and correct for systematic errors, and he repeated his observations many times in order to reduce statistical error. As a consequence, he knew how accurate his measurements were. For the first time, it was possible to know how much confidence to have in the data. Tycho's crowning achievements were a magnificent new map of the celestial sphere, with the positions of 777 stars placed more accurately than ever before, and a survey of all the planets' motions.

Tycho's approach to collecting planetary data was completely new. Ptolemaic theory had indicated that only certain observations were needed to fix the orbital calculations—those at opposition and conjunction in particular. (For example, if you believe the orbits are circles, then only three points determine the circumference of a circle.) However, Brahe had no preconceptions about the shapes of the orbits. Therefore, he measured them with an accuracy of four minutes of arc, at as many points around their orbits as he could, and not just at the few points deemed crucial by one theory or another. For these reasons, Tycho represents the first really significant advance of observational astronomy since the Greeks and Babylonians. This treasure was to be his legacy to Kepler. The accuracy of Brahe's data, together with Kepler's knowledge of how accurate it was, would enable Kepler to choose between different planetary theories.

Tycho Brahe had good reasons for disbelieving Copernicus, in addition to the usual physical arguments against the earth's motion. He had attempted to measure the annual parallax of stars and had failed; he, Tycho Brahe, who had constructed the best measuring instruments ever, still could not find a parallax. The earth must be at the center of the universe.

Tycho developed another argument against the immense Copernican universe based upon the size of the stars. No one knew precisely how big stellar images were because the stars twinkle. All one could do was guess at the size of the diffuse spot in the sky where the starlight comes from. Tycho estimated that a typical star might have a diameter of perhaps one minute of arc. What does this say about the actual diameter of the star? Tycho could not estimate how big stars were without knowing how far away they were. However, it was obvious that the stars could be much closer in the Ptolemaic system than in the Copernican; and this meant they did not have to be as big. The geometry is sketched in Fig. 4–3. Using his 1' estimate and a plausible distance to the Ptolemaic celestial sphere, the stars came out to be as big as the sun. On the other hand, using the minimum distance to the Copernican celestial sphere needed for the stars to show no annual parallax, the stars had to be at least as large as the earth's Copernican orbit around the sun! This was inconceivable to Tycho.

While Brahe had observational arguments against the earth's motion, the Copernican system did have several advantages. It did particularly well with the inner planets, Mercury and Venus, explaining

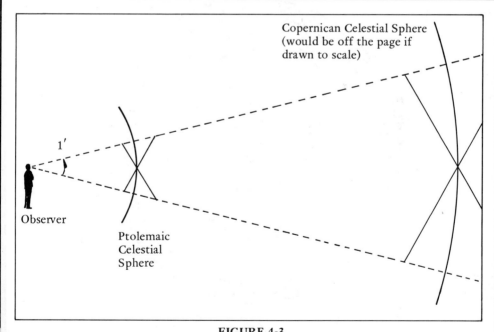

**FIGURE 4-3**
*Ptolemaic and Copernican Stellar Diameters, Deduced from the Absence of Parallax.*

why they never were far from the sun. Brahe revived a version of the old proposal of Herakleides: Mercury and Venus orbit the sun, while the sun, moon, and remaining planets orbit the earth. This kept earth fixed at the center of the universe. Tycho's system is shown in Fig. 4–4. While Tycho did not explain how there could be two centers of circular motion in the universe, his system was a reasonable compromise between the Ptolemaic and Copernican hypotheses. In the tense intellectual atmosphere of his times, when it was prudent to remain neutral on cosmological issues, the Tychonic system was seen as a viable alternative. In closing, we might ask ourselves: to what extent did the atmosphere of studied neutrality motivate Brahe to collect data with no particular theory in mind and to what extent was a period of neutrality necessary to the resolution of the tension between the Copernican and Ptolemaic hypotheses?

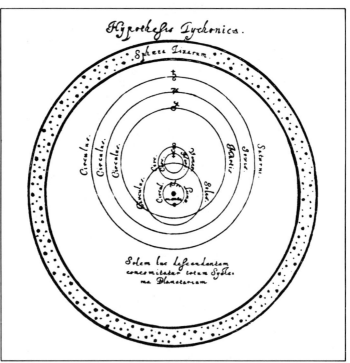

**FIGURE 4-4**
*The System of Tycho. Only the Sun Has the Center of Its Orbit at the Earth. The Other Planets Circle the Sun.*

## JOHANNES KEPLER AND THE COSMOGRAPHIC MYSTERY

One man had the imagination and extraordinary patience to extract from Brahe's mountain of observations a truly simple picture of the planets' motions. In the month of January in the year 1600, Johannes Kepler, then at Graz in Austria, set out for Prague. Here, for eighteen months he worked with whatever portions of Brahe's data he could cajole out of the Imperial Mathematician. Brahe had been reluctant to publish his new astronomical catalog without understanding it theoretically; Kepler would have remained merely an eccentric genius had his flights of speculative fancy not been confronted by Tycho's hard facts. From their brief and tempestuous encounter came a new cosmic order.

Kepler was born on December 27, 1571, in Weil-der-stadt, a small town near Stuttgart in southwest Germany. His family was poor and somewhat disreputable. (Years later, Kepler's mother was put on trial for being a witch; her then famous son saved her from being burned at the stake only at the last minute.) He was sent to school to be a clergyman, and eventually came to study theology and other subjects at the great Protestant University at Tübingen. There he studied astronomy under Michael Mästlin. Mästlin has a place in history in his own right, for he was one of the first astronomers to discuss the Copernican system seriously as a viable alternative to Ptolemy's. Mästlin also measured the parallax of the nova of 1572. His measurement was not nearly as accurate as Brahe's, but was good enough to show that the new star was beyond the moon. Mästlin awakened Kepler's interest in a heliocentric universe, and Kepler was convinced. Even as a student, Kepler publicly defended Copernicus. Kepler gave up his theological studies to teach mathematics at the provincial university of Graz, in Austria. While in the midst of an elementary geometry lecture, a new geometric conception of the planetary system burst upon him. He spent the year 1596 in intense labor working out his conception in his first book, the *Cosmographic Mystery.* This book is important not because his idea was correct; it was not. It did bring him to Tycho Brahe's attention and establish him as a formidable mathematical astronomer. Most important of all, it raised several powerful questions never before asked, which drove Kepler to work with fanatic intensity at astronomy all his life.

Kepler began the *Cosmographic Mystery* with the flat assertion that the Copernican system represented physical reality. This was one of the first affirmations of Copernicus by a professional astronomer. Then, Kepler posed the following three questions:

Why were there exactly six planets?

Was there a simple mathematical relationship between the radii of the different Copernican orbits?

Was there a simple mathematical relationship between the radius of a planetary orbit and its period?

These questions show that Kepler, like Pythagoras, always believed that deep mathematical harmony lay beneath the structure of the universe. "The ideas of quantities have been and are in God from eternity, they are God himself . . ."[2] Like Plato, Kepler at first believed that the planetary orbits should be explained using perfect geometrical objects. Circles and spheres had been Plato's examples of geometrical perfection. Kepler was also interested in another class of "perfect" geometrical figures, the regular polygons. A regular polygon is a plane figure that can have any number of straight sides, but the sides all have the same length. Moreover, all the angles between the sides are equal. Two familiar examples are the square and the equilateral triangle, which have four and three equal sides, respectively. The four angles of the square are each 90°, the three angles of the triangle are each 60°. A pentagon is a regular polygon with five equal sides.

One can inscribe a circle within a regular polygon, draw it inside the polygon so that it just touches the polygon in the middle of each of the sides. One can also circumscribe a circle outside the polygon so that each corner, or vertex, of the polygon lies on the circle. The inscribed and circumscribed circles will have the same center but different radii. What really excited Kepler was the fact that the ratio of the radii of the two circles is a fixed number, characteristic of the specific regular polygon between them. For example, the ratio of the circumscribed to inscribed radii is two for the triangle and the square root of two for the square.

According to Copernicus, the sun was near the center of all the planets' orbits, which seemed to be almost circular. Perhaps, thought Kepler, the radii of the orbits were determined by the condition that a regular polygon just fit between each pair of orbits. Figure 4–5 illustrates this idea. It shows two 2-planet solar systems and one 3-planet solar system constructed according to Kepler's hypothesis. Since Kepler knew of six planets, he had to find five polygons between their six circular orbits, and put them in the proper order.

Kepler worked feverishly to find a sequence of polygons that fit. The idea got off to a good start since the ratio of the radius of Saturn's orbit to Jupiter's is very close to 2, so that a triangle fits between them; and, as we saw in the previous chapter the ratio of the

Johannes Kepler and the Cosmographic Mystery

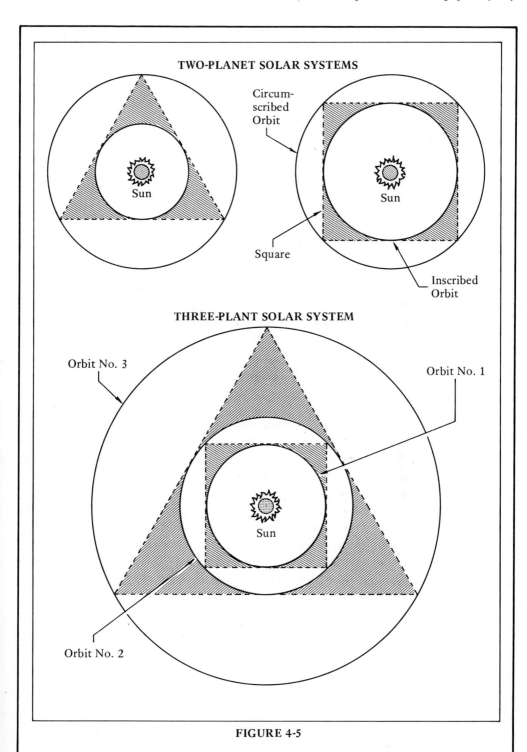

FIGURE 4-5

radius of the earth's orbit to that of Venus is the square root of 2, so that a square could fit between them. But gradually, it became clear to Kepler that the scheme could not work in detail. Then, he had a wonderful idea: Try it with solid figures instead of plane figures. A solid figure bounded by polygons is called a polyhedron. If all the polygons are identical and regular, and if all the angles that the surfaces of the polyhedron make with each other are identical, the solid is called a regular polyhedron. The cube and pyramid are familiar examples. Kepler's idea was to circumscribe and inscribe spheres about these regular solids, and hope that the radii of the spheres would correspond to the radii of the orbits. Why was this such a wonderful idea? Because an ancient theorem known to Plato and Eudoxus said that although you can construct a regular polygon with *any* number of sides, there exists only *five* possible regular solids. It just is not possible to make one with an arbitrary number of sides. (This is an impressive example of the power of mathematical reasoning, since the result is surprising and unanticipated. An elementary modern proof is shown in the chapter Appendix. Table 4.1 lists the five regular solids; they are drawn in Fig. 4-6.) This model explained something the regular polygon scheme, or any other, had been unable to do: Why there were exactly six planets! The six planetary spheres required only five regular solids between them.

In summary, Kepler revived the ancient Greek idea, lost since Ptolemy, that the heavens were governed by simple, harmonious geometrical laws. Also Kepler, unlike Ptolemy and Copernicus, believed that there had to be simple *numerical* relationships describing the orbits. Medieval scholars would have been satisfied just to pose such an elegant geometrical scheme. Kepler, living at the beginning of a new age, also required the planetary observations to fit his model. After first recounting triumphantly in the *Cosmographic Mystery* the beauty and symmetry of the five regular solids and their function in the design of the solar system, Kepler abruptly changed his style. He made a painstaking comparison of his model with the available observations. The fit

TABLE 4.1. *The Five Regular Solids.*

| No. of Sides | Shape of Each Side | Name |
| --- | --- | --- |
| 4 | triangle | tetrahedron |
| 6 | square | cube |
| 8 | triangle | octahedron |
| 12 | pentagon | dodecahedron |
| 20 | triangle | icosahedron |

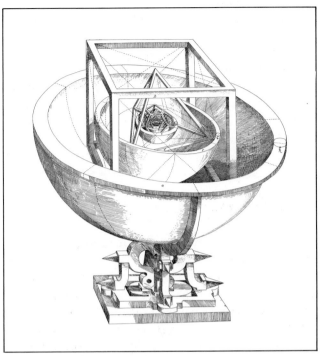

**FIGURE 4-6**
*Kepler's Drawing of His Model for the Solar System, Based on the Five Regular Solids.*

was not terribly good. Blaming poor data, Kepler maintained the hope that better observations would confirm the new natural law he had posed. Nonetheless, the ambiguity of his conclusion awakened in him a passion for more accurate theory and observation, a passion matched only by that of Tycho, whom he would soon seek out.

Kepler was convinced that understanding these simple numerical rules should lead him to the deeper laws governing the structure of the solar system. While Copernicus had been content to describe the solar system, Kepler wanted an explanation. Kepler's theory of the five regular solids seemed to explain why the planets are located precisely where they are. Was there a numerical rule that governed their speeds, or periods, as well? Copernicus had shown that the speeds of the planets decrease with distance from the sun. Why should this be so? Perhaps the sun, the center of the universe, could exert a force on the planets, a force that pushed them around their orbits. Perhaps, the more distant planets moved more slowly because the force weakened with distance from the sun. If the *same* force moved *all* the planets, then he should be able to find *one* numerical rule relating the size of an individual planet's orbit to its period of revolution. This rule should hold for all the planets.

Kepler had inverted the sense of causality in the universe. The cause of all motion had been at the outermost celestial sphere in Aristotle's universe; in Kepler's, it was the sun at the center. Kepler's conviction—there is a force in the sun, there is a force in the sun, there is a force in the sun—was first stated in the *Cosmographic Mystery;* it powered all his investigations.

## KEPLER'S WAR ON MARS

In 1600, Kepler journeyed to Prague and met Tycho. The two would have a brief, stormy relationship. The aristocratic, domineering Tycho and the mystical Kepler were not destined by temperament to get along. Moreover, Tycho had strong reservations about the Copernican system and Kepler's passionate espousal of Copernicus. Yet, a mutual fascination with astronomy brought them back together again and again. Kepler knew he must test his intuition against Brahe's data, and Tycho yearned for an explanation of it. When Kepler arrived in Prague, Mars was currently in opposition to the sun, and Tycho had prepared a table of all Mars' oppositions since 1580. Theories involving circles had encountered great difficulty with Mars. Tycho and his assistants had worked out a tentative new theory for Mars, but it was unsatisfactory. Tycho relieved one of his assistants of responsibility for Mars, and handed the problem to Kepler.

A few months later, Brahe died. Immediately, Tycho's assistants and relatives squabbled over the division of his wealth and expensive astronomical instruments. In all the commotion, Kepler quietly made off with the data. Kepler took Tycho's place as Imperial Mathematician to the court of Prague and settled down to work, free from the stress of arguments with Tycho and bolstered by Tycho's last wish: that Kepler use the data to prove the Tychonic plan of the universe. Kepler had boasted to Tycho that he would solve the orbit of Mars in eight days. It took him eight years; eight years of theoretical speculation, of frantic calculation, of patient sifting of data, and of glorious elation followed time and time again by black despair. Kepler came to call his extraordinary struggle, his "war on Mars."

In 1609, Galileo published his *Starry Messenger,* in which he disclosed the heavenly secrets revealed by the telescope. The book was written in breathless haste, in a simple style, and made Galileo a world celebrity, as we shall see in the next chapter. Kepler published his *Astronomia Nova—A New Astronomy, or a Physics of the Skies*—in 1609, too. It took him eight painful years to write, and was difficult to read; it established his reputation as an astronomer but not as a publicly recognized figure. In it he recounts his war on Mars, every blind lead, every false start, every speculation his fertile mind conjured up. The *Astronomia Nova* is a rich source of insight into the creative process in science; for Kepler never covered his tracks, as Newton would do some eighty years later.

Kepler began the *New Astronomy* by making three assertions aimed at purifying the Copernican hypothesis. Recall that while Copernicus' system in outline was one of perfectly circular orbits with the sun at the center, in detail the planets moved on complicated systems of epicycles, and the sun was not exactly at the center of anything. The center of Copernicus' universe was the center of the earth's orbit. If, as Kepler had speculated, a force from the sun moved the planets around their orbits, how the planets moved should depend on their distance from the sun, and not from the center of the earth's orbit. In using Brahe's data, he first recalculated all planetary distances, referring them to the sun, hoping that this would reduce the complexity of the orbits. Kepler relied only upon Tycho's data, because it was the only data whose accuracy he knew and trusted.

Next was the question of how fast the planets move around their orbits. Do they move with constant speed, or not? Suppose the planets turned out to have different distances from the sun at different points in their orbits? If the sun had a force, it should not only move the nearer planets faster than the farther ones as Copernicus had found; it should also move any of the planets faster when it is closer to the sun. Since for Kepler the physics of the sun's force and not the shapes of the planets' orbits was the important thing, his mind was open to the possibility of deviations from Plato's dogma of uniform circular motion. While he initially retained a circle for Mars' orbit, never again would uniform orbital speed appear in his thought.

Kepler's faith that the planets' orbits would be simple when their positions were referred to the sun was reinforced by his next realization. Because the ecliptic was a great circle, a curve that can be drawn in a plane, Copernicus knew that the earth's orbit around the sun had to lie in a single plane—called the plane of the ecliptic. What about Mars' orbit? It could not be in the plane of the ecliptic because Mars is observed sometimes above and sometimes below the ecliptic. Copernicus had tried to put Mars' orbit in a plane that always contained the earth. Since this plane had to move around the sun with the earth, the detailed description of Mars' motion was very complicated. Kepler, asserting that "the orbit of Mars was no business of the earth's", found (using Tycho's data) that Mars' orbit does indeed lie in a plane, but even more important, a plane that contains the sun and not the earth. Since the sun is motionless, Mars' orbital plane remains fixed. The orbital planes of earth and Mars are not parallel, but make a small angle (1° 51') where they intersect at the sun, as shown in Fig. 4-7. Upon making this discovery, Kepler remarked that Copernicus did not know how rich he was.

Kepler lost the first battle in the war on Mars. In the *Cosmographic Mystery,* he had tried a Platonic theory of the solar system; his first attack on Mars' orbit was Ptolemaic. Convinced that the planets' orbits must be simple, Kepler first assumed Mars' orbit must be a circle. Following the circular dogma more purely than either Copernicus or Ptolemy, he abondoned epicycles completely. How could the sun move Mars around an epicycle? It soon became clear, however, that the sun could not be at the circle's center; he tried putting it off to one side. Then, a speed law became important in order to calculate the variations in Mars' speed as it moved closer and farther from the sun. For this he used the Ptolemaic device of an equant. Kepler's combined eccentric-equant model for Mars' orbit is shown in Fig. 4-8. The equant point (marked "*Eq*") was not at the center, *C,* of the circle but opposite the center from the sun. Kepler imagined a line from the equant to Mars, rotating uniformly like the second hand of a clock, even though Mars did not stay at the same distance out along the clock hand. When Mars was at *M,* as in Fig. 4-8, it was as far from the equant point, *Eq,* as it ever got. It moved fastest when it was nearest the sun. When Mars was at *M',* furthest from the sun, it moved slowest. Moreover, if the imaginary clock hand rotated at a steady rate, it was easy to calculate Mars' speed at all times. By adjusting the distances of the sun and the equant point from the circle's center, and also the speed of the clock hand, Kepler tried to find an ever better agreement with Brahe's observations. After 70 separate attempts, Kepler came up with an orbit that deviated no more than 8' of arc (8/60 of a degree, or roughly one part in three thousand of a circle) from Brahe's data. He wrote:[3]

> If thou (dear reader) art bored with this wearisome method of calculation, take pity on me who had to go through with at least seventy repetitions of it, at a very great loss of time; nor wilst thou   be surprised that by now the fifth year is nearly past since I took on Mars...

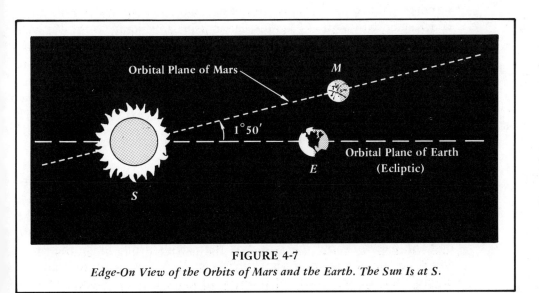

**FIGURE 4-7**
*Edge-On View of the Orbits of Mars and the Earth. The Sun Is at S.*

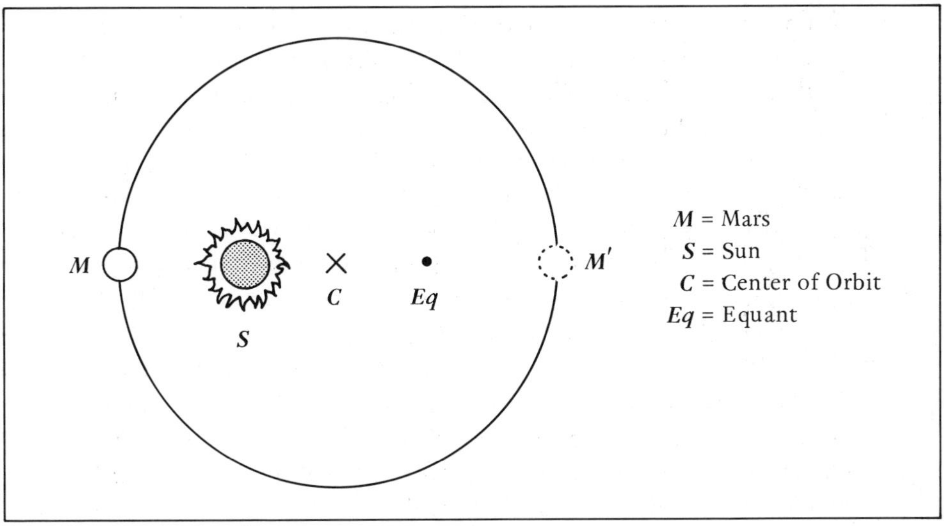

**FIGURE 4-8**
*Kepler's Ptolemaic Theory of Mars.*

Kepler, reasoning like a Ptolemaic astronomer, has arrived at a result better than all the generations of Ptolemaics had achieved. But then a surprising thing happened. Kepler threw out his solution. Kepler knew that Brahe's data were accurate to at least 4' of arc. There was no way he could come that close to the data with a circle. We can see now, said Kepler, why Ptolemy acquiesced in using an equant, for 8' was well within the limit of accuracy of his observations (10'); but to us Divine goodness has given us a most diligent observer in Tycho Brahe, and it is, therefore, right that we should with a grateful mind make use of this gift to find the true celestial motions. Even Copernicus should have been satisfied with Kepler's accuracy, but Kepler had to start again:[4]

> Who would have thought it possible? This hypothesis, which so closely agrees with the observed oppositions, is nevertheless false . . .

Kepler's next attempt was more systematic. He still believed the orbits had to be simple, but that the simplicity of the sun's force law apparently did not manifest itself in a circular orbit. This time he would allow himself no preconceptions about the shape of Mars' orbit. But first, he had to know the shape of the earth's orbit. Since the earth was a moving observation platform, its motion was mixed into all the observations of the apparent position of Mars. Copernicus had simply assumed that the earth's orbit was a perfect circle. Then, all the other planets turned out not to have the same center as the earth's orbit. Again, more Copernican than Copernicus, Kepler argued that the earth should not be a special planet, that quite possibly its orbit was similar to the other planets. This meant that he should not *assume* an orbit for the earth; he had to deduce it.

How can one find the earth's orbit when sitting on the earth? Kepler turned the problem around. He calculated the earth's orbit from the viewpoint of an observer on Mars. He supposed that Mars' orbit was roughly that described by Copernicus, but he did not assume it was a perfect circle or any other special shape. His only hypothesis was that Mars' orbit was closed, that Mars repeated the same orbit each time it went around the sun. Copernicus had calculated Mars' period of revolution around the sun to be 687 days. If the orbit were closed, Mars would return to exactly the same point in space every 687 days. Since the earth completes somewhat less than two full orbits in 687 days (see Table 3.2), it will arrive at different points in space every 687 days. Someone on Mars would see the earth against a different star background every 687 days. Similarly, observers on earth would see Mars from different vantage points. This permits the earth's distance from the sun to be calculated by triangulation. Figure 4-9 illustrates Kepler's procedure where, for convenience, we start the cycle of triangulation with earth at $E_1$ and Mars at $M$, in opposition. Then, 687 days later, Mars will again be at $M$, but the earth will be at another place, $E_2$.

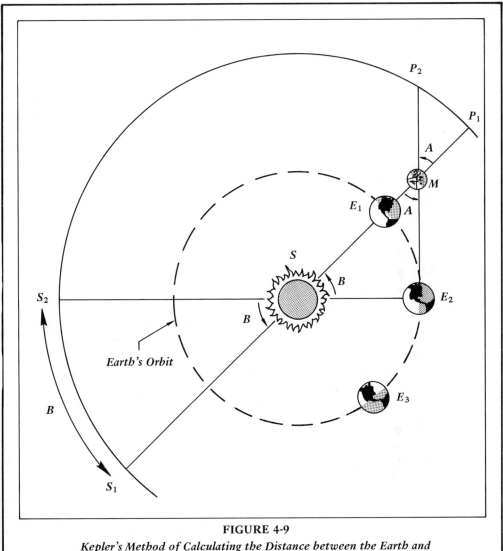

**FIGURE 4-9**
*Kepler's Method of Calculating the Distance between the Earth and the Sun along the Earth's Orbits by Triangulating on Mars.*

From Brahe's measurements Kepler knew the angle between $P_1$ and $P_2$ on the celestial sphere. If the celestial sphere were really very far away, then, the angle between $P_2$ and $P_1$ also would equal the angle $P_2 - M - P_1$, which we have labelled $A$ in Fig. 4-9. (The celestial sphere is drawn far too small in Fig. 4-9.) An elementary theorem of geometry states that angle $P_2 - M - P_1$ equals angle $S - M - E_2$, which we have also labelled $A$. Moreover, the angle, $E_2 - S - M$, called $B$ in Fig. 4-9, could also be calculated from the shift in the apparent position of the sun, $S$, against the background stars between times 1 and 2. Knowing $A$ and $B$, Kepler knew the shape of the triangle, $S - E_2 - M$; and, therefore, he knew the ratio of Mars' distance from the sun at $M$ to the earth's distance from the sun at $E_2$. 687 days later, the earth would arrive at $E_3$, where Kepler repeated the same calculation to find the ratio of the distance $S - E_3$ to $S - M$. By repeating this procedure many times, Kepler could calculate the shape of the earth's orbit. However, he only knew its size relative to the unknown distance from the sun to Mars, $S - M$.

To find $S - M$, Kepler had to determine Mars' orbit. Thus, he had to find both earth and Mars' orbit at the same time. Kepler calculated Mars' orbit by an inversion of the technique he used for earth. If he assumed the earth's orbit were closed and that he knew the earth's orbit exactly, he could use the same measurements to find Mars' orbit. For example, if the distance $S - E_2$ in Fig. 4-9 were known—from knowing the earth's orbit, it could be used as an arbitrary baseline. Then, knowing the angles $A$ and $B$ and, therefore, the ratio of $S - M$ to $S - E_2$; these values would yield one point on Mars' orbit. To find more points, Kepler applied the same procedure to other oppositions of Mars, when Mars was at other points in its orbit. In order for this procedure to work, Kepler had to alternate back and forth between earth and Mars, calculating and recalculating their orbits until he arrived at two orbits that were mutually consistent and agreed with all Tycho's data.

In fact, the above description is greatly oversimplified. Kepler's task was much more difficult. Brahe's data was not sufficiently accurate to distinguish between a circular orbit for the earth, with the sun slightly off center, and one of another shape. Thus, Kepler was not forced to abandon a simple circle for the earth. Moreover, Kepler needed speed laws for the earth's and Mars' motions, since he needed to calculate the earth's position at all times, and calculate Mars' position in between those points specified by Brahe's observations. Kepler tried several, each of which had the earth and Mars moving fastest when closest to the sun. We have already discussed his use of the equant. He also invented an "equal areas" speed law, which we will describe in detail shortly. Should he use an equal areas law for both earth and Mars, or an equant law for both, or an equant law for Mars and equal areas for earth, or *vice versa?* Thus, Kepler had not only to calculate two orbits but also to decide upon the appropriate speed law at the same time. Finally, he had to reconcile all his calculations and Brahe's observations with the fact that Mars' orbit lay in a plane intersecting the ecliptic at the sun at angle of 1° 51′.

Kepler found that Mars' orbit was clearly not a circle. It was some sort of oval figure. But what kind? Soon he realized that it was not just any oval, but the next best thing to a circle, an *ellipse*. An ellipse is an example of a "conic section," whose properties had been worked out by the late ancient Greeks. Imagine an ordinary cone, like an ice-cream cone, and slice it through with a plane. The intersection of the plane with the outside of the cone is a curved line called a conic section, a "cutting of the cone." If the cone is upright and the plane is horizontal, the curve is a circle. If the plane is tilted, the curve is an ellipse. Thus, a circle is a special case of an ellipse! Holding the cone fixed and upright, tilt the plane more and more. Eventually, it becomes parallel to the side of the cone, and the intersection is no longer a closed curve, but extends infinitely. This curve is also especially simple; it is a parabola. The circle and the ellipse are sketched in Fig. 4–10.

In an ellipse are two special points, called the foci (plural of focus). From any point on the ellipse, draw straight lines to both foci. The sum of the two distances is the same for *every* point on the ellipse. An ellipse is often defined in this way: choose two points, the foci, and then connect all the points having the property that the sum of the distances to the two foci is the same number. An ellipse is easy to draw with a string. Hold the string fixed at two points,

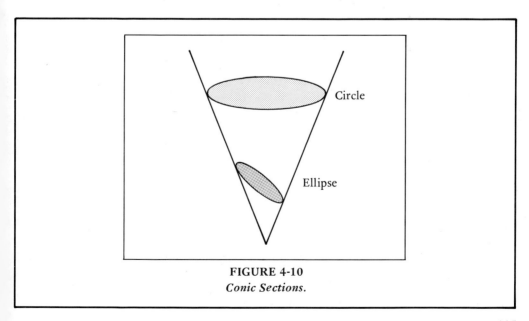

**FIGURE 4-10**
*Conic Sections.*

the foci, as in Fig. 4-11, and run the tip of a pencil around, keeping the string tight. The sum of the distances from the pencil to the two foci will always be the same, since it is the total length of the string, minus the distance between the foci.

Although not all ovals are ellipses, ellipses come in many shapes. Again, two special cases are important. First, suppose that the two foci are at the *same* point. This kind of ellipse has all its points the *same* distance from one fixed point, where both foci are. Of course, this is a circle, and the foci are at its center. In the opposite case, where the second focus gets farther and farther away from the first, the part of the ellipse near the first focus resembles a parabola.

Where was the sun? Now Kepler's conviction that the sun had a special role in determining Mars' motion was confirmed: the sun was at one focus of Mars' ellipse.

Kepler also had to find a speed law for Mars. He knew that Mars moves faster when it is closer to the sun. He even knew that the ratio of maximum to minimum speeds was the ratio of minimum to maximum distances. Precisely, this means the following: when Mars is closest to the sun, its position is called perihelion, so call the closest distance $R_p$ ($p$ for perihelion). When Mars is at the other end of its elliptical orbit, it is at aphelion, farthest from the sun. Call this distance $R_a$. $R_p / R_a$ is a number slightly smaller than one. Now, Mars' speeds at perihelion and aphelion are different. Let us call these speeds, $V_p$ and $V_a$. Then, $V_p$ is bigger than $V_a$. Mars moves faster when it comes

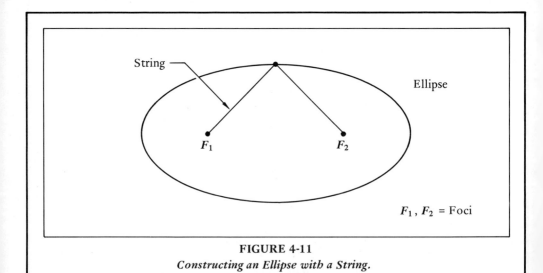

**FIGURE 4-11**
*Constructing an Ellipse with a String.*

closer to the sun. Kepler knew that the two speeds were inversely proportional to the distances, which means:

$$\frac{V_p}{V_a} = \frac{R_a}{R_p}$$

Kepler's equant system satisfied this rule. (An Appendix section to this chapter explains how this works.) But now, Kepler knew that Mars' orbit was not a circle at all. He tried the idea that the reciprocal speed rule worked everywhere around the orbit, not just at perihelion and aphelion. For example, if a planet's speed at aphelion is $V_a$, then, when the planet is 9/10 as far from the sun its speed is (10/9) × $V_a$. The rule worked reasonably well, but could not be made to reproduce Tycho Brahe's observations to the accuracy of four minutes of arc. Eventually, Kepler discarded it also.

The speed law Kepler finally found most convenient turned out to be exact for Mars' elliptical orbit. We have already referred to it as the equal areas law: A line drawn from the sun to Mars sweeps out equal areas in equal times. Precisely what this means is illustrated in Fig. 4–12. In Fig. 4–12, the two shaded regions have equal areas. Kepler's rule says that it takes Mars the same time to move from $A$ to $B$ as it does from $Q$ to $P$. Since the distance $AB$ has to be less than the distance $QP$ in order that the two areas be the same, Mars moves faster when it is nearer the sun. The difficulties Kepler faced and

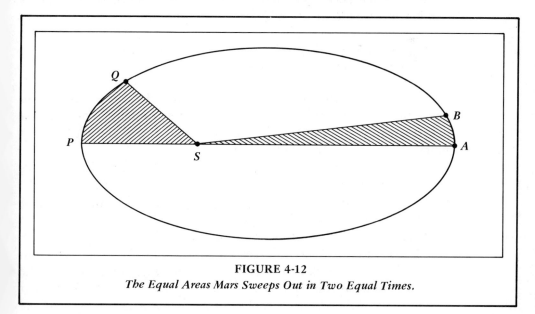

FIGURE 4-12
*The Equal Areas Mars Sweeps Out in Two Equal Times.*

overcame are exemplified by the fact (as we discuss in an Appendix section) that the equal area and the equant speed laws give the *same* result at aphelion and perihelion, and differ slightly only in-between.

### KEPLER'S LAWS

Kepler's interpretation of Copernicus demanded that the motion of all the planets, not just Mars, be determined in the same way by the sun. This remained to be proven. In subsequent years, Kepler showed that, within the accuracy of Tycho's data, the orbits of *all the planets* lie in planes. Like Mars, each of these planes passes through the sun and makes a small angle with the ecliptic. Moreover, all the planets obey the two other rules he found for Mars, which we now call Kepler's First and Second Laws:

> **KEPLER'S FIRST LAW:** The orbits of all the planets are ellipses, with the sun at one focus.

The fact that the sun is at a focus of all the ellipses more than compensated for the abandonment of Plato's circular dogma. The importance of harmony, symmetry, and order was as central to Kepler's thinking as it was to the Pythagorean Brotherhood.

> **KEPLER'S SECOND LAW:** A line drawn from the sun to each planet sweeps out equal areas in equal times.

With these two brief statements, equants, epicycles, deferents, and all other such devices vanished from planetary astronomy. For the first time, planetary positions could be calculated easily and with an accuracy that matched the observations.

Kepler also proved another surprising fact. The moon obeys Kepler's first two laws. The moon's orbit is also elliptical, with the earth at one focus, and the moon obeys the equal areas speed law. Not only the sun but also the earth can organize planetary motion. Even more important, both do it in the same way.

Kepler still had no rule relating the motion of one planet to another. His first two laws describe each planet separately. In each, the sun plays a crucial role, but that was all. Copernicus already knew that the farther a planet is from the sun the longer is its period and the slower is its speed. Is there a quantitative, numerical rule relating a planet's period, or speed, to its distance from the sun? Buried in Kepler's last important work, *Harmonice Mundi,* the "Harmonies of the World," and buried among pages about the music of the heavens lies what we call today Kepler's Third Law. It ties together the motions of the six planets. Historically, it is the most important of Kepler's three laws, for it led directly to Newton's law of universal gravitation. It is the solution to the puzzle of Table 3.2.

**KEPLER'S THIRD LAW:** The square of the period of revolution of a planet, divided by the cube of the radius of its orbit, is the same for all planets.

In mathematical notation, let $R$ be a planet's distance from the sun and $T$ be its period of revolution about the sun, as in Table 3.2. The combination:

$$\frac{T^2}{R^3}$$

is the *same number* for all six planets. Not only do all the planets' orbits have similar properties, they all share the common value of $T^2/R^3$. This was the strongest indication yet that the motions of all six planets have a common cause.

What is this important number, $T^2/R^3$? The numerical value of $T^2/R^3$ is not fundamental. It depends on the units we use to measure distances and times; i.e., whether we use feet, miles, or astronomical units for length and whether we use seconds, days, or years for time. Having made the choice, we can calculate the number, $T^2/R^3$, for any one planet. The content of Kepler's third law is that one obtains the same number for the other five planets. Another familiar form of Kepler's third law is:

$$T^2 = kR^3$$

where $k$ is some number that depends upon the units used to measure $R$ and $T$.

The calculation is especially easy if we agree to measure distances in astronomical units (A.U.) and periods of revolution in years. For the earth, $R = 1$ A.U. and $T = 1$ year; so, $T^2/R^3$ is $1^2/1^3 = 1/1 = 1$. Then, Kepler's third law says simply that for all the planets, $T^2/R^3 = 1$, or:

$$T^2 = R^3$$

It is easy to check that the rule is roughly right.* This fact alone might have satisfied Kepler. But, he discovered that if you do the arithmetic carefully, using accurate values for R and T, the third law comes out *exactly* right, as accuately as you can do the calculation.

In Table 4.2, the planets' distances from the sun ($R$) and their periods of revolution about the sun ($T$) are tabulated very precisely, so that Kepler's third law can be checked.

One should worry about a small point. Since the orbits are really ellipses, not circles, what is the right number to use for $R$ in Kepler's third law? Kepler's answer was that the right $R$ is the average of a planet's maximum and minimum distances from the sun. Table 4.2 records this average distance; if you use it for $R$ Kepler's third law is exactly right.

TABLE 4.2. *An Accurate Table of the Radii of the Planets' Orbits and Their Periods of Revolution.*

| Planet  | T (years) | R (A.U.) |
|---------|-----------|----------|
| Mercury | 0.241     | 0.387    |
| Venus   | 0.617     | 0.723    |
| Earth   | 1.000     | 1.000    |
| Mars    | 1.881     | 1.524    |
| Jupiter | 11.862    | 5.203    |
| Saturn  | 29.458    | 9.539    |

Finally, we can learn from Kepler's third law the rule relating a planet's speed to its distance, all in the approximation that they move at constant speed, $V$, along circles of radius $R$. Remember that a planet's speed is:

$$V = 2\pi \frac{R}{T}$$

We can solve this equation for $T$ to get:

$$T = 2\pi \frac{R}{V}$$

---

*Here are some approximate calculations you can almost do in your head. The point to check is that $R^3 = T^2$ for any planet where $R$ and $T$ are measured in A.U. and years.

Mercury's period is 88 days, about 1/4 year. From Table 4.2, we read that it is 0.39 A.U. from the sun; 4/10 A.U. is a good approximation. Therefore, $R^3 = (4/10)^3 = 4^3/10^3 = 64/1000 = 0.064$. $T^2 = (1/4)^2 = 1/16 = 0.0625$. The discrepancy is due to the approximations.

For Venus, $T = 225$ days, or 225/365 year, which is about 6/10 year; $R$ is 0.72 A.U., or about 7/10 A.U. Then, $R^3 = (7/10)^3 = 7^3/10^3 = 343/1000 = 0.343$, while $T^2 = (6/10)^2 = 6^2/10^2 = 36/100 = 0.36$.

Finally, we can check an outer planet, Jupiter. $R = 5.2$ A.U., while $T = 12$ years. Now, $12^2 = 144$, while $5^3 = 125$. Actually, $5.2^3$ is close to 141. Kepler's third law is really very good.

It follows squaring both sides that:

$$T^2 = 4\pi^2 \frac{R^2}{V^2}$$

Then, the universal constant in Kepler's third law, $T^2/R^3$, becomes:

$$\frac{T^2}{R^3} = \frac{(4\pi^2 R^2/V^2)}{R^3} = \frac{4\pi^2}{RV^2}$$

Therefore, $4\pi^2/RV^2$, and also $RV^2$ itself, is a universal constant; in A.U. and years, $T^2/R^3 = 1$, so:

$$\frac{4\pi^2}{RV^2} = 1$$

and so,

$$RV^2 = 4\pi^2$$

which can be written,

$$V^2 = 4\pi^2/R$$

The square of a planet's speed is inversely proportional to the planet's distance from the sun. From the more general form of Kepler's third law, $T^2 = kR^3$:

$$V^2 = \frac{4\pi^2}{kR}$$

Taking the square root, we get:

$$V = \frac{2\pi}{\sqrt{k}} \frac{1}{\sqrt{R}}$$

The speed decreases proportionally to $1/\sqrt{R}$. If one planet is four times as far away from the sun as another, it goes half as fast. We will need to know this fact when we come to study Newton. Kepler's third law is usually quoted as proportionality between $T^2$ and $R^3$. But, more of its simplicity appears when we rewrite it in terms of $V$ and $R$.

Kepler's insistence that physics rules the planets' motion had proven to be a powerful guide leading him through the labyrinth of his calculations to a simple description of the solar system. Thousands of years of planetary observations could be summarized in three sentences. Nonetheless, he went on to search fruitlessly for other mathematical laws in the heavens, for example, a connection between the planets' orbits and the rules of musical harmony. During his life, he also sought mathematical order in optics, the study of how light moves, which we will discuss in Chapter 8. He gave the first mathematical analysis of how a telescope works. He even had time to write a "science-fiction" story about a trip to the moon. He died in 1630, still wondering what the sun's force might be.

*Summary*

*Tycho Brahe revived the ancient tradition of making accurate astronomical observations. Brahe became famous by determining that the nova of 1672, and a comet five years later, had no observable parallax and were, therefore, beyond the moon, in direct contradiction to Aristotelian teaching. On an island given to him by the King of Denmark, he constructed the most accurate instruments ever made for observing the heavens.*

*He constructed an extremely accurate map of the stellar sphere, and made many measurements of the positions of the planets, recording their motions relative to the fixed stars, at many points around their orbits. His data was to provide Kepler the basis for his study of the laws of planetary motion. It is believed that the errors in his measurements were less than four minutes of arc.*

*Tycho proposed a compromise between the ancient geocentric and Copernicus' heliocentric model, one in which the earth remained at the center, circled by the sun, moon, and the outer planets while Mercury and Venus revolved about the sun.*

*In 1599, Tycho moved to Prague where the Emperor Rudolf II appointed him the Imperial Mathematician. In 1600, Johannes Kepler came to Prague to study Brahe's observations. Kepler had already proposed that by fitting snugly the surfaces of the five regular polyhedra between six concentric spheres one could explain not only the number of the planets, but also the sizes of their orbits. For eight years, Kepler studied the motions of the planets, especially Mars. He became Imperial Mathematician upon Brahe's death in 1601. Ultimately, he discovered three laws that completely describe the motions of the planets:*

1. *The orbits of all the planets are ellipses, with the sun at one focus.*
2. *A line drawn from the sun to a planet sweeps out equal areas in equal times.*
3. *If R is the average radius of a planet's orbit, and T is its period of revolution around the sun, the ratio, $T^2/R^3$, is the same for all six planets.*

*Another form of the third law is the statement that a planet's speed is inversely proportional to the square-root of its average distance from the sun.*

## Appendix to Chapter 4

### The Five Regular Solids

A polygon is a plane figure bounded by lines, and is "regular" if all the lines are the same length and all the angles between adjacent lines are the same. Regular polygons may have any number of sides, starting with three. The first two are the equilateral triangle and the square, which have three and four sides, respectively.

A polyhedron is a three-dimensional solid figure whose sides are planes. It is regular if all the planes are identical regular polygons, and all the corners look alike. The cube is a familiar example.

Unlike polygons, regular polyhedra cannot be made with any number of sides. There are only five different regular polyhedra. Most people find this result startling when they hear it for the first time. You can imagine how wonderful it must have seemed to Kepler, whose worship of numbers was so strong that he has often been described as a Pythagorean. Like Pythagoras, he felt that the physical world should reflect the wonders of mathematics, and gave this theorem a key place in his cosmology. The following proof is very elementary, although it was not known to the ancient Greeks:

Any polyhedron is characterized by its number of sides, edges, and vertices (corners). Let's assign them symbols:

$E$ = number of edges,
$V$ = number of vertices,
$S$ = number of sides.

For a cube, for example, $E = 12$, $V = 8$, and $S = 6$. Now, a relation exists among $E$, $V$, and $S$, which is satisfied by all polyhedra, regular or not. It is easiest to imagine constructing a regular solid out of bailing wire for the edges, and stretched balloons for the sides. Now imagine constructing a new vertex in the middle of one of the sides by adding some pieces of wire (e.g., see Fig. 4–13) and then by stretching the wire out to make a new solid, with one more vertex. We have created the same number of new sides as edges. But, one side was there before, so the difference, $S - E$, has decreased by one. The number of vertices, $V$, has increased by one, so we get the remarkable result that the difference, $V + S - E$, is unchanged! Now, if you think about it, you can convince yourself that any solid figure can be made from another with fewer vertices by this process; namely, adding a vertex, which does not change $V + S - E$, and then stretching the new figure to the desired shape, which obviously does not change $E$, $S$, or $V$. Therefore, $S + V - E$ is the same for all polyhedra, regular or not! You can discover

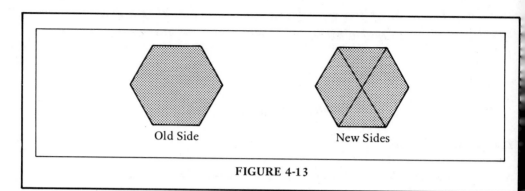

Old Side    New Sides

**FIGURE 4-13**

its value by looking at the most familiar shape, a cube: For a cube, $S + V - E = 6 + 8 - 12 = 2$. We have proved the following relation for *all* polyhedra:

$$S + V - E = 2$$

Check this for some other solid, like a pyramid. This is a theorem from a modern branch of mathematics called topology.

The rest is easy. If a polyhedron is regular, each side is a polygon with the same number of sides; and each vertex is identical to all the others, having the same number of lines meeting at it. Let us give these numbers names:

$n$ = number of edges of each side and
$r$ = number of edges meeting at each vertex.
For a cube, $n = 4$ and $r = 3$.

Now, the total of all the edges attached to all the vertices is $rV$. But, this counts each edge twice; therefore:

$$E = \frac{1}{2} rV$$

Similarly, the total number of edges of all the polygons is $nS$. But, this counts each edge twice also, so:

$$E = \frac{1}{2} nS$$

Next we solve the above equations for $V$ and $S$, to get:

$$V = \frac{2E}{r}$$

$$S = \frac{2E}{n}$$

These equations make sense only for regular polyhedrons. Now use the two equations immediately above to substitute for $V$ and $S$ in the general rule, $S + V - E = 2$:

Appendix to Chapter 4

$$\frac{2E}{r} + \frac{2E}{n} - E = 2$$

Divide each term by $2E$:

$$\frac{1}{r} + \frac{1}{n} - \frac{1}{2} = \frac{1}{E}$$

Add ½ to both sides:

$$\frac{1}{r} + \frac{1}{n} = \frac{1}{2} + \frac{1}{E}$$

The numbers $r$, $n$, and $E$ are all positive whole numbers. The right-hand side of the equation above is always greater than 1/2. Therefore, $1/r + 1/n$ must be greater than 1/2. Since $1/4 + 1/4 = 1/2$, we learn that if $r$ and $n$ are both greater than 3, $1/r + 1/n$ will be 1/2 or smaller, so the equation cannot be satisfied.

If $r = 3$, then:

$$\frac{1}{n} = \frac{1}{2} - \frac{1}{3} + \frac{1}{E} = \frac{1}{6} + \frac{1}{E}$$

Therefore, $1/n$ is greater than 1/6, so $n$ must be less than six. Similarly, if $n = 3$, $r$ must be less than six. Therefore, there can be only five possible cases, since by their geometrical meaning neither $r$ nor $n$ can be less than three:

| $r$ | $n$ | $E$ | $V$ | $S$ | Name | Shape of Sides |
|---|---|---|---|---|---|---|
| 3 | 3 | 6 | 4 | 4 | tetrahedron | triangle |
| 3 | 4 | 12 | 8 | 6 | cube | square |
| 4 | 3 | 12 | 6 | 8 | octahedron | triangle |
| 3 | 5 | 30 | 20 | 12 | dodecahedron | pentagon |
| 5 | 3 | 30 | 12 | 20 | icosahedron | triangle |

*More on Kepler's Speed Rule*

To see how difficult Kepler's problem was, let us understand how the equant device, as in Fig. 4-8, and the final form of the second law both agree with the observation that a planet's speed at perihelion, divided by its speed at aphelion, is the same as the ratio of the aphelion distance to the perihelion distance.

The equant model is illustrated in Fig. 4-8. The planet moves along its orbit so that a line to the equant point, $Eq$, sweeps out the same number of degrees each day.* Now, the equant point and the sun are chosen to be the *same* distance from the center. Therefore, when the planet is at aphelion, $A$, its distance to the sun is the aphelion distance; $R_a$, but its distance to $Eq$ is the perihelion distance, $R_p$. Similarly, when the planet is at perihelion, its distance to $Eq$ is $R_a$.

---

*The orbit is drawn as a circle in Fig. 4-8, but this is not necessary to the argument.

When the planet is near aphelion, how far does it move while a line drawn from the planet to $Eq$ moves through an angle of $1°$? In Fig. 4-14, $AB$ is the segment of the planet's orbit. The angle at $Eq$ is $1°$. The distance $EqA$ is $R_p$. Therefore, the arc $AB$ is 1/360 of the circumference of a circle whose radius is $R_p$, or:

$$AB = \frac{1}{360} \times 2\pi R_p$$

Let $T$ be the time the planet takes to go from $A$ to $B$. The planet's speed is, therefore:

$$V_a = \frac{2\pi R_p/360}{T}$$

Next we can do exactly the same calculation at perihelion. The distance from $P$ to $Q$ is $R_a$. So, when a line drawn from the planet to $Q$ sweeps out one degree, the planet moves a distance $2\pi R_a/360$. The time elapsed must also be $T$, since the imaginary line rotates at a uniform speed. The speed at perihelion is, therefore:

$$V_p = \frac{2\pi R_a/360}{T}$$

Divide the equation for $V_a$ by the equation for $V_p$. The factors, $2\pi / 360T$, drop out leaving:

$$\frac{V_a}{V_p} = \frac{R_p}{R_a}$$

This result is also a consequence of Kepler's second law. At perihelion, the earth is moving at right angles to the line $SP$ in Fig. 4-15. In Fig. 4-16, a small segment of the path near $P$ is enlarged.

Choose a small time interval, and call it $T$. In Fig. 4-16, the planet's true path during this time is $PE$, and the area that enters into Kepler's second law is bounded by the lines $PE$, $PS$, and $ES$. Now we make an approximation and calculate the area of the triangle $PFS$ instead. If $T$ is small, the approximation will be very good. Figure 4-16 greatly exaggerates the difference between $PE$ and $PF$.

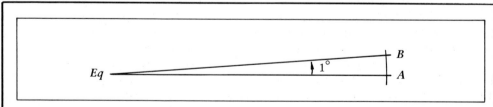

FIGURE 4-14

A Planets Motion through One Degree in the Equant Model.

Appendix to Chapter 4

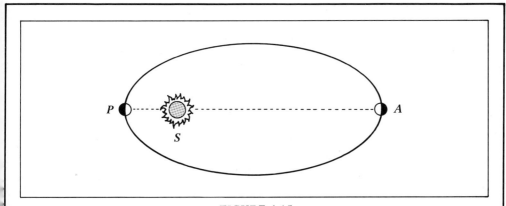

**FIGURE 4-15**
*A Planet's Elliptical Orbit, with Sun at One Focus.
P is Perihelion, A is Aphelion.*

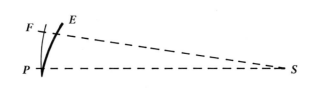

**FIGURE 4-16**
*A Small Segment of a Planet's Orbit Near Perihelion.*

The triangle *SPF* is a right-angled triangle. (Exactly at perihelion the planet is neither coming closer to the sun nor going away from it. This is possible only if its motion is exactly perpendicular to the line *PS*.) Its area is, therefore, 1/2 × *SP* × *PF*. Since the speed is $V_p$, the distance *PF* is given by $V_p T$; then:

$$\text{Area} = \frac{1}{2} R_p V_p T$$

Now we can repeat the calculation at aphelion. By an identical argument, the area swept out by a line from the planet to the sun in time *T* is:

$$\text{Area} = \frac{1}{2} R_a V_a T$$

By Kepler's second law, the two areas are equal:

$$\frac{1}{2} R_p V_p T = \frac{1}{2} R_a V_a T$$

137

Dividing $(1/2)\,T$ out from both sides, we get:

$$R_p V_p = R_a V_a$$

or,

$$\frac{V_a}{V_p} = \frac{R_p}{R_a}$$

## QUESTIONS

1. Since Kepler's time, three new planets have been discovered with the telescope. All are beyond Saturn, and are not bright enough to be seen with the naked eye. The following table lists their average distances from the sun.

   | Planet  | Distance (in Astronomical Units) |
   |---------|----------------------------------|
   | Uranus  | 19.18                            |
   | Neptune | 30.06                            |
   | Pluto   | 39.52                            |

   Use Kepler's third law to calculate their periods in years.

2. Comets are small, unusual objects that orbit the sun, obeying Kepler's three laws just as the planets do. They are probably made up of rocks and ice. Usually invisible, they become very bright when near the sun, and show a spectacular "tail" that stretches across the sky. Most famous is Halley's comet. The first record of its observation dates back to 240 B.C. Its orbit was determined by the astronomer Edmund Halley, who was a friend of Isaac Newton and preceded him as president of the Royal Society in London.

   Unlike the planets, the comets have orbits that are not almost circular, but usually are very elongated ellipses, The orbit of Halley's comet is very long and narrow, shaped like a cigar. Most of the time it moves very slowly, and is far from the sun and invisible. As Halley's comet approaches the sun, it grows brighter and brighter and moves faster and faster, according to Kepler's second law. For a few months, when it is closest to the sun, Halley's comet grows a splendid tail, stretching a vast distance across the sky (nobody knows exactly why, even today). Halley's comet is so bright that for a few weeks it is visible in the daytime.

   Halley's comet gets closer to the sun than Venus. Its nearest approach (perhelion) is 0.59 A.U. At its most distant point, it is farther than Neptune: 35.32 A.U. What is its period of revolution? Halley's comet was last seen in 1910. When will it reappear? Will you live to see it?

3. Kepler's third law applies also to the satellites of the earth, although Kepler did not know this, since in his day the earth had only one satellite, the moon. Today, in addition to the moon, many artificial satellites orbit around the earth. They all move in ellipses, with the center of the earth at one focus. A line drawn from the center of any satellite to the center of the earth sweeps out equal areas in equal times. Furthermore, let $R$ be the average of the satellites' maximum and minimum distance from the earth's center, and $T$ its period of revolution around the earth. Then, $T^2/R^3$ is the same number for all the earth's satellites in much the same way as for the sun's planets. However, the ratio $T^2/R^3$ for the earth's satellites is not the same number as $T^2/R^3$ for the planets. In other words, if we write Kepler's third law as $T^2 = kR^3$, then, this $k$ is not the same $k$ that appears in the law for the planets.

The earth is about 4000 miles in radius, and the moon's average distance from the earth's center is about 240,000 miles. Any low artificial satellite, just above the earth's surface, takes about 84 minutes to circle the earth. Use Kepler's third law to compute the moon's period of revolution about the earth.

4. The time interval between two successive new moons is about 29.5 days. This is a "month." If you did problem 3 correctly, you got a shorter period. What is the reason for the discrepancy?  *earths orbit*

5. For some purposes, it is useful to have an artificial satellite remain stationary above one point on the earth's surface. To do this, the satellite must be launched into orbit at just the right height so that its period will be twenty-four hours. What is that height?

6. Estimate the time it takes a spacecraft to go from the earth ($R$ = 1 A.U.) to Jupiter ($R$ = 5.2 A.U.). Imagine that the spacecraft's trajectory is a portion of an elongated elliptical orbit around the sun, whose perihelion is the radius of the earth's orbit and whose aphelion is the radius of Jupiter's orbit. (After the initial take off, the ship must coast on gravity to Jupiter.)

7. What is the ratio of the radius of a sphere circumscribed about a cube to the radius of an inscribed sphere? Is the ratio close to the ratio of the radii of the orbits of any two successive planets?

8. To what extent did Kepler actually *prove* that Copernicus was right?

9. Unlike the ancients, Kepler believed that using ever better observations would lead him to an ever simpler description of nature. While Tycho's data was the best available, it still had an uncertainty of roughly four minutes of arc. Given this uncertainty, how could Kepler be sure the planet's orbits are *exactly* ellipses?

10. You are the financial advisor to King Frederick of Denmark. Tycho Brahe has just come to him requesting more money for his research. In particular, Tycho wants to collect even more data on the orbits of the planets. You must advise the King on whether further investment in Tycho is wise. What will you tell him?

## REFERENCES

1. Dreyer, J. L. E. *Tycho Brahe* (Edinburgh, 1890), p. 27.
2. Kepler *Cosmographic-Mystery,* quoted by A. Koestler, *The Sleepwalkers* (Universal Library, New York, 1963).
3. Kepler *The New Astronomy,* quoted by Koestler, *The Sleepwalkers* (Universal Library, New York, 1963).
4. *Ibid.*

**GALILEO**
*(Photographed by Barry Donahue and by permission of the Houghton Library, Harvard University)*

CHAPTER

# 5

# GALILEO GALILEI

Kepler believed in the music of the spheres, and in later life tried to discover harmonies in the orbits of the planets. He thought that his difficulties might be due to his lack of understanding of some new discoveries in music—in particular, the appearance of polyphony—in the late Renaissance. He even read the recent book by Galilei on the new music, an influential work on harmony by a man whose compositions are still sometimes performed. Kepler was reading the wrong Galilei. The book Kepler was reading was by Galileo's father, Vincenzio Galilei, a cultured man who was active in the literary and artistic circles of Renaissance Florence and who provided a rich cultural and intellectual environment to nurture his son's many interests.

In 1564, Michelangelo died, and Shakespeare and Galileo Galilei were born. Galileo was the founder of modern mechanics and the great propagandist of the Copernican Revolution. At first a quiet and careful critic of Aristotelian physics, at the age of 43 he turned his telescope on the heavens and reported what he saw. Galileo wrote clearly and eloquently, so much so that today he is considered one of the founders of modern Italian prose style. He discovered the correct laws of motion for falling bodies and projectiles; he performed experiments where others had been content to speculate. His astronomical discoveries and his terrestrial mechanics were closely intertwined, and he emphasized the identity of terrestrial and celestial physical laws. He argued continuously and passionately against the ancient physics and cosmology of Aristotle. He was merciless in criticism of those who held what he considered ancient views. For his beliefs and behavior, he was put on trial by the Inquisition.

In 1581, Galileo entered the University of Pisa to study medicine, which was taught following Aristotle. In his classes, he soon developed a reputation for being obstinate and argumentative. He would remain so all his life. After two years, Galileo grew disenchanted with academic medicine and begged a family friend to tutor him in mathematics, which he had never studied before. He was a quick and apt student. But, it was not the search for abstract generalizations, which characterizes pure mathematics, that interested Galileo, so much as the preoccupation to be precise and coherent in discussing concrete problems. The elegant, economical proofs of mathematical theorems were for Galileo models of the way physical reasoning ought to be done. It ought to be as sequential, consistent, and logical as mathematics. Kepler's mystical flights of fancy were not for Galileo.

In the same year, 1583, Galileo made his first discovery in physics; it illustrates his attentive observation of every day events combined with sharp, physical insight. During a service at the Cathedral at Pisa, Galileo noticed the lamps hanging in the chandelier. They were all hung from ropes of the same length, but some were swinging through large arcs, others through small. Galileo noticed that, nevertheless, they all took the same time to complete one swing. This insight led to Galileo's first discovery in mechanics: The Law of the Pendulum. Notice how he formulated the question. He focused on the quantitative features of the pendulum's motion. He asked, "What is the mathematical relationship between the time a pendulum takes to swing back and forth, and the other numbers that describe the motion: the length of the string, the weight of the bob, and the size of the initial displacement?" Galileo's understanding of the pendulum eventually led to a great improvement in time-keeping, the pendulum clock.

In 1585, Galileo returned to Florence where he tutored occasional students, performed some minor researches, and tried to find a position. Finally, after four years, he got a three-year appointment to teach mathematics at Pisa; there he gave lectures on Euclid's geometry and Ptolemy's astronomy. It is not known exactly when he became a Copernican, but he certainly continued to lecture on Ptolemy's system long after he stopped believing in it.

At Pisa, Galileo studied mechanics, particularly the works of the medieval Parisian school whose ideas about "impetus" foreshadowed Galileo's own ideas about inertia. A legend, probably false, states that during this period Galileo dropped two balls of different weights from the Leaning Tower of Pisa, to test Aristotle's assertion that heavy objects fall faster than light ones. Both balls struck the ground nearly simultaneously. While this experiment had been performed many times before, Galileo continually referred to it as an example of the necessity of consulting nature rather than ancient authorities.

In 1592, Galileo moved to the university at Padua in the Venetian republic. His salary was only slightly better, but his prospects for advancement were good. Galileo's wit and charm gained him powerful and influential friends in nearby Venice. He also made many visits to the Venetian arsenal to discuss technical problems with the highly skilled artisans there. From them he must have learned useful techniques for fabricating experimental apparatus, as well as how cannonballs really do move through the air. From this practical problem, he would eventually abstract a general theory of projectile motion. Galileo said that the 18 years he spent at Padua were his happiest. They were scientifically productive. During this time he arrived at many of his great discoveries in mechanics, which he published only near the end of his life. At Padua, he began defending Copernicus, if in a low-key fashion. In 1598, he wrote Kepler to thank him for a copy of the *Cosmographic Mystery*, and admitted that in the last several years he had developed several arguments in favor of Copernicus. Although he still continued to lecture dutifully on Ptolemy, he did give three public lectures on the Nova of 1604 in which he discussed the Copernican theory.

From this brief summary of Galileo's career up to 1609, we can see the development of his mature ideas. They were, first, a belief that a successful description of motion was a description of its quantitative, mathematical characteristics; second, a resort to experiment and observation, rather than citation of authorities (principally Aristotle) to resolve scientific questions; and, finally, a growing conviction that the Copernican model of the universe might be correct.

### GALILEO'S ASTRONOMICAL DISCOVERIES

The year 1609 was a momentous one for astronomy. In Prague, Kepler was publishing his theory of the orbit of Mars in his *Astronomia Nova*. It probably contained the best work one could do with naked-eye observations. In Padua, astronomy was taking a new turn. Galileo was developing a machine to improve on the information man could obtain with his natural senses. It was a final, total departure from the methods of the ancients. The machine, of course, was the telescope, which had recently been invented in Holland. It was Galileo, though, who fully realized the telescope's scientific potential. Following Dutch descriptions, he manufactured one of his own. Here are his own words:[1]

> About ten months ago a report reached my ears that a certain Fleming had constructed a spyglass by means of which visible objects, though very distant from the eye of the observer, were distinctly seen as if nearby. Of this truly remarkable effect several experiences were related, to which some persons gave credence while others denied them. A few days later the report was confirmed to me in a letter from a noble Frenchman at Paris, Jacques Badovere, which caused me to apply myself wholeheartedly to inquire into the means by which I might arrive at the invention of a similar instrument. This I did shortly afterwards, my basis being the theory of refraction. First I prepared a tube of lead, at the ends of which I fitted two glass lenses, both plane on one side while on the other side one was spherically convex and the other concave. Then placing my eye near the concave lens I perceived objects satisfactorily large and near, for they appeared three times closer and nine times larger than when seen with the naked eye alone. Next I constructed another one, more accurate, which represented objects as enlarged more than sixty times. Finally, sparing neither labor nor expense, I succeeded in constructing for myself so excellent an instrument that objects seen by means of it appeared nearly one thousand times larger and over thirty times closer than when regarded with our natural vision.

Galileo was a sensation when he demonstrated his telescope to the Venetians in St. Mark's square. Venice, a great naval power, was easily convinced of the military importance of the telescope as a defense against invasion by the sea. The grateful republic offered Galileo a large salary and life tenure as professor at Padua, provided only that he discharge faithfully his teaching duties.

Galileo directed his instrument towards the sky, and made five great discoveries. First, that there are many more stars than appear to the unaided eye. The Milky Way is composed of stars, too faint to be seen individually with the naked eye. Second, that the moon has mountains just like the earth. Third, that the planet Jupiter has four moons, which form a miniature solar system. Fourth, that Venus has phases like the moon, as predicted by the Copernican theory. And fifth, that the sun has spots, which grow and diminish in intensity, and move slowly across the sun's disc. We will discuss each separately.

Galileo published his first discoveries in a short pamphlet, *Sidereus Nuncius,* or *The Starry Messenger* (1610). Galileo's style is worlds apart from Kepler's difficult, mystical *Astronomia Nova.* As the excerpts here show, Galileo communicated his mood of breathless excitement at seeing things never seen before. It took him a few years more to organize his argument clearly, but it is obvious that he considered each discovery a further refutation of the "Aristotelians". Here is the beginning of Galileo's book:

> Great indeed are the things which in this brief treatise I propose for observation and consideration by all students of nature. I say great, because of the excellence of the subject itself, the entirely unexpected and novel character of these things, and finally because of the instrument by means of which they have been revealed to our senses.
>
> Surely it is a great thing to increase the numerous host of fixed stars previously visible to the unaided vision, adding countless more which have never before been seen, exposing these plainly to the eye in numbers ten times exceeding the old and familiar stars.
>
> It is a very beautiful thing, and most gratifying to the sight, to behold the body of the moon, distant from us almost sixty earthly radii, as if it were no farther away than two such measures—so that its diameter appears almost thirty times larger, its surface nearly nine hundred times, and its volume twenty-seven thousand times as large as when viewed with the naked eye. In this way one may learn with all the certainty of sense evidence that the moon is not robed in a smooth and polished surface but is in fact rough and uneven, covered everywhere, just like the earth's surface, with huge prominences, deep valleys, and chasms.
>
> Again, it seems to me a matter of no small importance to have ended the dispute about the Milky Way by naming its nature manifest to the very senses as well as to the intellect. Similarly it will be a pleasant and elegant thing to demonstrate that the nature of those stars which astronomers have previously called "nebulous" is far different from what has been believed hitherto. But what surpasses all wonders by far, and what particularly moves us to seek the attention of all astronomers and philosophers, is the discovery of four wandering stars not known or observed by any man before us. Like Venus and Mercury, which have their own periods about the sun, these have theirs about a certain star that is conspicuous among those already known, which they sometimes precede and sometimes follow, without ever departing from it beyond certain limits. All these facts were discovered and observed by me not many days ago with the aid of a spyglass which I devised, after first being illuminated by divine grace.

When Galileo turned his telescope toward the moon, he saw that it was not smooth, as the philosophers had always taught, but rough, with mountains and valleys. The moon has many large craters. As the moon goes around the earth, the sun's rays strike its surface at a changing angle, and Galileo could see the shadow of a crater's walls slowly move across the flat bottom. The line dividing the light and dark portions of the moon is highly irregular. Beyond the line, in the dark portion, Galileo could see little points of light; he guessed correctly that these were mountains so high that their peaks were still in sunlight, while their bases were already in shadow. He even measured their heights, and found them comparable in size to mountains on the earth. (His method is explained in an Appendix section.)

These findings were in direct contradiction to Aristotelian and Scholastic doctrine, which held that the objects in the heavens, partaking of perfection, should be smooth, perfect, geometrical objects, polished spheres. Not only was the moon not a perfect sphere but, even worse, it was more or less like the earth. The fact that the height of the moon's mountains was comparable to the height of mountains on the earth suggested that at least in one way the moon, a celestial body, was not very different from the earth.

Next, Galileo turned his telescope to the fixed stars. Whereas the moon was enlarged by the telescope and the planets appeared as disks rather than points of light, the stars remained twinkling points without any measurable size. What really impressed Galileo was the vast number of stars his telescope enabled him to see. He wrote that:

> ... these are so numerous as almost to surpass belief. One may, in fact, see more of them than all the stars included among the first six magnitudes. The largest of these, which we may call stars of the seventh magnitude, or the first magnitude of invisible stars, appear through the telescope as larger and brighter than stars of the second magnitude when the latter are viewed with the naked eye. In order to give one or two proofs of their almost inconceivable number, I have adjoined pictures of two constellations. With these as samples, you may judge of all the others.
>
> In the first I had intended to depict the entire constellation of Orion, but I was overwhelmed by the vast quantity of stars and by limitations of time, so I have deferred this to another occasion. There are more than five hundred new stars distributed among the old ones within limits of one or two degrees of arc. Hence to the three stars in the Belt of Orion and the six in the Sword which were previously known, I have added eighty adjacent stars discovered recently, preserving the intervals between as exactly as I could. To distinguish the known or ancient stars, I have depicted them larger and have outlined them doubly; the other (invisible) stars I have drawn smaller and without the extra line. I have also preserved differences of magnitude as well as possible.
>
> In the second example I have depicted the six stars of Taurus known as the Pleiades (I say six, inasmuch as the seventh is hardly ever visible) which lie within very narrow limits in the sky. Near them are more than forty others, invisible, no one of which is much more than half a degree away from the original six. I have shown thirty-six of these in the diagram; as in the case of Orion I have preserved their intervals and magnitudes, as well as the distinction between old stars and new.
>
> Third, I have observed the nature and the material of the Milky Way.

*The Belt and Sword of Orion*

With the aid of the telescope this has been scrutinized so directly and with such ocular certainty that all the disputes which have vexed philosophers through so many ages have been resolved; and we are at last freed from wordy debates about it. The galaxy is, in fact, nothing but a congeries of innumerable stars grouped together in clusters. Upon whatever part of it the telescope is directed, a vast crowd of stars is immediately represented to view. Many of them are rather large and quite bright, while the number of smaller ones is quite beyond calculation.

But it is not only in the Milky Way that whitish clouds are seen; several patches of similar aspect shine with faint light here and there throughout the aether, and if the telescope is turned upon any of these it confronts us with a tight mass of stars. And what is even more remarkable, the stars which have been called "nebulous" by every astonomer up to this time turn out to be groups of very small stars arranged in a wonderful manner. Althoug!. each star separately escapes our sight on acccount of its smallness or the

*The Pleiades*

immense distance from us, the mingling of their rays gives rise to that gleam which was formerly believed to be some denser part of the aether that was capable of reflecting rays from the stars or from the sun.

The star-maps above are replicas of Galileo's own maps of Orion and the Pleiades, with the new stars added.

The universe was full of things never before suspected, suggesting that it was truly vast. This idea was unthinkable to most philosophers of his day. Did not all heavenly bodies have a purpose? Was it possible that the universe had *not* been created with humanity in mind? What were the new stars that one could not see *for*?

Most exciting of all was the fact that Jupiter had four satellites. Here is Galileo's blow-by-blow account of their discovery:

> On the seventh day of January in this present year, 1610, at the first hour of night, when I was viewing the heavenly bodies with a telescope, Jupiter presented itself to me; and because I have prepared a very excellent instrument for myself, I perceived (as I had not before, on account of the weakness of my previous instrument) that beside the planet there were three starlets, small indeed, but very bright. Though I believed them to be among the host of fixed stars, they aroused my curiousity somewhat by appearing to lie in an exact straight line parallel to the ecliptic, and by their being more splendid than others of size. Their arrangement with respect to Jupiter and each other was the following:

that is, there were two stars on the eastern side and one to the west. The most easterly star and the western one appeared larger than the other. I paid no attention to the distances between them and Jupiter, for at the outset I thought them to be fixed stars, as I have said. But returning to the same investigation on January eighth—led by what, I do not know—I found a very different arrangement. The three starlets were all to the west of Jupiter, closer together, and at equal intervals from one another as shown in the following sketch:

At this time, though I did not yet turn my attention to the way the stars had come together, I began to concern myself with the question how Jupiter could be east of all these stars when on the previous day it had been west of two of them. I commenced to wonder whether Jupiter was not moving eastward at that time, contrary to the computations of the astronomers, and had got in front of them by that motion. Hence it was with great interest that I awaited the next night. But I was disappointed in my hopes, for the sky was covered with clouds everywhere.

On the tenth of January, however, the stars appeared in this position with respect to Jupiter:

that is, there were but two of them both easterly, the third (as I supposed) being hidden behind Jupiter. As at first, they were in the same straight line with Jupiter and were arranged precisely in the line of the zodiac. Noticing this, and knowing that there was no way in which such alterations could be attributed to Jupiter's motion, yet being certain that these were still the same stars I had observed (in fact no other was to be found along the line of the zodiac for a long way on either side of Jupiter), my perplexity was now transformed into amazement. I was sure that the apparent changes belonged not to Jupiter but to the observed stars, and I resolved to pursue this investigation with greater care and attention.

And thus, on the eleventh of January, I saw the following disposition:

There were two stars, both to the east, and the central one being three times as far from Jupiter as from the one farther east. The latter star was nearly double the size of the former, whereas on the night before they had appeared approximately equal.

I had now decided beyond all question that there existed in the heavens three stars wandering about Jupiter as do Venus and Mercury about the sun, and this became plainer than daylight from observations on similar occasions which followed. Nor were there just three such stars; four wanderers complete their revolutions about jupiter, and of their alterations as observed more precisely later on we shall give a description here.

Here was a miniature Copernican system for all to see. Not only the sun and the earth but now also the planet Jupiter governed systems of satellites. And they all do it in the same way. The moon was not the only celestial body that did not orbit the sun. Galileo himself noticed that the nearer a satellite was to Jupiter the faster it moved. Later, when Kepler obtained a telescope and was able to observe the motions of Jupiter's satellites, he showed that they obeyed his three laws: (1) that their orbits were ellipses with Jupiter's center at one focus, (2) that each satellite sweeps out equal areas in equal times, and (3) most important, that the system obeyed Kepler's third law. The same laws that describe the sun's system of satellites (the six planets) also work for Jupiter's system of satellites* (Jupiter's four moons).

The *Starry Messenger* sold out immediately. The English Ambassador in Venice wrote to King James I that Galileo would be either the most famous man of all times, if he was proved right, or, otherwise, the most ridiculous. Kepler wrote Galileo a long letter of congratulation. Telescopes were in great demand. Those who could get one eagerly sought to confirm Galileo's discoveries. Jesuit teachers and scholars were particularly impressed, and translated the pamphlet into one language after another. Within five years, its contents were summarized in Chinese by a Jesuit missionary in Peking. Astronomers and mathematicians at the Roman College, the Pope's official scholars, were initially sceptical, but soon confirmed his observations. Galileo was invited to make a formal visit to Rome, where he was highly honored and had an audience with Pope Paul V. He was asked to be a member of the world's first scientific academy, the *Academia dei Lincei,* an honor that he treasured all his life.

But there was criticism, for Galileo's discoveries were disturbing. They contradicted two thousand years of accepted tradition. Many found it easier to believe that the new "planets" were mere illusions created by the telescope. Galileo was quick to reply that he would

---

*In this context Kepler's third law means simply that if $T$ is a satellite's period of revolution about Jupiter, and $R$ is its average distance from the planet's center, then, $T^2/R^3$ is the same number for all four moons. This number is not the same as $T^2/R^3$ for the planets, and is even a different number for the earth's moon.

agree if someone could produce a telescope that created four satellites about Jupiter, but not about every other star. One Francisco Sizzi argued against Jupiter's moons as follows:

> There are seven windows in the head, two nostrils, two eyes, two ears, and a mouth; so, in the heavens there are two favorable stars, two unpropitious, two luminaries, and Mercury alone, undecided and indifferent. From which and many other phenomena of nature, such as the seven metals, etc., which it were tedious to enumerate, we gather that the number of planets* is necessarily seven.

To an opponent who declared that the moon was covered with a smooth, transparent substance to make it perfectly round again so that Galileo's observations and the dogma of spherical perfection could both be maintained, Galileo sarcastically replied that he had an equal right to assume that such transparent material could make each mountain ten times as high as he had observed them. The head of the Roman College, Cardinal Robert Bellarmine, was also an officer of the Inquisition. Secretly, he began making inquiries into the theological and political significance of the novel discoveries, as well as of their flamboyant discoverer.

Galileo, meanwhile, quit the intellectual and political freedom of the Venetian Republic. He became Court Astronomer to Grand Duke Cosimo de Medici of Florence. Ultimately, the move proved to be his undoing; for Florence, unlike Venice, could not protect him from his enemies' political power. To obtain his position, Galileo flattered the Grand Duke by naming the newly discovered moons of Jupiter the "Medicean Planets." Part of Galileo's desire to return to Florence was probably simple nostalgia; it had been his childhood home. His public reason was that despite lifetime tenure and a good salary at Padua, its university atmosphere was intellectually conservative, and his teaching duties did not leave him enough time for his studies and research.

While parrying criticism and applying for a new job, Galileo had time for serious work. Galileo turned to two technical points that had influenced Brahe against the Copernican system. Recall that like most other astronomers Brahe thought that the stars and planets were a few minutes of arc in diameter, for this is how big they seem to the naked eye. Brahe's inability to measure parallax meant the parallax angle of even the nearest star was smaller than two minutes of arc. If Copernicus were right, this meant that the stars must be at least 2000 times farther than the sun. Now, suppose a star is 2000 astronomical units away, and appears to have a diameter of two minutes of arc. It was easy to compute that its diameter is about the radius of the earth's orbit. Not only were

---

*By "planet", Sizzi meant any heavenly body that is not a fixed star.

the stars very far away but also they were larger than the sun; indeed, large enough to fill up the entire orbit of the earth. This seemed ridiculous. In addition, it was known that Venus should have a particular cycle of changing phases in Copernicus' model, and these had never been observed.

Galileo weakened both objections by showing that the images of both stars and planets were smaller than people had thought previously. The planets, including Venus, present extended images in the telescope, and are never more than about one minute of arc in diameter. Venus' phases were not observable to the naked eye simply because the planet is too small. The stars did not increase in size at all when viewed with the telescope. Thus, they were certainly much smaller than Brahe claimed. Here is Galileo's description of his discovery:

> When stars are viewed by means of unaided natural vision, they present themselves to us not as of their simple (and so to speak physical) size, but as irradiated by a certain fulgor and as fringed with sparkling rays . . . [A telescope] . . . removes from the stars their adventitious and accidental rays, and then it enlarges their simple globes (if indeed the stars are naturally globular) so that they seem to be magnified in a lesser ratio than other objects . . . . Deserving of notice also is the difference between the appearances of the planets and of the fixed stars. The planets show their globes perfectly round and definitely bounded, looking like little moons, spherical and flooded all over with light; the fixed stars are never seen to be bounded by a circular periphery, but have rather the aspect of blazes whose rays vibrate about them and scintillate a great deal.

In his telescope, Venus appeared as a disk, rather than as a point of light. Like the moon, Venus showed changing phases as it went through its 584-day cycle. Therefore, Galileo could ask whether the cycle of phases agreed with that predicted by the Copernican model. During the sixty weeks it took Venus to move from greatest western to greatest eastern elongation, the planet appeared rather small and more than half full. During the remaining twenty weeks, which included the period of Venus' retrograde motion, the planet appeared in the telescope larger and as a crescent, like the moon when it is less than half full.

Galileo understood that his observations ruled out the Ptolemaic system. Figure 5-1 shows the earth and the sun, with Venus on its Ptolemaic epicycle. The center of Venus' epicycle is always lined up with the sun. Since, in this picture, Venus is always between us and the sun, it would always appear as a crescent. Figure 5-2 shows the situation in the Copernican model. The full phase occurs when Venus is beyond the sun; and the crescent phase, when Venus is between us and the sun,

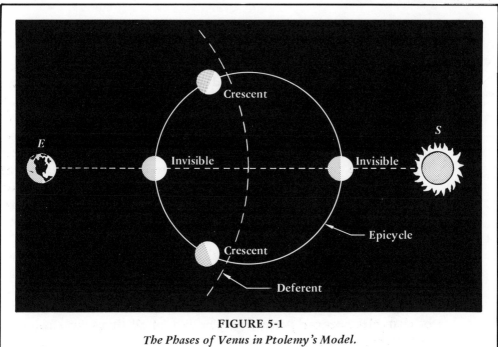

**FIGURE 5-1**
*The Phases of Venus in Ptolemy's Model.*

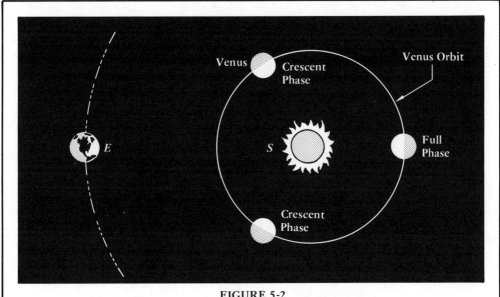

**FIGURE 5-2**
*The Phases of Venus in the Copernican Model.*

as Galileo observed. The planet appears exactly half-lit when it is at greatest elongation.

Although Galileo's observations argued against the Ptolemaic model, and its predecessors by Eudoxus and Aristotle, they in no way ruled out the compromise system of Tycho Brahe (see Fig. 4–4). Indeed, at any time the relative configuration of earth, sun, and Venus is the same in the models of Tycho Brahe and of Copernicus, so they both predict the same system of phases. Galileo did not seem aware of this fact, and began to insist that he had really proved that the earth moved. His serious opponents became not those who denied his observations or who made simpleminded philosophical arguments but men of learning, like Cardinal Bellarmine, who believed that Galileo was claiming more than was logically permissible.

Galileo published his new discoveries in 1612, in the *Letters on Sunspots,* which contained also his description of moving dark spots on the sun, and speculation on their nature. Here was another example of blemishes in what so recently had been the heavenly realm of changelessness and perfection. The sun, of all things, was not a perfect polished sphere.

Galileo was famous and controversial. His opponents began to mutter about heresy. Galileo took them on, and wrote an open *Letter to the Grand Duchess Christina,* stating his views on the proper relation between the authority of religion and natural science. On the subject of the proper use of the Bible, Galileo quotes a famous epigram: The Holy Spirit, he says, "teaches us how to go to heaven, not how the heavens go." The "letter" was an eloquent plea for intellectual freedom. It also warned the Church not to insist too strongly on traditional interpretations of Biblical revelation; for they would be proven wrong. Copernicanism and the new physics would triumph, and organized religion ought to be on the winning side if it is to survive.

The Church, however, was busy re-establishing its authority after the severe challenge of the Reformation. Bellarmine's official opinion was that it should not lightly change its traditional teaching that the earth was the center of the universe. If the Copernican theory were proven, the Church would have to reinterpret some Biblical passages, such as the one where Joshua commands the sun to stand still for a few hours; such a reinterpretation was not yet necessary, since Galileo and the Copernicans had not proved their case. It was legitimate to discuss Copernicanism as a fictitious but convenient mathematical model for astronomical predictions, but in 1616 it was proclaimed a heresy to assert that the sun was *truly* at the center of the universe.

Copernicus' *De Revolutionibus* and many lesser pro-Copernican works were put on the Index of Prohibited Books. Although Galileo's books were not prohibited, the warning was unmistakable. Galileo was admonished to abandon his heretical views, and he promised to refrain from teaching or advocating them publicly. However, what was to be his great work, the *Dialogue on the Two Great World Systems,* soon took shape in his mind. This was Galileo's long-promised, full length work: a debate between advocates of the Ptolemaic and Copernican world views. Ostensibly a balanced debate, in fact, the "Aristotelian" objections to Copernicus were systematically demolished, and in Italian rather than Latin, so that laymen all over Italy could read it.

First came the astronomical discoveries that we have been discussing; then, a detailed and fascinating discussion of scientific methodology and an elaboration of the arguments in the *Letter to the Grand Duchess,* refuting the religious and philosophical objections to the idea that the earth moves. Once more, and at greater length, Galileo affirmed that observation, or experiment, together with logical, mathematical argument—not the citation of Aristotle—is the way to discover the truth. Preconceived ideas about beauty and perfection should have no place in science.

What is new in the dialogue is the refutation of the arguments against the earth's motion. It is a clear statement of what Einstein called the Galilean principle of relativity, which simply states that, contrary to the beliefs of Aristotle and his successors, motion cannot be detected, there is no experiment one can do on the earth to test whether or not it is moving; and, therefore, the physical arguments against the earth's motion have no basis. Here Galileo relied on a great discovery he had made about motion in general; most of the rest of this chapter will be a discussion of this point. The *Dialogue* was finished in 1630, but had understandable difficulties in being published. Not the least of its troubles was that Galileo's enemies spread the rumor that Simplicio, the advocate of Aristotle in the *Dialogue,* was a caricature of the reigning Pope, Urban VII. The *Dialogue* finally appeared in print in 1632.

The next year, at the age of seventy, Galileo was summoned to Rome and tried by the Inquisition for publishing heretical opinions. After three months of grueling interrogation, he was sentenced to house imprisonment and the recitation of penetential psalms. The most famous part of Galileo's punishment was his formal "abjuration", or forswearing, of his belief in the Copernican system. Threatened with torture and more vigorous punishments, Galileo was compelled to kneel down in public and recite the following:[2]

> I, Galileo, son of the late Vincenzio Galilei of Florence, my age being seventy years, having been called personally to judgement and kneeling

before your Eminences, Most Reverend Cardinals, general Inquisitors against heretical depravity in the entire Christian dominion, and having before my eyes the sacred Gospels, which I touch with my own hands, do swear that I have always believed, do now believe, and with God's aid shall believe hereafter all that is taught and preached by the Holy Catholic and Apostolic Church. But because, after I had received a precept which was lawfully given to me that I must wholly forsake the false opinion that the sun is the center of the world and moves not, and that the earth is not the center of the world and moves, and that I might not hold, defend, or teach the said false doctrine in any manner, either orally or in writing, and after I had been notified that the said teaching is contrary to the Holy Scripture, I wrote and published a book in which the said condemned doctrine was treated, and gave very effective reasons in favor of it without suggesting any solution, I am by this Holy Office judged vehemently suspect of heresy; that is, of having held and believed that the sun is the center of the world and immovable, and that the earth is not its center and moves.

Therefore, wishing to remove from the minds of your Eminenences and of every true Christian this vehement suspicion justly cast upon me, with sincere heart and unfeigned faith I do abjure, damn, and detest the said error, heresy, and sect contrary to the Holy Church. And I do swear for the future that I shall never again speak or assert, orally or in writing, such things as might bring me under similar suspicion; but if I should know any heretic or person suspected of heresy I shall denounce him to this Holy Office or to the Inquisitor or governor of the place where I shall find him . . .

A famous tradition holds that after reciting this oath, Galileo kicked the ground and muttered under his breath, *"Eppur si muove"* (But it does move).

His book was added to those of Copernicus on the Index of Prohibited Books. But, since the Reformation, the Church's sway was not universal, even in Europe. The *Dialogue* was easily smuggled out of Italy, soon published in Protestant Holland, and widely circulated. The judgement and sentence against Galileo served mainly to dramatize his point of view.

Under house arrest in Siena, in his seventies, and gradually losing his eyesight, Galileo returned to his early studies of the laws of motion, and wrote his great textbook on mechanics: "Dialogues Concerning the Two New Sciences". The book constructed the foundation of the modern science of motion. It explained clearly the idea of inertia, the law of falling bodies, the principle of compound motion, and, finally, the principle of relativity. This new book was also smuggled out of Italy and printed in Holland in 1638. Galileo died in 1642, and was buried in the Church of Santa Croce in Florence, where one can still see his tomb, just across the aisle from Michelangelo's.

## THE SCIENCE OF MOTION—INERTIA

In the "Two New Sciences", Galileo summarized in systematic and sequential fashion not only his own unrecorded researches but also a whole medieval tradition of criticism of Aristotelian mechanics. Ever since Aristotle's works had been rediscovered in the West in the 12th century, the scholastic world made extensive commentaries upon his mechanics. Two schools, Oxford and Paris, had taken the lead. The difficult points in Aristotelian theory had been explored, criticized, and reformulated in the hope of strengthening the Master's intellectual edifice. There were attempts at Oxford in the early fourteenth century to make the description of motion more quantitative. The Paris School, under the leadership of Jean Buridan and Nicole Oresme, developed the conception of *impetus* (an important prototype of inertia) to account for the obvious fact that bodies do, in fact, persist in unnatural horizontal motion even after they have lost contact with their moving agent. Nonetheless, medieval science never succeeded in codifying its insights in a new mechanics. Many of the arguments and ideas Galileo presented were not new. Much of the work Galileo himself did had been done before. What was new was Galileo's realization that the old criticisms of Aristotle constituted a condemnation. Galileo concentrated upon the weak points of Aristotelian theory; the systematic resolution of these weak points led to a new mechanics. This inversion of viewpoint, dressing old ideas in new clothes, was Galileo's creative contribution.

Galileo's law of inertia stemmed from his criticism of the Aristotelian description of horizontal motions. The actual motions we observe around us are too complicated for either Aristotle or Galileo to have described exactly. Both, therefore, tried to analyze particularly simple examples, hoping to derive from them general insights applicable to all motions. However, the Aristotelians and Galileo emphasized different idealizations of actual motions. To the Aristotelians, horizontal motion was unnatural, or "violent". Objects move horizontally only when pushed; when left to themselves they stop. This is everyday experience. Try pushing a piece of furniture across the floor. The idea that some motions were "unnatural" only made sense in a world with the earth fixed at the center; for then any deviation from straightline motion towards the center of the universe could be regarded as unnatural. Furthermore, the words, "to stop moving", had an absolute meaning. On the other hand, in the Copernican universe the earth was unquestionably moving. Is a chair's motion relative to an earth itself moving in an orbit about the center of the universe really violent and unnatural?

Furthermore, when you stop pushing the chair, it comes to rest relative to an earth that is moving: What do the words, "to stop moving", mean?

The Aristotelians knew that even after you stop pushing a chair across the floor it takes a short time for the chair to come to rest relative to the floor. While this was unimportant to them, to Galileo it was one of the most important features of the motion. The chair is *rubbing* against the floor, and this friction stops it. What counted was the friction resulting from the *relative* motion of chair and floor.

Galileo then asked a question Aristotle had ruled meaningless: What would happen if there were no friction at all? Even though we encounter in everyday life no situations where there is no friction, some motions exist where there is *very little* friction. Galileo studied these, and from them guessed what frictionless motion might be like. His ideal simple example of motion was more nearly represented by a steel ball rolling on a hard surface, or a hockey puck on ice. Both persist in horizontal straightline motion long after they have been pushed, and only slowly come to rest. If the friction were even weaker, they should persist in motion longer. If there were no friction, then bodies would persist in uniform straightline motion. In fact, Galileo imagined motion in a vacuum, where there is no gravity, friction, or air resistance. An object at rest will remain at rest. An object in motion will remain moving in a straight line at constant speed. This is Galileo's law of "inertia". When we say a body has inertia, we mean that, if left to its own devices, it would remain at rest or in straightline motion at constant speed. It is *deviations* from uniform straightline "inertial" motion, not the motion itself, that physics must explain.

Galileo presented a simple argument for the principle of inertia, illustrated by Fig. 5-3. As a ball rolls down an inclined plane, its speed increases. Tilt the inclined plane the other way and start the ball

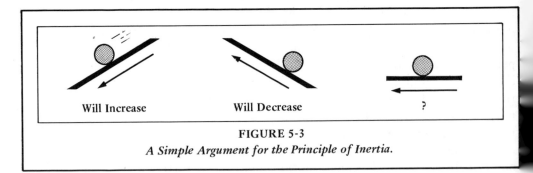

**FIGURE 5-3**
*A Simple Argument for the Principle of Inertia.*

rolling in the same direction with the same speed as before. The ball will slow down as it goes upward. Everybody will agree to these facts. Now, what happens if the surface is exactly horizontal? Since the ball does not speed up, and does not slow down, it must continue with exactly the same speed.

## SPEED AND ACCELERATION

By imagining the ideal case of no friction, Galileo arrived at a conception of horizontal motion sufficiently simple that he could describe it mathematically. However, in his experiments, objects did not move forever in straight lines at constant speeds. There were at least two reasons for this: friction and gravity. Galileo had the insight to understand that friction was complicated and would elude easy mathematical description, but that gravity was simple. It was crucial to separate the two effects. He studied the motions of falling bodies using experiments in which friction and air resistance could be minimized, and he neglected these effects altogether in his mathematical description of motion under gravity. The question Galileo asked and answered was: If an object is moving vertically, solely under the influence of gravity, can one calculate how far it has gone and how fast it is going at any time?

Aristotle had taught that the speed of a falling object never changed, and that heavier objects fall faster than light ones. From experiment Galileo knew that in the absence of air resistance (which in practice meant for heavy enough objects), everything falls at the same rate regardless of its weight. Furthermore, an object's speed increases as it falls. But much debate focused on exactly how the speed changed. Did it increase simply with distance (constant $v/d$), or with time (constant $v/t$)? In discussing inertia, we had to consider only unchanging, uniform speeds. Now we are about to describe the speed of a falling object, which increases as time passes. Galileo was one of the first to do this properly. To understand his work, let us study more exactly what we mean by speed and an important related concept, acceleration, the rate of change of speed.

Speed is simply how far an object goes, divided by the time it takes to go that distance. For example, if something is moving at 10 feet per second, after one second it will have gone 10 feet; after two seconds, 20 feet, and after 1000 seconds, 10,000 feet. These facts can be written as a mathematical formula. Let $t$ stand for the time after the motion has started, and $d$ for the distance the object has gone; then:

$$d = 10t$$

If you know $t$, the formula tells you how to calculate $d$. Of course, a similar formula works for any speed. Let the letter $v$ (for "velocity") stand for an object's speed. Then, the distance it has gone after moving for a time $t$ is:

$$d = vt$$

which is the same as:

$$v = \frac{d}{t}$$

Remember, these formulas describe motions with *constant* speed.

Another way of representing speed will be useful. Figure 5-4 is a graph from which you can read the distance an object moves at any time. As $t$ changes, so does $d$. When $t$ is 0, $d$ is also 0. At time $t_1$, the distance is $d_1$, and so forth. The times, $t_1, t_2, t_3, \ldots t_n$, can be arbitrary. The solid line is the graph of the motion, and because the motion is at a constant speed the graph is a straight line. Now, suppose we had only the graph in Fig. 5-4. Suppose, further, that $t_1$, $t_2$, and $t_3$ are 1 second, 3 seconds, and 4 seconds, respectively; and $d_1$, $d_2$, and $d_3$ are 10 feet, 30 feet, and 40 feet, respectively. Then, the speed between $t_1$ and $t_2$ is the distance gone, $d_2 - d_1$, divided by the time elapsed, $t_2 - t_1$:

$$v = \frac{d_2 - d_1}{t_2 - t_1} = \frac{30 - 10}{3 - 1} = \frac{20}{2} = 10$$

The speed between $t_1$ and $t_3$ is:

$$v = \frac{d_3 - d_1}{t_3 - t_1} = \frac{40 - 10}{4 - 1} = \frac{30}{3} = 10$$

Both answers are the same because $v$ is a constant and does not change.

So far, all this seems obvious. But, let us apply it to a case where the motion changes. Figure 5-5 might represent a journey. In Fig. 5-5, the traveller goes at a constant speed until $t_1$, stops until $t_2$, goes at a slower speed between $t_2$ and $t_3$, then speeds up, and then at $t_4$, stops again. The speed on each straight segment is still defined as before; for example:

$$\frac{d_3 - d_2}{t_3 - t_2}$$

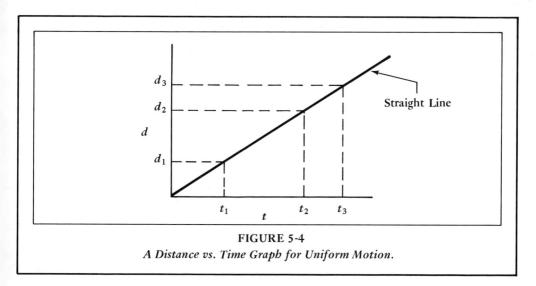

**FIGURE 5-4**
*A Distance vs. Time Graph for Uniform Motion.*

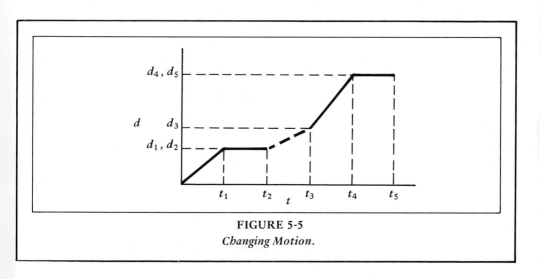

**FIGURE 5-5**
*Changing Motion.*

is the speed between $t_2$ and $t_3$. But, it is not equal to:

$$\frac{d_2 - d_1}{t_2 - t_1}$$

Now, what is the meaning, say, of:

$$\frac{d_5 - d_1}{t_5 - t_1}$$

It is the *average speed* between $t_5$ and $t_1$. This is not a conclusion. We are simply defining the words, "average speed." Here is an example. Suppose the value of the distances and times in Fig. 5-5 are as follows:

$$t_1 = 1 \text{ (sec)} \quad d_1 = 10 \text{ (ft)}$$
$$t_2 = 2 \quad d_2 = 10$$
$$t_3 = 3 \quad d_3 = 15$$
$$t_4 = 4 \quad d_4 = 35$$
$$t_5 = 5 \quad d_5 = 35$$

Between $t_2$ and $t_3$, the speed is:

$$\frac{d_3 - d_2}{t_3 - t_2} = \frac{15 - 10}{3 - 2} = \frac{5}{1} = 5 \text{ (ft per sec)}$$

The average speed between $t_1$ and $t_5$ is:

$$\frac{d_5 - d_1}{t_5 - t_1} = \frac{35 - 10}{5 - 1} = \frac{25}{4} = 6.25 \text{ (ft per sec)}$$

A real journey, of course, is not as jerky as in the graph of Fig. 5-5. Figure 5-6 is a more realistic example.

Even though the speed is continuously changing (no portion of the graph in Fig. 5-6 is a straight line), it is still useful to say that:

$$\frac{d_2 - d_1}{t_2 - t_1}$$

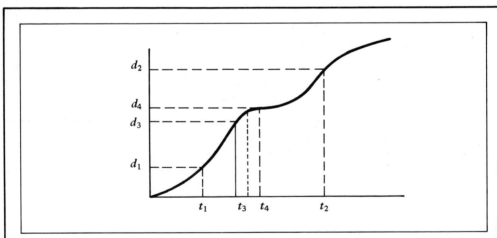

**FIGURE 5-6**
*Continuously Changing Motion.*

is the average speed between $t_1$ and $t_2$. However, the notion of average speed is not completely adequate to describe a motion like that in Fig. 5-6. Galileo knew he had to give a meaning to the speed *at any instant* of time, $t$. We do this today as follows. Suppose the instant of interest, $t$, is between $t_1$ and $t_2$. Calculate the average speed between these two times as above. Next, take a shorter interval between these two times, say, from $t_3$ to $t_4$. Calculate the average speed during that shorter time interval:

$$\frac{d_4 - d_3}{t_4 - t_3}$$

As we take the time interval, $t_4 - t_3$, smaller and smaller, the average speed comes closer and closer to what we call the instantaneous speed at time $t$. The reason this works is that a little piece of the curved graph looks almost like a straight line. If $t_4 - t_3$ is small enough, we can get a very good approximation to a straight line.

Next, we can make a different sort of graph. Instead of graphing the distance an object has gone at time $t$, we can plot its instantaneous speed, $v$, against time. For example, if an object moves at a constant speed of 100 feet per second, the graph of its speed would look like the graph in Fig. 5-7.

The graph in Fig. 5-7 is a horizontal line because the speed does not change. Figure 5-8 is a more interesting example; the one Galileo found applies to falling bodies. The speed starts at zero and increases in a regular way. The graph in Fig. 5-8 is a straight line. This is a very special way in which speed may change as time progresses. Since speed is the rate of change of distance, and a straight line distance-versus-time graph means *constant* speed, a speed-versus-time graph that is a straight line, like Fig. 5-8, also means a constant something. The something, by direct analogy, is the rate of change of speed, which is called *acceleration*.

We define acceleration precisely as follows: Suppose $v_2$ and $v_1$

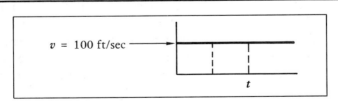

FIGURE 5-7
*The Speed of an Object Moving at a Constant Speed of 100 Feet Per Second.*

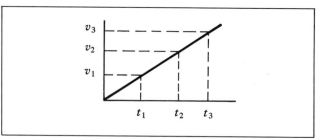

**FIGURE 5-8**
*The Speed of an Object with Constant Acceleration.*

are the instantaneous speeds of an object at times $t_2$ and $t_1$. Then, the average acceleration, $a$, between $t_1$ and $t_2$ is:

$$a = \frac{v_2 - v_1}{t_2 - t_1}$$

In the case illustrated by Fig. 5-7, the speed is constant so that $v_2 - v_1 = 0$, and the acceleration is zero. In Fig. 5-8, the acceleration is not zero, because the speed is increasing. Since the graph is a straight line, the average acceleration between any two times is the same:

$$a = \frac{v_2 - v_1}{t_2 - t_1} = \frac{v_3 - v_2}{t_3 - t_2}$$

It is possible to imagine motions with continuously changing accelerations. Then, we make a distinction between the average acceleration during some time interval, and the instantaneous acceleration at some instant of time. To find the instantaneous acceleration at time $t$, simply consider a time interval that contains $t$ and calculate the average acceleration during that interval. If the time interval becomes very small, this average acceleration is very close to the instantaneous acceleration. The idea is analogous to the definition of instantaneous speed, with "acceleration" replacing "speed", and "speed" replacing "distance". Compare the two sentences: "Speed is the rate of change of distance"; and, "Acceleration is the rate of change of speed." For the time being, constant accelerations are all we will have to talk about.

The law of inertia can now be rephrased in an elegant way:

> For horizontal motions, the acceleration is zero, when there is no friction.

Now we notice something important about Galileo's law of inertia. Not only does an object's *speed* not change, but it keeps on moving in the same *direction*. When we describe an object's motion, we must not only say how fast it is going, but also in what *direction* it is moving. *Any* change in inertial motion we will call an acceleration—either changes in speed in the same direction, or changes in direction that preserve the same speed, or, in general, both together. Acceleration also has a direction. The notion that acceleration includes changes of direction of motion will be important to us in the next chapter when we discuss Newton's theory of the planets' motions, which continuously change the direction in which they move.

## FALLING BODIES

As Kepler had done in studying the planets' orbits, Galileo also searched for a simple mathematical rule describing falling bodies. If the orbits of the planets were not circles, Kepler found they were the next simplest mathematical curves, ellipses. Although bodies do not fall with constant speed, Galileo showed they do the next simplest thing: They fall with constant acceleration. This is how he discovered it.

Dropping balls of different weights from the Leaning Tower of Pisa could not be an accurate experiment, since the balls fall so fast. In principle, Galileo might have dropped a ball from the top of the first flight of stairs and timed it; then measured the time for a ball dropped from the second flight to reach the ground, and so on. But these times were all much too short for him to measure; even a ball dropped from the top of the tower strikes the ground in less than four seconds. In order to make more accurate measurements, he had to stretch out the time each experiment took. He had to "dilute" gravity. It occurred to him to roll balls down inclined planes. Since they rolled slowly, air resistance would be very small. Friction would also be small, since the ball and inclined plane only touch each other at one point. He presumed that as he made the inclined plane steeper and steeper, the rate of fall would come closer and closer to free fall. He measured the rate of fall for a succession of steeper planes, and assumed he could simply extrapolate to the rate of a vertical plane, to get the law for a freely falling body.

Galileo was the first to establish the central role of *time* in the description of motion. He, therefore, needed to measure time accurately. At first, he simply counted the number of beats of his pulse between the time he let the ball go and the time it arrived at the bottom of the inclined plane. But, this was inaccurate. Then, he devised a water clock—using a faucet that dripped water at a constant rate into a jar. He removed his finger from the faucet upon letting the ball go, replaced it when the ball arrived at the bottom of the inclined plane, and then measured the height of the water in the jar. Later, he turned to a pendulum clock, counting the number of pendulum swings. Thus, Galileo realized, as Einstein would several centuries later, that time had to be defined by one experiment before the motion could be defined by another. There is no "absolute" way of knowing the time.

Galileo measured the distance an object falls during each successive second, and discovered that these distances were given by succesively increasing odd numbers. Moreover, this simple rule is true for a ball rolling down an inclined plane at *any* angle. Refer to Fig. 5-9. Galileo discovered that, whatever the inclination of the plane, if the distance the ball went during the first second was taken as the unit of distance, then, during the next second it fell three units; during the third second, it fell five units, and so forth. The only difference the steepness of the plane made was the size of the unit, the distance the ball went during the first second.

If this is true for a ball rolling down an inclined plane, whatever the angle, it ought to be true for a falling body as well. During the first second it goes a certain number of feet. During each succeeding second it falls that number of feet, multiplied by 3, 5, 7, 9, and so forth. Pythagoras was right after all. Nature's fundamental laws are elegant

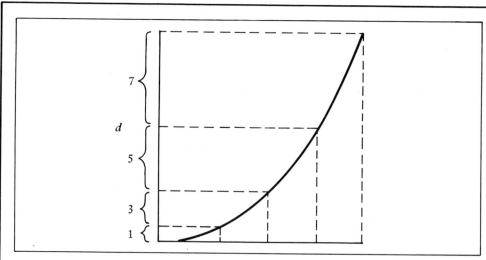

FIGURE 5-9
*Graph of Distence Fallen Vs. Time Ellapsed for a Real Object Rolling Down an Inclined Plane.*

games with numbers. Galileo soon realized that this formulation of the law of falling bodies was a roundabout way to recognize a simple fact. Let $d_1, d_2, d_3, d_4$, etc., signify the distance the falling object has gone after one, two, three, four, etc., seconds. We take as the basic unit of length the distance fallen during the first second, $d_1 = 1$. Then, we can find the remaining distances by simple addition:

$$d_1 = 1$$
$$d_2 = 1 + 3 = 4 \qquad\qquad = 2 \times 2 = 2^2$$
$$d_3 = 1 + 3 + 5 = 9 \qquad\qquad = 3 \times 3 = 3^2$$
$$d_4 = 1 + 3 + 5 + 7 = 16 = 4 \times 4 = 4^2$$

The sums are perfect squares! It is a simple theorem of mathematics that the sum of the first $n$ odd numbers is $n^2$.

Here, then, is Galileo's Law of Falling Bodies: The distance an object falls after a time, $t$, is some number multiplied by $t^2$. The number is almost exactly 16, if time is measured in seconds and distance in feet. Then, the distance, $d$, an object falls after a time, $t$, is:

$$d = 16t^2$$

Thus, we can tabulate the distance fallen at different times.

TABLE 5.1. Galileo's Law of Falling Bodies in Tabular Form.

| $t$ (sec) | $d$ (ft) |
|---|---|
| 0 | 0 |
| 1 | 16 |
| 2 | 64 |
| 3 | 144 |
| 4 | 256 |
| 5 | 400 |

Figure 5–10 is a distance vs. time graph of this sort of motion. Figure 5–10 is simply a graph of the formula, $d = 16t^2$. Like Kepler's laws, Galileo's Law of Falling Bodies is simple: Bodies do not fall at constant speeds, which would give $d = vt$; however, the rule has only $t^2$, the square of the time, in it, and is still not very complicated.

Galileo's law has an even greater hidden simplicity, which becomes apparent if we try to deduce the *speed* of a falling body, instead of the distance it has gone. Since the graph in Fig. 5–10 is not a straight

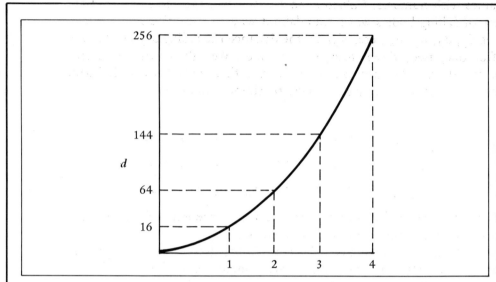

**FIGURE 5-10**
*Galileo's Law of Falling Bodies in Graphical Form.*

line, the speed of a falling body is not a constant; it increases as time passes. The average speed during the first second is 16 ft per sec. During the next second, the average speed is 48 feet per second. During the third second, the average speed is 80 feet per second. What is the instantaneous speed one second after the object is dropped? First, we must find the average speed between zero and two seconds; it is:

$$\frac{64 - 0}{2 - 0} = \frac{64}{2} = 32$$

To get a better approximation to the instantaneous speed at one second, we compute the average speed between ½ second and 1½ seconds. When $t = ½$, $d = 16(½)^2 = 16/4 = 4$. When $t = 3/2$, $d = 16 \times (3/2)^2$, which is $16 \times 9/4 = 36$. The average speed between $t = ½$ and $t = 3/2$ seconds is, therefore:

$$\frac{36 - 4}{3/2 - ½} = \frac{32}{1} = 32$$

the same average speed we computed between $t = 0$ and $t = 2$ seconds
This result is not a coincidence. If the rule for the distance is $d = 16t^2$, then, if we calculate the average speed during a time interval that contains the time, $t = 1$ second, we will always get 32 feet per second *provided* the beginning and end of the chosen time

interval are equally spaced from the time, one second. Therefore, 32 feet per second is the correct instantaneous speed at $t = 1$ second.

Now, we want to compute the *instantaneous* speed at any arbitrary time. We give this arbitrary time an algebraic name, $t$. We must calculate the average speed over a short time interval around $t$, and then imagine the direction of the interval becoming smaller and smaller. The average speed over a very short time interval more nearly approximates the instantaneous speed as the interval diminishes. Let us put $t$ in the middle of a time interval whose duration is $2x$. That is, the time $t_1$, when the interval begins, is $t_1 = t - x$; and the interval ends at $t_2 = t + x$. Now, according to Galileo's law of falling bodies, the distance $d_1$ at the beginning of the interval, $t_1$, is:

$$d_1 = 16t_1^2$$

and at the end, $t_2$, is:

$$d_2 = 16t_2^2$$

The average speed, $v$, between $t_1$ and $t_2$ is, therefore:

$$v = \frac{d_2 - d_1}{t_2 - t_1} = \frac{16t_2^2 - 16t_1^2}{t_2 - t_1} = 16\frac{t_2^2 - t_1^2}{t_2 - t_1}$$

Now, we know that $t_2^2 - t_1^2$ factors into $(t_2 + t_1)(t_2 - t_1)$ so that the factor $t_2 - t_1$ appears both in the numerator and denominator and can be cancelled:

$$v = 16\frac{(t_2 + t_1)(t_2 - t_1)}{(t_2 - t_1)} = 16(t_2 + t_1)$$

Substitute $t_2 = t + x$ and $t_1 = t - x$ into the formula above:

$$v = 16(t + x + t - x) = 16(2t) = 32t$$

If we choose $t = 1$ second, this formula reproduces the speed we already obtained, $v = 32$ ft/sec.

Something interesting happened. The length of the time interval, $x$, completely disappeared from our calculation of $v$. This means the answer, $v = 32t$, is valid for any time interval, no matter how small; $v = 32t$ must, therefore, be the instantaneous speed of the arbitrary time, $t$, after the ball is dropped. We have discovered the "trick" necessary to calculate the instantaneous speed in this case, without doing an infinite number of calculations.

The rule, $v = 32t$, is a very simple prescription for the way speed increases with time. Since the graph of speed vs. time is a straight line, as in Fig. 5-8, the average acceleration between any two instants is the same as between any other two, 32 feet per second per second.* This number, 32, is the *instantaneous* acceleration of the falling body at any time.

A simple way to state the law of falling bodies is: A falling body always accelerates at the rate of 32 feet per second per second. This particular acceleration is a "natural constant", called the acceleration of gravity. It is usually represented by the symbol, *g*. In terms of *g*, the distance fallen, speed, and acceleration of a body dropped from rest are:

$$d = \tfrac{1}{2} g t^2$$

$$v = gt$$

$$a = g$$

We have shown that these three equations all mean the same thing.

### COMPOUND MOTION

Galileo's laws can be summarized by saying that (ignoring friction, air resistance, and so forth) an object in purely horizontal motion has zero acceleration, while an object in purely vertical motion has downward acceleration, *g*.

Galileo demanded an exact description of the motion of a projectile, rather than a qualitative explanation such as the *impetus* theory provided. Just as in the case of falling bodies, he proposed a rule that described where a projectile is at any time, *t*.

Galileo's insight was to describe the motion of a projectile—or any object flying through the air solely under the influence of gravity—by describing *separately* its vertical motion and its horizontal motion, but, nevertheless, thinking of them as taking place simultaneously. By separating in his mind these two types of motion, Galileo could see that his law of inertia could explain how a cannonball can keep moving horizontally after it has left the cannon.

How can one describe the motion of an object that has both vertical and horizontal motion? A simple example would be a cannonball fired from a cannon at the top of a castle wall. To the Aristotelians even the simple facts that the cannonball travels horizontally and then hits the ground far from the castle wall was hard to explain. What

---

*Since speed is the rate of change of distance, it is measured in feet per second. Similarly since acceleration is the rate of change of speed, it is measured in feet per second per second.

pushed the cannonball along after it leaves the cannon? In the middle ages, the conceptions of *impetus* had been developed to answer this question. *Impetus,* which flowed into the cannonball upon firing the cannon, temporarily made unnatural horizontal motion natural. Gradually, the *impetus* would leave the cannonball; and when it was all gone the ball would revert back to its natural motion—falling in straight lines towards the center of the universe. When the cannonball possessed impetus, it had a different nature than when it did not. Therefore, it could exhibit two different kinds of natural motion, but only at different times. Since it was logically impossible for an object to have two different natures at the same time, the cannonball could not have the two different natural motions, horizontal and vertical, at the same time.

To start with a simple example, suppose a cannonball leaves the cannon with an initial horizontal speed of 100 ft/sec. We will have to introduce some new notation. The ball's position at any time is given by two numbers, the horizontal distance it has moved and the vertical distance it has fallen. Let us keep symbol $d$ for the vertical distance, and use $x$ for the horizontal distance.

Figure 5-11 shows the path such an object will take until it strikes the ground (after which Galileo's laws no longer apply, of course). Initially its speed is 100, and its direction is horizontal. As it moves, both the speed and the direction will change. Instead of studying its speed and direction, it is simpler to talk about two speeds, a vertical speed and a horizontal speed. Call them $v_x$ and $v_d$; $v_x$ is the rate of change of $x$, and $v_d$ is the rate of change of $d$. If we know $x$ and $d$ at all times, then we have described completely the motion of

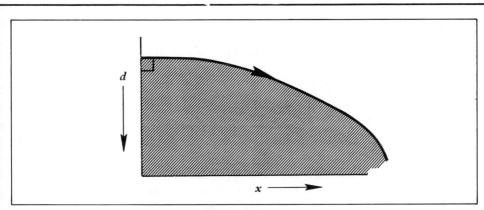

**FIGURE 5-11**
*The Path of a Projectile Whose Initial Direction Is Horizontal.*

the ball. The horizontal speed never changes, even though the ball is also falling:

$$v_x = 100$$

so that

$$x = 100t$$

The vertical speed obeys the law of falling bodies, even though the ball also has horizontal motion:

$$v_d = 32t$$

and,

$$d = 16t^2$$

There is nothing special about the speed 100 ft/sec. If the initial horizontal speed is any number, call it $v_x$, the horizontal speed will continue to be $v_x$. The horizontal distance at any time will be:

$$x = v_x t$$

while,

$$d = \tfrac{1}{2} g t^2$$

is still true.

The important point is that the horizontal motion obeys Galileo's law of inertia even if the object is falling, and the vertical motion obeys his law of falling bodies even though the object is moving horizontally. This *compound* motion can still be completely described by the statements that the horizontal acceleration is zero, and the vertical acceleration is $g$. This means that if a ball is thrown horizontally, as in Fig. 5-11, and at the same instant another ball is simply dropped, they will both strike the ground at the same time. It is easy to calculate what this time is. Suppose the two balls were released from the height above the ground, $h$. They both fall according to the rate, $d = \tfrac{1}{2}gt^2$, and hit the ground when $d = h$, at the time given by $h = \tfrac{1}{2}gt^2$. Taking the square root, we find that $t = \sqrt{2h/g}$. How far does the thrown ball move horizontally before it hits the ground? Since its horizontal distance obeys $x = v_x t$, we can substitute $t = \sqrt{2h/g}$ to find that it moves the distance $v_x \sqrt{2h/g}$. For example, if the ball is thrown horizontally with speed $v_x = 10$ ft/sec from height $h = 64$ ft, it will take $\sqrt{2h/g} = \sqrt{2 \times 64 / 32} = \sqrt{2 \times 2} = \sqrt{4} = 2$ sec to hit the ground; and it will travel the horizontal distance $v_x \sqrt{2h/g} = v_x \times 2 = 20$ ft.

What is the geometrical shape of the curve in Fig. 5-11? Since we know both $d$ and $x$ at any time, it is not difficult to find out what

$d$ is when the ball has gone a horizontal distance $x$. Mathematically, simply think of the two equations:

$$d = \tfrac{1}{2} g t^2$$

and,

$$x = v_x t$$

as simultaneous equations for $d$, $x$, and $t$; and eliminate $t$. The answer is:

$$d = \frac{g x^2}{2 v_x^2}$$

This is the equation for a parabola, which is a conic section—a special, limiting case of an ellipse. In practice, if an ellipse is very elongated, and if you look at the sharply curved piece of it near one focus, the piece is almost indistinguishable from a segment of a parabola. Is it possible that Galileo's parabolas on earth are related to Kepler's ellipses in heaven? Galileo saw a different cosmological implication of the principle of compounded horizontal and vertical motions. From his laws of motion, he could extract a general mechanical principle that refuted the Aristotelian physical objections to the earth's motion.

## GALILEO'S PRINCIPLE OF RELATIVITY

Motion had an absolute meaning for Aristotle. A body was at rest when it did not move relative to the center of the universe. And, any motion relative to the center required an explanation. Galileo's law of inertia denied absolute motion; for it said that bodies would continue indefinitely in uniform straightline motion in the absence of friction or other forces. Inertial motion required no "explanation." Galileo realized that it is only possible to define "relative" motions; for example, the motion of the cannonball in the section above, relative to observers standing still on the ground. Suppose someone could run along at 100 ft/sec remaining under the cannonball. Would he not see the cannonball fall straight down, accelerating according to Galileo's law? If, as Copernicus claimed, the earth were not at the center of the universe, how could one define absolute rest, as Aristotle had tried to? In fact, Galileo's law of inertia implied that there was no fundamental difference between rest and uniform straight-line motion. Imagine a ball rolling along a table at 1 ft/sec. Next, let the ball be fixed on the table, but imagine walking by it at 1 ft/sec. In both cases, the observer measures speeds of 1 ft/sec relative to him. What is the difference between these two situations? Galileo's relativity principle addresses precisely this question of relative motion. It states that any two observers, in uniform straight-line motion relative to one another, will find that bodies obey the same laws of physics.

We may deduce Galileo's relativity principle by applying his laws of inertia and compound motion to the simplest possible dynamical experiment—dropping a ball and measuring its acceleration under gravity. Imagine a "laboratory" that may slide over a frictionless frozen pond. Once set in a uniform straight-line motion relative to the pond, the laboratory will continue to move the same way. Inside, an observer performs Galileo's experiment of dropping a ball and measuring how far it has gone after a certain time. Of course, he discovers the law of falling bodies, $d = \frac{1}{2}gt^2$.

Now, imagine that another observer, outside the "laboratory", is looking in and recording the result of the same experiment. If the laboratory is at rest relative to the pond, both observers see the ball fall straight down, following Galileo's law of falling bodies; and both agree on the time the ball hits the floor. But, suppose the laboratory slides across the frozen ice in a straight line, with some speed, $v$. According to the observer outside the moving laboratory, the ball, after it is released, moves horizontally with the speed, $v$, of the sliding laboratory. This observer verifies the rule for compound motions: the ball continues to move horizontally with the laboratory's speed, $v$, and simultaneously falls downward with speed $gt$, where $t$ is the length of time since the object was released.

How does the experiment appear to an observer in the moving laboratory? Relative to him, the downward motion still obeys the law of falling bodies: $d = \frac{1}{2}gt^2$. But horizontally, he is moving with the same speed as the projectile. Even though the moving observer might know that he is inside a room sliding across the ice and even though he might know that he and the projectile are moving with the room, all that he can see is that relative to him the projectile is not moving horizontally at all; it is simply falling straight down.

Which observer is really moving? Both observers agree that Galileo's laws hold. The "moving" observer simply records motion according to the simple law of falling bodies. The observer "at rest" observes parabolic compound motion, according to Galileo's law for projectile motion, which we have just studied. The observer whom we have labelled as "in motion" could describe the experiment another way. He could say that he is at rest, and when he dropped the ball it fell to his feet according to Galileo's law. The other observer is moving and that is why he saw something more complicated. Which observer is right?

Evidently, there is no way to distinguish rest from uniform straight-line motion using Galileo's law of falling bodies.

Using his principle of relativity and mechanical laws, Galileo was able to describe the general motion of any projectile. (By projectile we mean any object moving solely under the influence of gravity.) If a ball were thrown upward, he could calculate how high it would go, when it would reach the top of its path, and when it would hit the ground. He could answer the same questions even if the ball were thrown in an arbitrary direction. These solutions are discussed in an Appendix section. Galileo even computed artillery tables, calculating where cannonballs shot at various angles would hit the ground, the first time anyone had ever done so from general theoretical principles.

The reader may have noticed that when riding in a vehicle moving at constant speed, it is impossible to tell how fast one is moving without referring to something outside. This is the principle of relativity at work. Our example of a laboratory sliding on an ice pond suggests that the principle seems to be a consequence of Galileo's way of combining the law of falling bodies with the principle of inertia. Galileo then turned the argument around, asserting that the principle of relativity was a universal, general law of nature and that detailed laws, like his mechanical ones, must conform to it. His own example is more colorful than ours:[3]

> Shut yourself and a friend below deck in the largest room of a great ship, and have there some flies, butterflies, and similar small flying animals; take along also a large vessel of water with little fish inside it; fit up also a tall vase that shall drip water into another narrow-necked receptacle below. Now, with the ship at rest, observe diligently how those little flying animals go in all directions; you will see the fish wandering indifferently to every part of the vessel, and the falling drops will enter into the receptacle placed below. . . . When you have observed these things, set the ship moving with any speed you like (so long as the motion is uniform and not variable); you will perceive not the slightest change in any of the things named, nor will you be able to determine whether the ship moves or stands still by events pertaining to your person. . . . And if you should ask me the reason for all these effects, I shall tell you now: "Because the general motion of the ship is communicated to the air and everything else contained in it, and is not contrary to their natural tendencies, but is indelibly conserved in them."

## RELATIVITY AND THE EARTH'S MOTION

The Principle of Relativity refuted the ancient physical arguments against the earth's motion although, of course, it did not prove that the earth is moving. For example, Galileo knew that the earth's radius is 4000 miles and that a point on its surface rotates once every 24 hours according to Copernicus. It was not hard to calculate that if Aristotelian physics were correct but the earth was, nevertheless, rotating, then during the few seconds it takes a ball dropped from the Leaning Tower of Pisa to hit the ground the ball would land several hundred yards to the east of the tower. However, the ball falls straight down, as observed, because the ball preserves the earth's motion, following the law of inertia. No force is required to keep the ball moving with the earth. Because no one had made an accurate determination of the earth's distance from the sun, Galileo did not know exactly how fast the earth was moving around the sun. However, this motion was certainly at least as fast as that due to the earth's rotation, and would lead to an even larger displacement of a dropped ball in Aristotelean physics. That such large displacements are predicted, but none observed, had once been a proof that the earth does not move; Galileo saw that their absence would be consistent with the principle of relativity if the earth did move. Thus, there were, he claimed, no experiments that could detect the earth's motion.

Once again, Galileo dared to do the unthinkable. He claimed that results of experiments on earth could be applied to motions in the heavens. Nonetheless, he was very careful in stating to what extent his experimental results applied to the motion of the earth. For, according to Copernicus, the earth is not in uniform straight-line motion. It moves in a circle about the sun, and it rotates on its axis once a day. Galileo and his laboratory actually were executing a very complicated path through the heavens while he was performing his experiments. All he was certain about was the results of his small-scale experiments. All he could say was that inertial motion was in a straight line on a small scale; and since a small piece of a circle is almost a straight line he thought his laws were valid on a small enough scale. This was enough to refute the Aristotelean arguments against the earth's motion, which were also based on small-scale experiments. However, on the heavenly scale, his experiments did not rule out the possibility that inertial motion might be in a circle. In fact, Galileo never did relinquish the idea of circular inertia for heavenly motions. It remained for Isaac Newton and his contemporaries to realize that the question, "Why *doesn't* the earth move in a straight line?", would have a very interesting answer.

In concluding, let us summarize Galileo's contribution to the style of science. Much of his life was a struggle to define a proper scientific method. He was concerned with how scientific statements should be defined and proven, with the criteria of scientific truth. For

Galileo, a proper physical statement can be tested according to *only* two criteria. A theory must be logically consistent internally. Mathematics is the means by which the elements of theory are defined, clarified, and logically ordered. The language of science should be unambiguous, in contrast with the rich but imprecise language of everyday life. Secondly, agreement with experiment is the ultimate arbiter among theoretical possibilities. Galileo admitted no intellectual authority in science other than these: logical consistency and experimental agreement. He ridiculed the criterion of "perfection" imposed upon science by scholastic, and ultimately Greek, philosophy. "Perfection for what specific purpose?" he asked. The imposition of philosophical, theological, or political values upon the methods or conclusions of science could only interfere with the direct and simple thought necessary for the accurate formulation of scientific ideas. Galileo divorced mechanics from all considerations of the nature of moving bodies. Their color, shape, texture, composition, and, surprisingly, even their weight were all irrelevant to the description of their motion: all that mattered was their position and the way their position changed with time.

Galileo's austere criteria implied a necessary limitation of his objectives. Kepler's search for cosmic harmony, however pleasing personally, was nevertheless meaningless in Galileo's terms. It was a quest for an imprecisely defined goal. It is no accident that in Galileo we find the first appearance of truly abstract theoretical reasoning in physics. However unattainable in reality and however they violate common sense and experience, Galileo's idealized dynamical principles can be used simply and efficiently. This gave them extraordinary power as instruments of thought. What Galileo sacrificed of Aristotle's demand for direct perception of the multiplicity of nature, he gained in clarity, accuracy, and efficiency of thought. Similarly, experimental science was for him not a passive monitoring of the rich complexity of nature, but a direct intervention into nature to construct an idealized, experimental environment, one which best approximates the abstractions embodied in theory. He preferred to think with complete mathematical accuracy about simple dynamical processes—uniform motion parallel to the surface of the earth or uniform gravitational acceleration—rather than lose clarity with more realistic problems. His program for mechanics was a new beginning; others would carry on.

## Summary

*Galileo Galilei was both an important early advocate of the Copernican idea that the earth is one of several planets orbiting the sun and also* the founder of modern mechanics. He was the first to examine the sky systematically with the telescope. Although with modern hindsight we can say that his astronomical discoveries were not proofs of the Copernican model, to Galileo they were strong arguments against the traditional, geocentric model since they refuted Aristotle's idea of the perfection of the heavens, which had been elaborated by theologians and philosophers. At any rate, his discoveries were spectacular, whatever their interpretation.

Galileo discovered that the Milky Way was, in fact, a collection of faint stars and that many stars could be seen in the telescope that were too faint to be visible to the naked eye. The surface of the moon was not smooth but made of mountains and valleys, as was evident by the changing pattern of shadows cast by the sun. In the telescope the stars appeared to be points of light, refuting the idea that they appeared so large that, in a Copernican world, they would fill up the earth's orbit.

The planets were not points of light, but displayed disks. Jupiter, the largest, was surrounded by four faint "stars" moving around the planet; evidently Jupiter had four moons. Venus showed a sequence of phases, like the moon, consistent with the Copernican model but not easy to explain in a geocentric model. The sun had spots on its surface, which rotated once in 28 days.

Galileo studied the motion of projectiles, i.e., objects moving under the influence of "gravity" only. In the absence of other effects (friction, or air resistance), an object on a horizontal surface continues moving with an unchanging speed and direction. This is the principle of inertia, as abstracted from real phenomena by Galileo. Vertical motion is more complicated. The distance d an object falls in time t is given by:

$$d = \tfrac{1}{2}gt^2$$

where g, the acceleration due to gravity, is *32 feet per second per second*. This means, as can be shown by calculation, that an object's downward speed increases simply with time: the speed v is:

$$v = gt$$

An even more succinct statement of Galileo's law of falling bodies is the statement that their acceleration is always g. When an object has both horizontal and vertical motion, its downward acceleration is g; its horizontal acceleration is zero. The law of falling bodies and the law of inertia are obeyed separately.

A consequence of all this is the principle of relativity. Absolute motion or rest does not exist. Precisely, if one person is moving uniformly relative to another, there is no way to tell which one is "really moving". The laws of physics are the same to both. Galileo claimed that the principle of relativity refuted all the "physical" arguments of the Aristoteleans against the motion of the earth.

Galileo led a dramatic life. His telescopic discoveries, reported in the brief Sidereus Nuncius, *which was published in 1610 when he was 46 years old, made him famous. He quickly developed a talent for argument and rhetoric. After several shorter works, he published in 1630 the* Dialogues on the Two Great World Systems, *an elaborate discussion of the relative merits of the geocentric and the heliocentric models of the universe. In it the modern "scientific method" is stated clearly. Experiment and observation, plus mathematical reasoning, are our only sources of knowledge about nature; against them "authority" has no place. For arguing for the motion of the earth, Galileo was tried and imprisoned by the Inquisition.*

*Under house arrest in his seventies, Galileo wrote the* Dialogues on the Two New Sciences, *explaining his work on mechanics, the principle of inertia, the law of falling bodies, and much else.*

# Appendix Chapter 5

### The Height of the Mountains on the Moon

Here is a good example of the way Galileo applied mathematical thinking to nature. Through his telescope, he carefully observed the portion of the moon's surface near the "terminator"—a space-age word for the line dividing the sunlit portion of the moon from that part in shadow. Figure 5-12 gives an exaggerated view:

Figure 5-12 shows the moon half-lit, i.e., at "quarter" phase. Most of the surface beyond the terminator is dark, but in a telescope little points of light can be seen occasionally. Galileo guessed that these were lunar mountains high enough that their peaks were still in sunlight even though their bases were already in shadow. Figure 5-13 is an expanded diagram of the geometry.

Galileo measured the distance, $MT$, between the terminator and a point of light in the dark portion. One can assume that the center of the point of light, just before it goes out, is the peak of the mountain just barely poking itself into the sunlight. Then, the triangle, $CTM$ in Fig. 5-12, is a right triangle with the right angle at $T$.

What is $h$, the height of the mountain? We solved this geometrical problem in Chapter 2 when discussing the size of the earth. The answer is:

$$MT = 2rh$$

181

Galileo Galilei

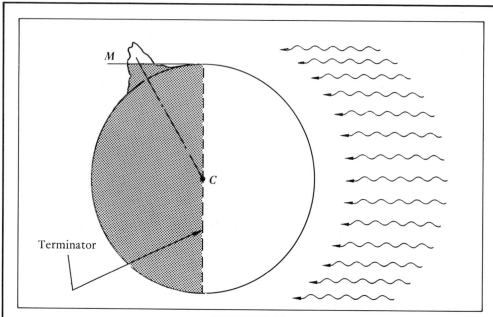

**FIGURE 5-12**
*The Illumination of a Mountain on the Moon.*

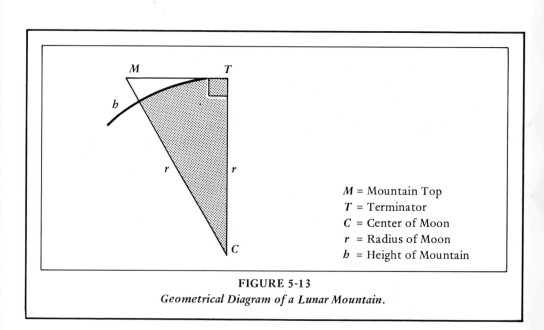

M = Mountain Top
T = Terminator
C = Center of Moon
r = Radius of Moon
h = Height of Mountain

**FIGURE 5-13**
*Geometrical Diagram of a Lunar Mountain.*

Solving for $h$, we get:

$$h = \frac{MT}{2r}$$

As before, the answer is formally an approximation, correct only insofar as the height of the mountain is small compared to the diameter of the moon.

## Kepler's Third Law Applied to Jupiter's Moons

It was easy for Kepler to measure the periods of each of Jupiter's four brightest satellites once he had a telescope. It was probably a bit harder to be sure of the distances because the angle the satellites orbit makes with the line of sight is unknown. Fortunately, however, their orbits are almost edge-on, seen from the earth, so it is sufficient to measure their apparent distance when they are the farthest from Jupiter. Kepler could only measure the angle between Jupiter and a satellite. To know how many miles it was at the time, he would have had to know how far away Jupiter is. To test his third law, he did not care about this difficulty, and neither do we since we are interested in the relative value of $T^2/R^3$ for the four satellites. So, we will list distances in minutes of arc, and will not care how many miles one minute of arc is at Jupiter's distance from the earth.

You can test how well the law works for the Jupiter system from the following table:

| Satellite | Distance | Period |         |
|-----------|----------|--------|---------|
| Io        | 2.3'     | 1 day  | 18 hours |
| Europa    | 3.7'     | 3 days | 13 hours |
| Ganymede  | 5.9'     | 7 days | 4 hours  |
| Callisto  | 10.3'    | 16 days | 18 hours |

Io and Europa are about the size of our moon; Ganymede and Callisto are much bigger. Ganymede and Io are the brightest.

## Another Proof That $d = \frac{1}{2}gt^2$ Is the Same As $v = gt$

In the text we showed that if the distance a body falls, starting from rest at time $t = 0$, is given by the equation, $d = \frac{1}{2}gt^2$, then, the speed at time $t$ is $v = gt$; and the acceleration is always $g$. Galileo actually assumed the acceleration is always $g$, and made the argument backward. Suppose we know that the object falls with constant acceleration, $g$. From our study of speed-time graphs, we know that this means that the graph of the speed vs. time is a straight line, and that the equation is $v = gt$.

What is the *average* speed between time 0 and time $t$? It is the distance divided by the time:

$$v_{average} = \frac{d}{t}$$

## Galileo Galilei

Since at the beginning of the motion the speed was zero, and at the end the speed is $v$, the average speed is $\frac{1}{2}v$.

The distance is the average speed multiplied by the time:

$$d = v_{average}\, t = \tfrac{1}{2}vt = (\tfrac{1}{2})(gt)t = \tfrac{1}{2}gt^2$$

which was our result obtained using a completely different method.

### The General Motion of Projectiles

With the principle of relativity and the law of falling bodies as our guides, we may deduce the general motion of a projectile, by which we mean simply an object moving solely under the influence of gravity; and we can apply our knowledge to solve all sorts of problems.

We begin with the motion of an object that has an initial vertical speed. Suppose at time $t = 0$ a ball is not released at rest, but rather thrown down, say at 10 feet per second. Now, we apply the principle of relativity. Imagine an observer who is moving downward at constant speed of 10 ft/sec and just passes the falling ball at time zero. If the idea is hard to visualize, try surrounding the falling ball with an elevator moving steadily downward at 10 ft/sec, containing an imaginary mad scientist equipped with a ruler and a stop watch. The point is that the moving observer sees the ball at rest at time zero. Relative to him, therefore, the ball must seem to fall according to $d = 16t^2$. To an observer on the ground, the total distance the ball has fallen is not just the distance it has fallen relative to the observer in the elevator, of course. To this distance he must add the distance the elevator has moved down; since it is going at 10 ft/sec this distance is $10t$. The total distance the ball has gone since time zero is, therefore:

$$d = 16t^2 + 10t$$

More generally, suppose an object is thrown downward at time $t = 0$ with any speed, call it $v_0$. (The notation is suggested by the idea of a speed, $v$, at a time zero.) By the argument above, but with 10 replaced by $v_0$:

$$d = \tfrac{1}{2}gt^2 + v_0 t$$

Galileo also put this answer another way. There is also a vertical principle of inertia. To find the motion of a body that is already in motion, simply add the motion due to the law of falling bodies, no matter what direction the initial motion is.

Since we have our formula for the distance fallen, $d$, at any time, $t$, we can also find the instantaneous downward speed at any time. The only hard part of the argument is how to treat the piece of the formula for $d$ that contains $t^2$, but we have already done that. The answer is:

$$v = v_0 + gt$$

The initial speed remains, and the speed imparted by gravity is simply added to it. Figure 5-14 is a graph of the speed.

The acceleration is easy to calculate. Since the graph in Fig. 5-14 is a straight

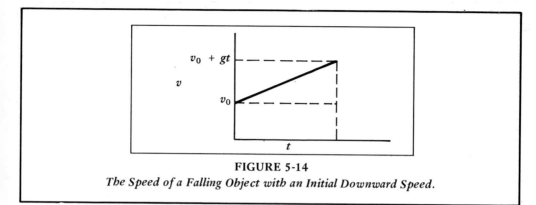

**FIGURE 5-14**
*The Speed of a Falling Object with an Initial Downward Speed.*

line, the acceleration is still a constant, and the instantaneous acceleration at any instant is the same as the average acceleration between any two times. Of course, the answer is still $g$. In fact, we could have simply stated the rule by saying that the downward acceleration is $g$, and worked backwards to find $v$ and $d$.

A much more interesting problem is the motion of an object whose initial speed is upward, like a ball thrown straight up. How high does it go? How long does it take to fall back to the ground? We can use our same formulas. (But, to get them we have to imagine an elevator going up!) Instead of going through the whole argument again, we can use a simple trick, which illustrates the power of mathematical formulas to describe a more general situation than was originally intended. Suppose, at time zero, an object has an upward speed of 10 ft/sec. What is its downward speed? Its downward speed has a simple definition in terms of where the object is at different times. The downward speed is a negative number, $-10$ feet per second. Now, we have just solved the problem of an object moving vertically with a known initial speed, $v_0$. The answer is:

$$d = v_0 t + \tfrac{1}{2} g t^2$$

and,

$$v = v_0 + gt$$

If $v_0$ is negative, then, at time zero it was moving up. Figure 5-15 is a graph of the speed for that motion. Now, we can solve all sorts of problems. For example, how high does the object go? We do it this way: At any time, its speed is $v = v_0 + gt$, where $v_0$ is a negative number. The speed, $v$, is a negative number at first; then, as the object stops going up and starts coming down, $v$ becomes a positive number. At the top, $v = 0$. Let us put that in. At the time when the object at the top:

$$0 = v_0 + gt$$

This equation is easy to solve for $t$:

$$t = \frac{-v_0}{g}$$

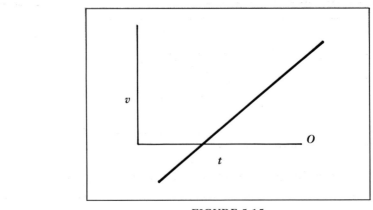

**FIGURE 5-15**
*The Speed of an Object with an Initial Upward Motion.*

Do not be worried by the minus sign. Since $v_0$ is negative, $t$ is positive, as it has to be.

How high does it go? At any time:

$$d = v_0 t + \tfrac{1}{2} g t^2$$

All we have to do is put for $t$ the value we just calculated:

$$d = v_0 \left(\frac{-v_0}{g}\right) + \frac{1}{2} g \left(\frac{-v_0}{g}\right)^2$$

$$= \frac{-v_0^2}{g} + \frac{1}{2} g \left(\frac{v_0}{g}\right)^2$$

$$= \frac{v_0^2}{g}\left(-1 + \frac{1}{2}\right) = -\left(\frac{1}{2}\right)\frac{v_0^2}{g}$$

Of course, this value for $d$ is also negative—the highest point is *above* the starting point.

As an example, suppose you throw a ball upward at a speed of 32 feet per second. To use our formulas, set $v_0 = -32$. How high does the ball go?

$$d = -\frac{1}{2}\frac{(32)^2}{32} = -\frac{1}{2} \times 32 = -16 \text{ feet}$$

The ball goes up 16 feet.

Finally, we come to *projectile motion*, a ball thrown, or shot, at an angle. Its horizontal position is $x$, its vertical position is $d$. There is an initial horizontal speed, $v_x$, and an initial vertical speed, $v_0$, which will be negative if the angle is up. We know the answer; the horizontal and vertical motions are compounded independently:

$$x = v_x t$$

$$d = v_0 t + \tfrac{1}{2} g t^2$$

To *prove* this result from the principle of relativity, it is necessary to create one of our imaginary observers, but this time the observer is riding in a monorail or funicular whose angle is the angle of the original motion, and with just the right speed so that at the time zero, he sees the projectile at rest.

Projectile motion is the most general case treated by Galilean mechanics. Although Galileo took pains to confirm the result by experiment, he showed that it could all be deduced from two rules. One, the law of falling bodies, he had discovered by experiment. The second, the principle of relativity, he knew in his soul was true, for it alone could refute the arguments against the earth's motion and the Copernican system.

At the other end of the argument, projectile motion can be summarized by two simple rules. First, the horizontal acceleration is always zero; and second, the vertical acceleration is always $g$ (32 ft/sec/sec). What is the shape of the projectile's path in the general case? We have to solve the simultaneous equations:

$$x = v_x t$$

$$d = v_0 t + \tfrac{1}{2} g t^2$$

and eliminate $t$. The answer is the formula:

$$d = v_0 \frac{x}{v_x} + \tfrac{1}{2} g \left( \frac{x^2}{v_x^2} \right)$$

One can make a graph of $d$ vs. $x$ using the above formula. The resulting curve is a picture of the actual path the projectile will take, and is *always* a parabola. The curve in Fig. 5-11 was just a special case of this general result.

We have displayed here just a taste of what can be discovered with Galileo's mechanics using only the principle of relativity and the law of falling bodies. Some of the problems that follow at the end of this chapter can give you some further idea of the possibilities of Galileo's mechanics.

## QUESTIONS

1. The Leaning Tower of Pisa is 170 feet high. If a heavy ball is dropped from the top, with what speed will it hit the ground? Convert your answer to miles per hour. [1 foot/sec = (3600/5280) miles/hour]

2. A good baseball pitcher can throw a ball horizontally at 130 ft/sec. Suppose the ball leaves his hand 5 feet above the ground. Ignoring air resistance, compute how far it will go before it hits the ground.

3. Suppose an object is moving at an angle so that it has both a horizontal speed, $v_x$, and a vertical speed, $v_d$. Use Pythagoras' theorem to find a formula for its speed along the direction it is moving.

4. Suppose a cannonball is fired at an angle of $45°$ so that its initial horizontal and vertical speeds are the same. Call these speeds $v_0$. Use the result of problem 3 to compute, in terms of $v_0$, the actual speed with which the ball leaves the cannon.

5. How far has the cannonball in question 4 travelled horizontally when it is at the top of its trajectory (in terms of $v_0$)? How far has it gone when it strikes the ground? In this way get a formula, in terms of the initial speeds, $v_0$, for the range of a cannon placed at a $45°$ angle.

6. Two automobiles are speeding down a one-lane road on a collision course. One is travelling west at 35 miles per hour, the other east at 45 miles per hour. What is their relative speed? A pilot in an airplane overhead flying east at 400 miles per hour observes the impending crash. What are the speeds of the two automobiles relative to the airplane?

7. Why must an automobile driver keep his foot on the accelerator to maintain a constant speed?

8. Many people today say that Galileo's astronomical discoveries and mechanics *proved* that the Copernican hypothesis was correct. Did he really do this and, if so, how?

9. Suppose you are in an elevator in the Empire State Building, going up steadily at the rate of 10 feet per second. You release a ball from a height of four feet off the elevator's floor. What acceleration do you observe for the ball? When will it hit the floor of the elevator?

10. Suppose, after performing the experiment in the previous questions, somebody cuts all the cables connecting the elevator to the building. You pick up the ball and repeat the experiment, releasing it from 4 feet above the floor. What acceleration do you now measure for the ball? When will it hit the floor of the elevator?

11. An automobile, starting from rest, increases its speed with a constant acceleration of 10 feet per second per second. How fast is it going after three seconds? How far has it gone after three seconds? How long will it take to get up to 60 miles per hour?

12. Zeno, a fifth century B.C. Greek philosopher, proposed a series of paradoxes about motion. One went as follows. Achilles runs a race with a tortoise, but, to be fair, lets the tortoise start a distance ahead of himself. Before Achilles can catch up to the tortoise, he must get to the tortoise's starting line, and the tortoise is ahead. When Achilles gets to the place the tortoise was when Achilles got to the tortoise's starting line, the tortoise is still ahead. By the time he covers the distance separating them, the tortoise is still ahead. Achilles can

never catch up! How do you think Galileo would have resolved Zeno's paradox?

13. We stated that when Galileo rolled a ball down an inclined plane he found the acceleration to be constant, some number, $d$. The speed is $v = at$ and, therefore, the distance, $d$, along the inclined plane is $d = \frac{1}{2} at^2$. The acceleration, $a$, depends on the steepness of the incline, and, as Galileo discovered, it does so in a simple way. In the illustration, $c$ is the length of the inclined plane, $b$ is the horizontal distance below the plane.

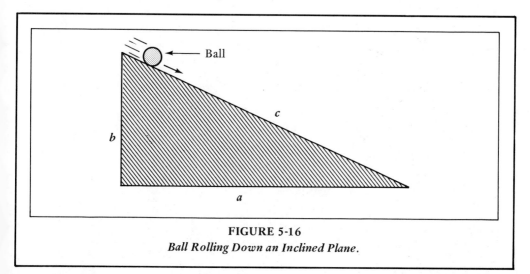

**FIGURE 5-16**
*Ball Rolling Down an Inclined Plane.*

Galileo discovered that the acceleration is always $a = g'(b/c)$. The number $g'$ is a universal constant, close to $g$, but not exactly equal to $g$.

(a) Write a formula for the ball's speed at any time, $t$, in terms of $g'$, $b$, and $c$. (b) Write a formula for the time, $t$, when the ball reaches the bottom.

Insert this time into your general formula for speed to get a formula for the speed at the bottom. (If your answer is correct, the speed at the bottom will depend only on $g'$ and the height, $b$, but not on the length, $c$, of the plane. The speed depends only on the vertical distance fallen, not on how long it takes to fall!)

### REFERENCES

1. This and the quotations below, with drawings, are from *Discoveries and Opinions of Galileo,* copyright © 1957 by Stillman Drake. Reprinted by permission of Doubleday and Company, Inc.
2. From *Galileo Galilei; A Biography and Inquiry into His Philosophy of Science,* by Ludovico Greymonat.
3. Galileo Galilei, *Dialogues on the Two Great World Systems.*

**ISAAC NEWTON**
*(Photographed by Barry Donahue and by permission of the Houghton Library, Harvard University)*

# CHAPTER 6

# HEAVEN AND EARTH IN ONE FRAMEWORK: SIR ISAAC NEWTON

### THE SEVENTEENTH CENTURY IN SCIENCE

Galileo and Kepler, who began the great seventeenth century in science, set the style for the work to come. Both believed that accurate measurements permit clearer expression of physical law, whereas accuracy had seemed previously only to complicate theory. Both also sought precise mathematical theories. Kepler, believing the sun moves the planets, purified Copernicanism of all remaining Ptolemaic devices, and so produced a simple unified description of the solar system. Galileo willingly sacrificed reality, imagining motion in a vacuum—unattainable, yet so simple he could describe it mathematically. He performed experiment after experiment attempting to approximate ever more closely the ideal vacuum state that existed only in his mind, and proclaimed the results of his experiments a better foundation for mechanics than twenty centuries of Aristotelian thought.

The number of scientists increased rapidly. In France, René Descartes invented "analytic geometry"—the method of graphs. The problems suggested by Descartes' method led directly to the calculus. In Germany, the philosopher Gottfried Leibnitz, independently of Newton, worked out the calculus—the key to understanding all but the simplest accelerated motions. In Holland, Willebrord Snell revised Ptolemy's optics and extended Kepler's, to produce the exact mathematical law of refraction. Christian Huygens proposed the wave theory of light and greatly extended Galileo's mechanics. In England and France, many telescopes were manufactured and systematic astronomical observations were made. Tycho Brahe's measurements were easily surpassed. The speed of light and size of the solar system were measured for the first time. In England, Robert Boyle discovered how gases like air behave when they expand and contract. Boyle was the father of modern chemistry; his book, *The Skeptical Chymist*, argued against Aristotle's four elements and suggested that there were many more. Just as Galileo had turned his new telescope to the heavens and reported the immensity of the universe, the Englishman Robert Hooke described the world of the very small seen under the microscope. Hooke also arrived at some features of the law of gravitation independently of Newton. In Italy, Galileo's student Torricelli measured the weight of air with a mercury barometer.

Galileo had been a member of the first scientific society, the *Accademia dei Lincei*. The year 1645 found a group of Englishmen gathering informally to discuss the new science; by the 1660s they had obtained a charter from the King to become the Royal Society of London. In 1666, the French Academy of Sciences had its first meeting. People needed to discuss their work regularly, away from the too-traditional atmosphere of the universities. Besides, they needed a place to do their experiments. Hooke was commissioned to perform one experiment each week before the members of the Royal Society, a task he performed with great energy for more than thirty years. The scientific societies sponsored the publication of their member's work. The secretary of the Royal Society, Robert Oldenburg, maintained an extensive correspondence, communicating the most recent developments all over Europe. The scientific societies soon defined the standards and style of scientific work.

## SEVENTEENTH CENTURY ASTRONOMY AND MECHANICS

Galileo's *Starry Messenger* molded a climate of thought whose central question was no longer, "Is Copernicus right?" but, "Since Copernicus may well be right, what is wrong with our physics that we cannot explain his system?" Copernicus had shown that the solar system had unified geometry, and Kepler's laws strongly suggested that the solar system must have a unified physics. If, as he claimed, a force in the sun pushed the planets around their orbits, the sun's force acted the same way throughout the solar system. But, what was the sun's force and how did it act?

With the moon having mountains like the earth and the earth a planet moving through the heavens, what was the difference between heavenly bodies and the earth? Aristotle's total separation of heavenly and earthly physics was unnecessary. Since the daily rotation of the earth removed the need to attach the stars to a crystalline sphere, a universe filled with celestial material was also questionable. People became comfortable with Galileo's idealization of motion in a vacuum. They even invented air pumps to create high vacuum conditions in the laboratory. The absence of air, a vacuum, seemed no less real than the air itself. Perhaps, even, the motion of planets through empty space was the one place where Galileo's frictionless mechanics might really work. The planets' motions should really be described by the same laws as the motion of balls rolling down inclined planes. A "mechanistic philosophy" developed, which attempted to explain *all* phenomena of nature as the action of tiny, invisible particles, each moving through the vacuum obeying the laws of mechanics.

Nonetheless, serious questions remained unanswered. While Galileo's law of inertia weakened the ancient objections to the earth's motion, he had been uncertain whether inertial motion on a large scale—over distances comparable to the size of the earth—was in straight lines or circular. He leaned towards circular inertial motion. Descartes maintained that even in the heavens, inertial motion was in straight lines. The new "natural" motion became uniform straight-line motion; it was accelerations (changes in speed or direction) that required some sort of "force" to explain them. Much discussion focused on what the forces might be and how to describe them mathematically, but no one doubted any more when they were needed. The almost circular elliptical planetary motions, considered natural by the Greeks, suddenly demanded an explanation. Why did the planets not fly off into space in inertial straight-line orbits? Perhaps Kepler had not been exactly right; a force was not needed to push the planets around, a force was needed to keep them in their orbits. What force was this? Was it a celestial force? Or, could we learn about it from experiments on earth?

With the earth adrift about the sun, what about the old, comfortable explanation for gravity that all heavy matter seeks its natural place at the center of the universe? Galileo's laws still did not say *why* bodies fell. People took up an idea of Copernicus' to explain why there are many concentrations of heavy material (planets) orbiting the center of the universe. It was "natural" for heavy bodies near each other to attract one another, to be pulled together. Perhaps this was why the earth was formed in the first place, why it did not fly apart, and why things fall. But, could one prove that heavy bodies would fall, for example, on Jupiter? Would they obey Galileo's law of falling bodies?

One of the most pressing difficulties was mathematical. Most problems of physics involved continuously changing instantaneous speed. Exact solutions were still intractable mathematically. Galileo had restricted himself to the next simplest case after uniform motion—uniform acceleration under gravity, and that had been a great triumph. Ptolemy had approximated the nonuniform planetary motions by adding up many uniform circular motions. Even Kepler's second law avoided the question of general nonuniform motion, since something—the area swept out by the line from the sun to a planet—advanced steadily in time. Yet Kepler's work made it even clearer that further understanding of the solar system depended upon first describing nonuniform motion mathematically.

One man, Isaac Newton, tied together the three threads of our story and of seventeenth century physics: First, the principle of inertia, stated by Galileo and clarified by Descartes; second, Galileo's law of falling bodies; and third, the geometry of the solar system, perceived by Copernicus, propagandized by Galileo, and described mathematically by Kepler. Newton, along with Leibnitz, also originated the basic principles of the calculus, necessary for nonuniform motion. To solve one of these problems, he had to solve them all; and not the least was the mathematical one. Newton meditated upon them all in secrecy for twenty years, until he could present the world a complete system of terrestrial and celestial mechanics—his *Principia Mathematica Philosophiae Naturalis,* the mathematical principles of natural philosophy.

Newton was born on Christmas day, 1642, the year of Galileo's death, in Woolsthorpe in rural England. His father had died before he was born, and his mother soon remarried and moved away, leaving the lonely boy in the care of a relative. After he was famous, his neighbors managed to recall the many toys and models he constructed as a child—the first signs of the manual dexterity that would make him the greatest experimental physicist of his age. Stories multiplied that even then he was so absorbed in thought that he neglected his farm chores. Since he was clearly not to be a gentleman farmer, he was permitted to enter Trinity College, Cambridge, in 1661. Again, legends abound that he taught himself Kepler's *Optics,* and skipped over Euclid's *Geometry* to read Descartes' *Analytic Geometry.* It is certain that he met Isaac Barrow, scientist, preacher, the first Lucasian professor of mathematics, and the first man to recognize and encourage Newton's astounding talents.

### NEWTON'S MIRACULOUS YEAR

In the summer of 1665, the Great Plague struck London, decimating the population. Fearing a similar calamity at Cambridge, the authorities closed the university. Newton returned home to Woolsthorpe and in enforced yet not unwelcome isolation meditated upon what he had so recently learned at Cambridge. He began working upon many of his greatest discoveries in Woolsthorpe. Much later he wrote:

> In the same year* I began to think of gravity as extending to the orb of the moon, and having found out how to estimate the force with which a globe revolving within a sphere presses the surface of the sphere, from Kepler's rule of the periodical times of the planets being in a sesquialterate proportion of the distances from the centers of their orbs** I deduced that the forces which keep the planets in their orbs must be reciprocally as the squares of their distances from the centers about which they revolve: and thereby compared the force requisite to keep the moon in her orb with the force of gravity at the surface of the earth, and found them to agree pretty neatly. All this was in the two plague years of 1665 and 1666, for in those days I was in the prime of my age for invention, and minded Mathematics and Philosophy more than at any time since.

Newton returned to Cambridge when the university reopened in 1667 and stayed until 1692. It is not clear that he told anyone what he had done during the plague years. However, we will reconstruct in modern terms what he said he learned during the miraculous year, 1666, from his discussion in the *Principia*.

Newton first constructed a conceptual argument proving that Galileo's laws of mechanics apply to the heavens. He returned to Galileo's cannonball fired in a vacuum from a mountaintop, but he put the mountaintop on a round earth. He imagined firing the cannonball faster and faster so that it moves further and further around the earth before gravity pulls it to the ground. Eventually it should continue right around the earth. If it arrived back at the cannon with the same speed it started with, it would make another circuit. The cannonball would orbit around the earth, just like the moon. In fact, imagine firing the cannon from higher and higher mountains, including one as high as the moon's orbit. What would be the difference between the cannonball and the moon in orbit? Would the numbers work out so that the same explanation could apply to the cannonball and the moon? Figure 6-1 is the sketch Newton made to illustrate his argument.

In later life, Newton recounted that he had been sitting under an apple tree at Woolsthorpe wondering how the moon could stay in orbit around the earth, when an apple fell. Was it gravity, he wondered, that pulled on the apple, his imaginary orbiting cannonball, and on the moon? While many felt generally that terrestrial physics should apply to the heavens, this was the first suggestion that specific laws,

---
*Newton is referring to the year 1666.
**Newton means Kepler's third law here.

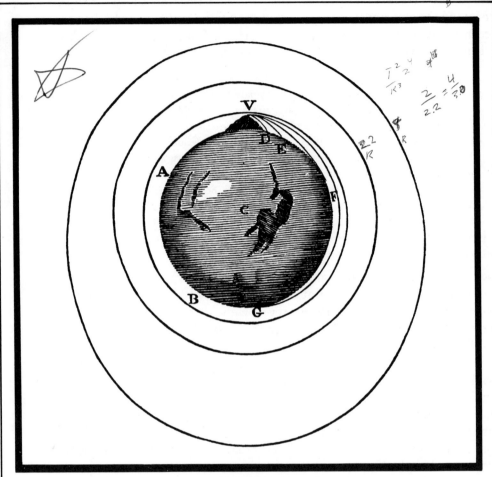

**FIGURE 6-1**

*Newton's Thought Experiment. Imagine firing a cannon from a high mountain, V. If the initial horizontal speed of the cannon ball were not very large, it would fall to earth at D. Its trajectory would be almost exactly the parabolic one originally described by Galileo. If the initial horizontal speed is increased, the ball would fall to earth at E, F, and G. One could imagine a ball fired so fast that it would even arrive back at V. It would then be in orbit, just like the Moon. Of course, we must imagine that there is no air resistance.*

Galileo's laws of inertia and acceleration under gravity, should apply to a celestial body, the moon. And if the moon, why not all others?

Let us analyze Newton's thought experiment in more detail. According to Galileo, a cannonball moving over a small part of the earth maintains a constant horizontal speed while uniformly accelerating downward. The Greeks had known that dropped objects always fall straight down towards the center of the earth. Thus, Galileo had really demonstrated that falling bodies always accelerate towards the center of the earth. Newton's cannonball moving around the spherical earth must, therefore, always accelerate towards the center of the earth, too. It must achieve a delicate balance between flying off into space in inertial straight-line motion and accelerating towards the center of the earth. This balance just keeps it moving in a circle. The same analysis should apply to the planets orbiting the sun and the moon orbiting the earth—to all orbits.

Suppose an object, any object, is moving around a circular orbit. It is, therefore, accelerating. Kepler's equal areas law, when applied to a circular orbit, says that the speed around the orbit does not change. In this case, the acceleration is a change in *direction* of motion. (Elliptical orbits are much more difficult to analyze, since a planet changes both its speed and direction of motion.) Newton had to answer two questions very precisely. What is the direction of the acceleration, and how big is the acceleration? The answer to the first was obvious: just as the acceleration of the cannonball was towards the center of the earth (which was also the center of its orbit) so should the acceleration in any circular orbit be directed towards the center. We repeat: *an acceleration towards the center is required.* The second question was more difficult to answer. Let us rephrase it more precisely. If a body moves with constant speed, $v$, in a circular orbit of radius $R$, how big is the acceleration towards the center required to keep the body in orbit? Newton needed to answer this question to prove that the moon remains in orbit because it is *falling* due to the earth's gravity. He needed to calculate the acceleration required to keep the moon in orbit and then compare it with the acceleration of gravity measured near the earth's surface, $g = 32$ ft/sec/sec. Newton knew a cannonball in orbit near the earth must have the same acceleration as a ball that is dropped, or an apple that falls from a tree, $g = 32$ ft/sec/sec. But the moon is very far away, where the acceleration of gravity had not been measured. How could he be certain that it would be the same far away from the earth as it is near the earth's surface?

We now calculate the acceleration required to keep the moon in circular orbit. Let us call this unknown acceleration $a$. Newton treated the moon's orbit as an example of compound motion. The moon moves in inertial straight-line motion and simultaneously accelerates towards the earth. Following Galileo, he treated each motion separately. Figure 6-2 shows the moon, the earth, the path $M_1 M_2$ the moon would follow if there were no gravity, and its path if it had no speed and

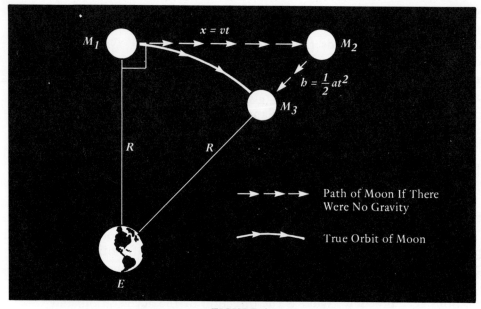

**FIGURE 6-2**

*The Moon's Orbit as an Example of Compound Motions. Imagine the moon's true orbit, $M_1M_3$, to consist of two separate motions, straight-line inertial motion along $M_1M_2$ obeying Galileo's mathematical relation, $x = vt$, and uniform acceleration toward the center of the earth, along $M_2M_3$, obeying Galileo's law, $h = \frac{1}{2}at^2$. The acceleration, $a$, towards the center of the earth is unknown at the distance of the moon and needs to be calculated. If the time interval, $t$, is very short, then, $a$ turns out to be $v^2/R$, where $R$ is the radius of the moon's orbit. The moon's orbit has been assumed to be a perfect circle in this simple example.*

simply fell towards the earth, $M_1E$, and finally, its true orbit, $M_1M_3$. Along the path $M_1M_2$, the moon moves in a straight line at constant speed, $v$. It covers the distance $x = vt$ during a short time interval, $t$. During the same interval, $t$, the moon accelerates towards the earth so that it falls the distance $h = \frac{1}{2}at^2$, according to Galileo's formula for accelerated motion. We calculate the unknown acceleration, $a$, by requiring that the distance $h$ the moon falls be precisely enough to bring the moon back to its circular orbit.

We find distance $h$ the same way Galileo found the height of the mountains on the moon or the Greeks found the radius of the earth by watching ships sail away from them. In Fig. 6-2, the triangle $EM_1M_2$ is a right triangle, with a 90° angle at $M_1$. Therefore, we can use Pythagoras' theorem for its sides. The side $EM_1$ has length $R$, the radius of the moon's orbit; the side $EM_2$ has length $R + h$; and the side $M_1M_2$ has length $x$; therefore:

$$(R + h)^2 = R^2 + x^2$$

Expanding out $(R + h)^2$, we find:

$$(R + h)^2 = R^2 + 2hR + h^2 = R^2 + x^2$$

Our aim is to find the *instantaneous* acceleration. Therefore, we imagine time interval $t$ to be very small. Similarly, the distance $h$ the moon falls is also very small when $t$ is small; since $h$ is much smaller than $2R$ we may ignore $h^2$ compared to $2Rh$. To a very good approximation, then:

$$(R + h)^2 = R^2 + 2hR$$

This approximation improves as the time interval gets smaller and smaller. Thus, we have:

$$R^2 + 2hR = R^2 + x^2$$

Or, cancelling the $R^2$:

$$2hR = x^2$$

Rearranging terms:

$$h = \frac{x^2}{2R}$$

as before.

Thus far, we have only applied Pythagoras' theorem to the geometry of Fig. 6-2. Now, we require the moon to obey Galileo's laws of mechanics, but with the acceleration, $a$, unknown. Into the formula, $h = x^2/2R$, we substitute $vt$ for $x$ and $½at^2$ for $h$ to get:

$$½at^2 = \frac{(vt)^2}{2R} = \frac{v^2t^2}{2R}$$

Note that the ½'s and even more importantly, the time factors, $t^2$, cancel, so that acceleration $a$ required to keep the moon in orbit is:

$$a = \frac{v^2}{R}$$

Since the $t^2$ terms cancelled, the acceleration, $a$, does not depend upon the length of time interval, $t$. Thus, $a = v^2/R$ is correct for an *arbitrarily short* time interval. It is, therefore, the instantaneous acceleration.

Since the radius of the moon's orbit, $R$, had been known since ancient times to be about 60 times the radius of the earth, and its period, $T$, to be about 28 days, Newton could calculate the moon's speed using the formula $v = 2\pi R/T$. Then, he could calculate its acceleration, $a$, as follows:

$$a = \frac{v^2}{R} = \left(\frac{2\pi R}{T}\right)^2 \left(\frac{1}{R}\right) = \frac{4\pi^2 R^2}{T^2 R} = \frac{4\pi^2 R}{T^2}$$

The size of $a$ is not Galileo's $g$ but only about 1/3600 of $g$, or about 0.009 feet per second per second. The moon accelerates much more slowly towards the earth than the falling apple. Thus, at this point, it did not appear that the earth's gravity kept the moon in orbit, at least not with Galileo's acceleration, $g$.

The rule, $a = v^2/R$, applies not only to the moon, but also to any body in a circular orbit. It is a theorem of geometry, not just about the moon. Newton also could calculate the accelerations of the planets towards the sun, if he approximated their orbits by circles with the formula $a = v^2/R$, where $v$ is now a *planet's* speed and $R$ is its distance from the sun. Each planet would have a particular acceleration towards the sun. Perhaps these accelerations would depend in a regular way upon distance from the sun. Perhaps Newton could prove Kepler's speculation that the sun's force weakens with distance from the sun. He could have gone to a table of planetary positions and laboriously computed their speeds and distances from the sun and then calculated their accelerations. Kepler's third law saved him this trouble, for the observations could be summarized by the simple formula, $T^2 = R^3$, where $T$ is measured in earth years and $R$ in astronomical units. If he used these units in the formula, $a = 4\pi^2 R/T^2$, he could substitute in $T^2 = R^3$ directly to find:

$$a = 4\pi^2 R/T^2 = 4\pi^2 R/R^3 = 4\pi^2/R^2 \text{ (astronomical units/year/year)}$$

Thus, the planets' accelerations decrease in a simple way with distance from the sun—as the square of the distance.

The Galilean satellites of Jupiter also obey Kepler's three laws. Since, in particular, they obey Kepler's third law, Jupiter exerts a force on its satellites that decreases as the square of the distance from Jupiter's center. Now, what about the moon? Suppose the earth had two moons instead of one. Wouldn't the two moons obey Kepler's third law too? Since our one moon has an elliptical orbit and obeys the equal areas law, it is plausible that the third law would be obeyed if there were two moons. But, then their accelerations would have to decrease as the square of the distance from the center of the earth

Now, it was obvious what to do. Galileo had stood on the surface of the earth, one earth radius from the earth's center, and measured the acceleration of gravity, $g$. Newton had argued this is also the acceleration of the imaginary cannonball in an orbit skimming the earth. The moon is 60 times as far away from the center of the earth as the earth's surface. If it is gravity that accelerates the moon, and if, as for the planets, that acceleration also decreases as the square of the distance, the acceleration of the moon should be $(1/60)^2$ of $g$. But, $(1/60)^2 g$ is $1/3600$ of $g$, precisely the lunar acceleration Newton had calculated. As Newton said, the numbers "agree pretty neatly."

Now see what Newton could conclude. The force that causes apples to fall, gravity, also keeps the moon in orbit. Kepler's "force in the sun" was again gravity, but instead of pushing the planets around the sky, it pulls them back to their orbits. Jupiter's gravity pulls on its satellites also. A ball dropped on Jupiter would obey Galileo's law of falling bodies (with a different $g$). Observe how far the scientific revolution against Aristotelian thought started by Galileo had progressed against Aristotelian thought. Newton could propose to violate the once perfect and sacred heavens with an artificial satellite made of heavy terrestrial material. Newton could even compute the orbital period of an artificial satellite of the earth even though he could not make one. If the satellite's orbit is close to the earth's surface, its orbital radius, $R$, is approximately 4000 miles, the radius of the earth. It is accelerated downward toward the center of the earth with acceleration $g$, since $g$ is the acceleration of objects near the surface. Thus, we know that:

$$g = \frac{v^2}{R}$$

or $v^2 = Rg$; taking the square root, we obtain:

$$v = \sqrt{Rg}$$

For a circular orbit, the speed $v$ is related to the period $T$ by $v = 2\pi R/T$, which may be rearranged to give:

$$T = \frac{2\pi R}{v}$$

Then, substituting $v = \sqrt{Rg}$, we find:

$$T = \frac{2\pi R}{v} = \frac{2\pi R}{\sqrt{Rg}} = 2\pi \sqrt{\frac{R}{g}}$$

Now, $g$ is 32 in units of ft/sec/sec, so we must convert $R = 4000$ miles to feet by multiplying by 5280, the number of feet in a mile, to make the units consistent. Thus, $R = 21{,}100{,}000$ feet. Then, substituting:

$$T = 2\pi \sqrt{\frac{R}{g}} = 2\pi \sqrt{\frac{21{,}100{,}000}{32}} = 5100 \text{ sec}$$

Dividing through by 60, we see that 5100 seconds is approximately 85 minutes.

When an artificial satellite is in orbit, the earth has two satellites, two "moons". Do the artificial satellite and the moon obey Kepler's third law? We need only calculate $T^2/R^3$ for both. For the satellite, $T = 5100$ seconds and $R = 21{,}100{,}000$ feet so that:

$$T^2/R^3 = (5100)^2 / (21{,}100{,}000)^3 = 0.000{,}000{,}000{,}000{,}0027$$

For the moon, $T = 656$ hours $= 2{,}360{,}000$ seconds and $R = 240{,}000$ miles $= 1{,}167{,}000{,}000$ feet, so:

$$T^2/R^3 = (2{,}360{,}000)(1{,}267{,}000{,}000)^3 = 0.000{,}000{,}000{,}000{,}027$$

It is no coincidence that the same number can be calculated in two such different ways. Previously, we showed that Kepler's third law implied that the sun's gravitational acceleration decreased as $1/R^2$ from the sun, and Jupiter's as $1/R^2$ from Jupiter. Now, we have worked the problem in reverse, showing that Kepler's third law would be obeyed by the moon and an artificial satellite if the earth's acceleration of gravity decreases as $1/R^2$.

Newton did not disclose the conclusions he reached during the plague year of 1666 until very much later. Many explanations have been offered as to why he was so secretive; for example, that he was shy and unsociable, and hated criticism. However, no one could have been more aware than Newton that his discoveries of 1666 were very incomplete. To begin with, the orbits of the moon and planets were ellipses, not circles. Newton demanded a complete and accurate solution for the planet's orbits. Would the change to ellipses affect his conclusions? Second, Newton had assumed that the acceleration of gravity decreases as $1/R^2$ from the *center of the earth*. What was so special about the center? It was altogether too Aristotelian to assume that a special point, the center, exerts an attraction on the moon. It was more likely that the *material* in the earth attracted the *material* of the moon. If so, it appeared that the earth acted as if all its material were concentrated at its center. But why? A third problem was to determine exactly how widespread gravitational attraction really was. How did the material in the earth attract the material in the moon? If the earth were not a sphere but some other shape, would it still attract the moon the same way? Did the moon also attract the earth? Did the moon also exert a gravitational force? Did all celestial bodies have gravity? Which bodies had gravity, and which not? Finally, Newton sought to include all he had learned about gravitation in a more general theory of motion that would include other effects such as friction and electrical attraction. A *theory* of motion was not enough; he would not be satisfied until he had developed the mathematics necessary to describe nonuniform motions, so that his theory would predict observations exactly. The answers to these questions had to await the further maturation of Newton's thought, and so, he pondered in silence for twenty years.

## THE *PRINCIPIA MATHEMATICA*

Newton returned to Cambridge in 1667, where he would remain until 1692. He left Cambridge only a few times during these twenty-five years. He did not continue to work much on mechanics, but turned his attention also to pure mathematics and to experiments in optics. His mathematical work so impressed his teacher, Barrow, who wanted to return to teaching theology, that Barrow resigned his professorial chair in favor of Newton. Thus, Newton, at age 27 the second Lucasian professor of mathematics, could look forward to the quiet life of scholarly contemplation he so deeply yearned for. In 1672, Newton published his first scientific paper in the *Proceedings of the Royal Society of London*, on the results of his optical experiments. It immediately embroiled him in controversy and made him a lifelong enemy. Robert Hooke complained that, while Newton's work said how light behaved, Newton did not state what light is. Enraged at such obvious misunderstanding of his basic philosophy that science was limited to describing nature, Newton retreated into further isolation. He published very little of his work.

Edmond Halley, the astronomer after whom Halley's comet is named, earned a place in history by goading Newton out of his isolation. Halley also had concluded that for circular orbits and constant speed he could show, using Kepler's third law, that the force of attraction between sun and planets decreases as the square of the distance from the sun. The difficulty was to show that the true elliptical orbits lead to the same result. The architect Christopher Wren mediated a scientific wager between Halley and Robert Hooke, who had reached a conclusion similar to Halley's. Here is Halley's own description:[1]

> I met with Sir Christopher Wren and Mr. Hooke, and falling in discourse about it, Mr. Hooke affirmed that upon that principle* all the laws of the celestial motions were to be demonstrated, and that he himself had done it. I declared the ill success of my own attempts; and Sir Christopher, to encourage the enquiry said, that he would give Mr. Hooke, or me, two months time, to bring him a convincing demonstration thereof; and besides the honour, he of us, that did it, should have from him a present of a book of 40 shillings.

Neither man won the book. After several more months of wrestling with the problem, Halley journeyed to Cambridge to visit Newton. Perhaps Newton might have something interesting to say, even though he had published precious little. John Conduitt, later Newton's successor at the Mint and husband of Newton's favorite niece, recounted the fateful meeting long after the fact. What would be the orbit of a planet, Halley had asked, if its acceleration were always toward the sun and inversely proportional to the square of its distance from the sun?

> Newton immediately replied, an *ellipse*. Struck with joy and amazement, Halley asked him how he knew it. "Why," replied he, "I have calculated it" and being asked for the calculation he could not find it...

Newton had mislaid the solution to the central scientific problem of his age! Halley perceived what he would have to do. He urged Newton to reconstruct the solution. This fortunately had more than the expected effect, for Newton in so doing made an error that he did not find for several months. This so infuriated him that he set about working on mechanics seriously. Soon he completed and sent to Halley a short work on mechanics, one so impressive that Halley went to Cambridge again to urge Newton to write a complete treatise on mechanics. There followed 18 months of excruciatingly hard labor; 18 months in which it is said that Newton often neglected to eat or sleep; in which he occasionally remained motionless all day, lost in thought; these 18 months resulted in the *Principia Mathematica Philosophiae Naturalis*—the Mathematical Principles of Natural Philosophy. While Newton had thought through the problems of mechanics and gravitation in 1666, he had not created a complete mathematical theory of the planets' motions with the exactitude only he perceived possible. Most of the demonstrations in the *Principia* were created during this 18 month period—definitions, laws, and theorems, the law of universal gravitation, the explanation of Kepler's three laws, the theory of motion in vacuum, the theory of motion in fluids (a forerunner of modern aerodynamics), the theory of waves in fluids, the study of the tides and of comets, a theory describing the small deviations of the earth's shape from

---

*By principle is meant the $1/R^2$ law.

perfect sphere, an exact calculation of the precession of the equinoxes, and many other things. Singlehandedly Newton demonstrated that science need not be qualitative, as it was for Aristotle, or mathematically precise only about ideal situations, as it was for Galileo, but that it could describe God's real universe with great precision.

## NEWTON'S THREE LAWS

So far, we have studied three problems in mechanics: uniform straight-line motion, free fall under gravity, and circular orbital motion. In 1666, Newton realized that free fall under gravity was intimately connected with the problem of the planets' orbits. Halley, Hooke, and Huygens soon reached this conclusion, too. A fourth problem also interested seventeenth century scientists, that of collisions between particles. With their "mechanistic philosophy", they had tried to understand the universe in terms of Galileo's balls moving in a vacuum. Descartes had imagined that all the phenomena of nature were produced by a swarm of tiny, invisible particles flying through a vacuum. The planets were carried around the sun by giant swirls of these particles. Given this point of view, then, it became important to ask what happens when any two particles collide. Let us rephrase this question more carefully: Consider two hard spherical balls that collide and bounce off one another. Before the collision, each ball is in a state of uniform, straight-line motion; and after the collision, each ball is in a different state of uniform, straight-line motion. Given the speed, direction of motion, and weight of each ball before the collision, could one calculate their speeds and directions after the collision? We will return to collisions shortly.

Newton remained unsatisfied until he could formulate a complete mechanics that would apply not only to the four problems above but also at least in principle to all conceivable problems involving motion. It is a mark of Newton's genius that of all the possible statements about motion, he recognized that three and only three completely define a logically consistent framework within which all problems of motion can be analyzed quantitatively. These are Newton's three laws. They define what quantities describing motion it is important to know and calculate. But understanding what is important to calculate is different from actually calculating. Newton's ambition was to solve mechanical problems exactly; i.e., to say in detail how the planets move or how balls collide. This required the invention of calculus. We will not discuss the calculus here, but using our understanding of straight-line motion, free fall and orbital motions, and collisions, we will discuss Newton's three laws, which must apply to all these situations.

### Newton's First Law

Newton's first law is simply Galileo's law of inertia, and so we state it without further ado:

> Every object remains in its state of rest, or of uniform motion in a straight line, unless it is compelled to change that state by forces impressed upon it.

Certain key words are contained in Newton's definition of the principle of inertia. The first is "every"—Galileo's principle of inertia should apply to *all* bodies whether they are square or round, orange or blue, light or heavy, and celestial or terrestrial. The second key word is the "or" between "state of rest" and "of uniform motion". This is the statement of Galilean relativity—no experiment can distinguish between rest and uniform motion. They are equivalent states, and both are called inertial states. The third key word is the "unless" in "unless it is compelled to change that state . . . :" for this defines when forces are needed. Uniform straight-line motion is natural; it requires no explanation. A force changes the state of rest or of uniform straight-line speed. In other words, forces produce *accelerations*. When two balls collide, they exert forces on each other that change them from one inertial state to another. In this case, it is easy to see what exerts the force and when it is exerted. Each ball exerts a force on the other during that short interval of time when they are in contact. The earth also exerts a gravitational force on the moon, and the sun on the planets. We know there is a force on the moon and planets because they are not moving in straight lines. Instead of acting during a brief instant, like the forces between coliding balls, the gravitational force acts continuously. Newton knew there were other forces in nature. His own experiments suggested electrical forces were very powerful, as were the forces holding matter together. He also suspected that matter could exert a force on light. Nonetheless, not enough was known experimentally about these other forces to permit the same careful mathematical analysis he could bring to collisions, free fall, and orbital motion.

### Newton's Second Law

Newton's first law states when a force is needed, but it certainly does not define the mathematical quantity, force; it does not say *how much* force is required to produce a given change of inertial motion. For one thing, it was clear that heavier bodies tend to resist changes in inertial motion more than light ones. How could Newton make quantitative the tendency for bodies to persist in straight-line, uniform motion? How could he measure their inertia?

Let us consider a simple example. A baseball weighing one pound and a cannonball weighing one hundred pounds both move in straight lines at one hundred miles per hour. Suppose both are brought to rest in one second. It is obvious that a larger force is required to stop the cannonball. Similarly, a larger force should be required to stop a two-hundred mile per hour cannonball in one second than one moving at one hundred miles per hour. Suppose the cannonball moved at one mile per hour; and the baseball at one-hundred. How much force is required to bring each to rest in one second? Is a larger force needed to stop the 200 mile/hour cannonball in two seconds than the 100 mile/hour cannonball in one second?

To make the motion of force quantitative, Newton first had to define precisely the "quantity of inertial motion", which is called *momentum*. The above examples make it clear that the faster a body moves and the heavier it is, the more momentum it should have. With these facts in mind, momentum is defined as the product of the mass $M$ and the velocity $v$.

$$momentum = Mv$$

We will discuss shortly how the mass of a body is measured. For now, it is enough to know that its mass is measured by its weight. In other words, the 100 pound cannonball has 100 times the mass of the one pound baseball. When they both travel at the same speed, the cannonball has 100 times the momentum. When the cannonball moves at 1 mile/hour and the baseball at 100 miles/hour, they have the same momentum. The cannonball moving at 200 miles/hour has twice the momentum of the 100 mile/hour cannonball. Finally, one other thing is important: momentum, like speed and acceleration, has a direction—the direction of motion.

With a definition of momentum, Newton then could define force precisely. His second law, slightly paraphrased, states:

> The rate of change with time of momentum equals the applied force and is in the direction of that force.

Again, Newton's second law contains certain key phrases. The first is the "rate of change with time of momentum." It is not the *total* change of momentum, but *how fast* the momentum is changing that measures the force applied to a body. This important distinction makes it possible for Newton's definition of force to apply both to collisions and to free fall and orbital motion. When two balls collide, they exert a force on each other for a short period of time. Their mutual interaction produces a large *total* change of each ball's momentum. It is easy to measure the total change of momentum of each ball: subtract the ball's speed before the collision, $v_{before}$, from its speed, $v_{after}$, and multiply by the mass. Thus, the total change in momentum is $M(v_{after} - v_{before})$. But, it is difficult to measure *how fast* the ball's momenta are changing during the collision because the duration of the collision is so short. If collisions were the only problem Newton wanted to solve, he might have been tempted to define the force as the total change in momentum. However, this definition did not apply to falling bodies. As Galileo showed, a falling body's velocity does not increase all at once, but continuously, obeying the rule, $v = gt$. Since Newton defined momentum as mass $M$ times velocity $v$ and since for a falling body $v = gt$, the momentum of a falling body is $Mgt$, and increases continuously. Recall that the rate of change of speed of a falling body is $g$, Galileo's acceleration of gravity. What is the rate of change of momentum? If an object's mass never changes, rate of change of momentum = (*mass*) × (*rate of change of speed*). Thus, the rate of change of momentum of a falling body of mass $M$ is $Mg$. The earth exerts a force on a body of magnitude $Mg$.

The second key phrase in Newton's second law is "in the direction of that applied force." Forces have a magnitude *and* a direction. To catch a baseball, a force must be directed opposite to the ball's direction of motion. To increase the ball's speed, a force must be directed along the ball's direction of motion. The direction of the earth's gravitational force is downwards, towards the center of the earth. Thus, the horizontal velocity of Galileo's cannonballs is constant because the earth exerts no horizontal force, only a downward gravitational force. Only the downward velocity increases with time. Similarly, the directions of the earth's gravitational force on the moon, and the sun's on the planets, are towards the centers of their orbits, because the moon and planets are accelerated towards the center.

### Mass and Weight

It was not enough for Newton to realize that a quantity called mass was needed to measure an object's resistance to acceleration; he had to define how mass could be measured. He knew that mass had something to do with weight. With his laws of motion and with Galileo's law of falling bodies, he could define both weight and mass precisely.

An object's weight is simply the force the earth's gravity exerts on it at the surface of the earth. Galileo's law of falling bodies states that objects accelerate at the rate $g$. Thus, the force the earth exerts on the falling body, the weight $W$, is according to Newton's second law, $Mg$. However, an object's weight cannot be measured by dropping a ball and measuring its acceleration. For all objects have the same acceleration, $g$, regardless of their weight.

A separate experiment is needed to measure weight. Many different kinds of scales have been invented to measure weight. They are all based on a very simple principle: exert a force of known magnitude in the direction opposite to gravity. When the known force and gravity are equal in size (but opposite in direction), there will be no total force on the object. It will no longer accelerate, but remain at rest or at constant speed. By measuring the size of the force opposing gravity, we measure also the earth's gravitational force.

A very simple scale is shown in Fig. 6-3. The mass to be weighed compresses a spring. Springs have one important property. When we push on a spring, it pushes back. In other words, a compressed spring exerts a force in the direction opposing its compression. The more the spring is compressed, the larger the force it exerts. By measuring how much the spring is compressed, we measure the size of the force compressing it. An object's weight is just the force of gravity on it; and it can be measured by observing how far it must compress the spring of a scale so that the upward force of the scale and the downward force of gravity just balance. The fact that $g$ is a universal constant, 32 feet per second per second, is a remarkable physical fact. An object's weight as measured on a scale is a universal number, $g$, times is mass; but mass also measures its inertia, its resistance to acceleration by all forces. Newton had no explanagion for this fact, and carried out many experiments to test whether inertial mass and gravitational mass were the same.

Of course, the acceleration of objects due to gravity in other parts of the universe is not $g$, as Newton already understood. Near the surface of the moon, or of Jupiter, an object accelerates at a rate different from $g$. Thus, a hundred-pound person on the earth's surface will "weigh" only about 16 pounds on the moon's surface. But still, all objects will fall down on the moon with the same acceleration (a value different from $g$).

Weight and mass both measure the amount of matter in an object. Cut the object exactly in half, and each half will weigh half as

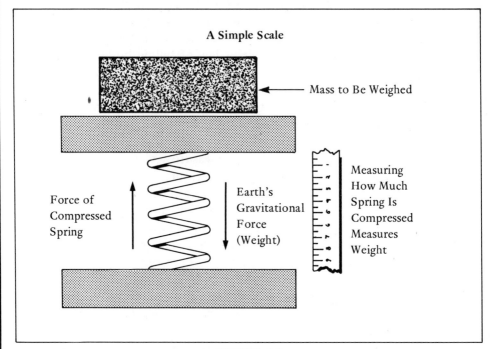

**FIGURE 6-3**

*A Simple Scale. The earth's gravity pulls the mass downwards with the force Mg. The spring, when compressed, exerts a force upward. The more the spring is compressed, the larger the force it exerts. When the spring is compressed enough, its upward force just compensates gravity's downward force. Since the mass experiences no net force, it remains at rest. Measuring how much the spring is compressed measures the weight.*

much as the whole body; the spring will be compressed half as much. Put two identical objects on the scale, and together they will weigh twice as much as either one separately. Since the weight measures the mass, the mass—the resistance to acceleration—also depends upon the amount of matter in the object.

Newton made one more important assumption. The mass of a body does not depend upon its state of motion. A moving ball has the same amount of matter in it as when it is at rest. Thus, its mass would also be the same. Its "resistance to acceleration" is the same whether it is sitting still or moving very rapidly. The same force is required to accelerate the ball from 0 to 5 miles per hour in one second as it is to accelerate the same ball from 1000 to 1005 miles per hour in one second, or from 1,000,000 to 1,000,005 miles per hour in a second.

In most cases, then, the mass simply does not change. This means there is a simpler form of Newton's second law:

*force* = *rate of change of momentum*

= *rate of change of (mass × speed)*

But the mass does not change, so:

*force* = *mass × rate of change of speed*

Since the rate of change of speed is the acceleration, *force* = *mass × acceleration*, which we write in symbols as:

$$F = Ma$$

Newton expressed his second law in terms of the more general definition, *force = rate of change of momentum*, to include those few special cases where one knows the mass of a body is changing. For example, rockets expel their exhaust gases, and this decreases their mass. However, the simpler formula, $F = Ma$, applies to most cases.

### Newton's Third Law

Newton's first two laws had been proposed in various forms by Galileo, Hooke, and Huygens. Newton's third law is completely original, and makes the laws of mechanics logically complete. It states a basic property of forces. Forces can only come from *interactions* of material bodies—one body acting on another. Newton's third law rules impossible Aristotle's explanation of gravity, in which a geometric point, the center of the universe, exerts a "force" on all heavy matter causing it to fall towards the center to form the earth. Newton's third law also states that in every interaction, the two interacting bodies exert upon one another forces equal in magnitude but opposite in direction.

René Descartes had postulated that God created the universe, giving it a certain quantity of motion that never changed thereafter. In Newton's language this meant that if one took all the masses in the universe, multiplied them by their velocities to get their momenta, and then added up all their momenta to get the total momentum of the universe; the total momentum of the universe would never change. The momentum of the universe is "conserved."

Newton's third law is equivalent to conservation of momentum in *each* interaction between masses in the universe. For example, imagine a universe with only two balls in it, and so only one possible interaction—their collision. Let the two balls have masses of $M_1$ and $M_2$, and call their speeds *before* the collision, $u_1$ and $u_2$. We label their speeds *after* the collision by $v_1$ and $v_2$. Suppose we restrict ourselves at first to a special kind of collision; one where the balls move in one straight line, both before and after the collision. Let us call their speeds positive if they move from left to right, and negative if they move in the opposite direction. The momentum of ball (1) before the collision is $M_1 u_1$, $M_2 u_2$ is ball (2)'s momentum, and the total momentum beforehand is:

$$total\ momentum\ before = M_1 u_1 + M_2 u_2$$

similarly,

$$total\ momentum\ after = M_1 v_1 + M_2 v_2$$

The law of conservation of momentum simply states that:

$$M_1 u_1 + M_2 u_2 = M_1 v_1 + M_2 v_2$$

We can rearrange the above equation so that all terms involving ball (1) are on the left-hand side, and ball (2) on the right-hand side:

$$M_1(u_1 - v_1) = -M_2(u_2 - v_2)$$

This is an interesting form of the equation for conservation of momentum because $M_1(u_1 - v_1)$ is the change of the momentum of ball (1) during the collision, and $M_2(u_2 - v_2)$ is the change of the momentum of ball (2). Thus, conservation of momentum implies that ball (2)'s change of momentum is equal in size and opposite in direction to ball (1)'s change of momentum. (Recall that a minus sign in the speed means that a ball moves from right to left rather than left to right. Thus, the minus sign above means that the change in momentum is also in the opposite direction.)

According to Newton's second law, the force equals the rate of change of momentum. In a collision, each ball feels the other's force for the same length of time. One way for the total changes in momentum to be equal and opposite is for the mutual forces to be equal and opposite at every instant of time during the collision. Thus, using this special problem, we have arrived at Newton's third law, which applies to all problems of motion:

> To every action, there is an equal and opposite reaction.

Unlike Newton's expression of his first two laws, his wording of the third is not particularly illuminating. We rephrase it in the terms we have used so far.

Two interacting bodies exert mutual forces on each other that are equal in magnitude and opposite in direction.

Newton's third law and the law of conservation of momentum are entirely equivalent. Each implies the other.

Newton's third law applies to all problems of mechanics, including more general collisions than the one we discussed above, where the balls are not restricted to move in one straight line but can fly off in any direction. If you compute the change of momentum in each direction separately (say up and down, north and south, and east and west), changes in momentum in *each* direction are equal and opposite.

Newton's third law even means that when you throw a ball to the ground, the earth should move! (It is, after all, only a collision between a very small and a very big ball.) Why do we not feel the earth move? Suppose the ball weighs 1 pound, and moves at 100 mph when it hits the ground. Its momentum before the collision is, therefore, $Mv = 100$. After the collision, the earth acquires this momentum. What is the change in the earth's speed? Let $M_e$ be the mass of the earth, and $v_e$ its change in speed; then:

$$M_e v_e = 100$$

or,

$$v_e = \frac{100}{M_e}$$

But, the mass of the earth is $M_e$ = 10,000,000,000,000,000,000,000,000 pounds so that the change in the earth's speed is $v_e$ = 1/100,000,000,000,000,000,000,000/ miles/hour. The earth hardly moves because its mass is so large. In fact, the tiny change in speed, $v_e$, is unmeasurable; there is no reason why Newton's third law ought not to apply to this collision.

A mutual force exists not only in collisions, where we can see when and how it is applied, but also in gravitational interactions. If the sun exerts a gravitational force on Jupiter, and if Newton's third law is correct, Jupiter exerts an equal and opposite gravitational force on the sun. Even more interesting, if the earth exerts a gravitational force on a falling ball, the ball must exert a gravitational force on the earth. The earth must "fall" upwards towards the ball! Again, we cannot measure the earth's tiny acceleration because the earth's mass is so large.

## NEWTON'S SYSTEM OF THE WORLD

Newton devoted Book I of the *Principia Mathematica* to the motion of bodies in a vacuum subject to specified forces. Here he stated his three laws of motion. Book II considered the motion of bodies in resistive media. Newton showed that his three laws can describe air resistance. He described air resistance on a falling ball as another force on the ball, opposing gravity and trying to slow the ball down. The faster the ball moves through the air, the larger the force of air resistance. If a ball were dropped from a great height, its speed would at first obey Galileo's law, $v = gt$, because the force from air resistance is very small compared to gravity when the speed is small. However, as $v$ increases with time, the force of air resistance eventually just balances that of gravity. Then, there is no net force on the ball, and it does not accelerate but continues to fall at a constant speed (just as Aristotle had said). When one drops a heavy ball, it hits the ground before deviations from Galileo's law are noticeable. However, light objects like raindrops or snowflakes attain their final constant speed very quickly. That is why raindrops do not hit you at hundreds of miles per hour, even though they have fallen miles. Thus, Book II completed Galileo's program for mechanics—first to understand motion in a vacuum, and then to learn how to include air resistance. Book II also systematically demolished Descartes' speculation that giant whirls of microscopic particles carried the planets around the sun. Kepler had been right, in a sense; there is a force in the sun, a force that acts on the planets *through the vacuum.*

In Book III, Newton discussed this force. Gravity, the only force of nature adequately documented experimentally, was the crucial test of his mechanics. But, it was more than that. It was a new organizing principle for the universe. With it, he could calculate in detail the motions of the moon and planets, comets, the tides, and the precession of the equinoxes. For the first time since the Greek cosmos had been shattered, the universe seemed understandable. We can sense the exhilaration Newton must have felt when he wrote the introduction to Book III:[2]

> In the preceding books, I have laid down the principles of philosophy; principles not philosophical but mathematical . . . These principles are the laws and conditions of certain motions, and powers or forces; but, lest they should have appeared of themselves dry and barren, I have illustrated them here and there . . . giving an account of such things as are of more general nature . . . such as the density and resistance of bodies, spaces void of all bodies, and the motion of light and sounds. It remains that, from the same principles, I now demonstrate the frame of the *System of the World.*"

## UNIVERSAL GRAVITATION

Newton first returned to the question he had left unanswered twenty years earlier: "Which bodies exerted gravitational forces and which did not?" Since the planets obeyed Kepler's laws, the sun exerted a gravitational force on them. Jupiter exerted a gravitational force on its moons, and the earth exerted a gravitational force on the moon. Since the earth and Jupiter had gravity, it stood to reason that the other planets had gravity too. Huygens had recently discovered that Saturn had satellites that obeyed Kepler's laws: thus, Saturn, too, had a force of gravity that decreased as the square of the distance from its center. From his third law, Newton could deduce that the moon had its own gravity. If the earth exerted a gravitational force on the moon, the moon must exert a gravitational force equal in strength and opposite in direction on the earth. Similarly, the sun exerted a gravitational force on the planets; they in turn exerted a gravitational force on the sun. Thus, at least the planets, their moons, and the sun exerted mutual gravitational forces upon each other.

Did gravity happen only between one large body and one small body? Newton's astonishing answer to this question was no. A mutual force of gravity is exerted between *any* two masses, regardless of size. Let us see how Newton reached this conclusion. First of all, all bodies falling to earth have the same acceleration, $g$, regardless of their mass. The *force* the earth exerts on them is, therefore, given by their mass times acceleration, or *Mg*, from the second law of motion. This is the weight. An object twice as big will have twice the force on it, but accelerate at the same rate, $g$. Divide the ball in half and each half will have half the force on it; divide into tenths and each part will have one tenth the force, and so on. By entirely similar reasoning, the gravitational force exerted by the falling ball on the earth should also be proportional to the earth's mass. This meant that if you imagine splitting the *earth's* mass up into many small pieces and adding all their forces upon the falling ball the total force will be the same as if the earth's total mass interacted with the falling ball. In Newton's words, "the force of gravity towards any whole planet arises from, and is compounded of, the force of gravity towards all its parts." The problem could be done in reverse. You could imagine splitting the *ball* up into many arbitrarily small masses. Thus, Newton argued that each piece of the falling ball exerts a gravitational force on each piece of the earth.

Similarly, each piece of the earth exerts a gravitational force on every other piece of the earth. However, the mutual forces between the pieces of the earth do not accelerate the earth as a whole, because the mutual forces are equal and opposite. In sum, gravity is a universal force of interaction between *any* two bodies.

> There is a power of Gravity pertaining to all bodies, proportional to the several quantities of matter which they contain.*

It remained to describe the force of gravity mathematically. Let us rephrase this question more precisely. Given two objects of mass $M_1$ and $M_2$ and the distance, $R$, between them, how big is the mutual force of gravitation between them and in what direction is it pointed?

The gravitational force is clearly one of attraction. It is directed along the line joining the two bodies, pulling them together. How does the size of the force, $F$, depend upon the two masses $M_1$ and $M_2$, and how does it depend upon $R$? Earlier we argued that the larger an object's mass, $M$, the larger the earth's force, $W$, which is the weight. By entirely similar reasoning, the force it exerts on the earth is proportional to the earth's mass. Since the two "action and reaction" forces are equal in size according to Newton's third law, we conclude that the mutual gravitational force should be proportional to both masses at the same time. The formula for the mutual force should contain the ball's mass and the earth's mass multiplied together. This should be true for any two masses. We can express the fact that $F$ is proportional to both $M_1$ and $M_2$ mathematically as follows:

*Gravitational force*, $F$ = *some other numbers* $\times M_1 M_2$

Since gravitation is universal, $M_1$ and $M_2$ denote any two masses, and $F$ the force between them. In the case of the falling ball, $M_1$ could be the mass of the earth, $M_2$ the mass of the ball, and $F$ the force of the earth's gravity.

"Some other numbers" describe how the gravitational force depends upon the distance, $R$, between the two masses. Newton knew that the gravitational forces between the sun and planets, earth and moon, Jupiter and its satellites, and Saturn and its satellites decreased as the square of the distance. If the law of gravity were to apply to all interactions between bodies, the force would have to decrease as $1/R^2$ for all. Thus, finally, we arrive at the mathematical form of Newton's law of universal gravitation:

$$F = \frac{GM_1 M_2}{R^2}$$

$G$ is a number, called "Newton's constant of gravitation." We have encountered a similar constant before when we wrote Kepler's third law in the form, $T^2 = kR^3$, where $k$ was Kepler's constant. There, $k$ was needed to make consistent the units in which we measure $T$ and $R$.

---

*Another way to express this is as proportional to their masses.

A natural system of units was known in that problem: if we measured $R$ in A.U. and $T$ in earth years, then $k = 1$ and Kepler's law became $T^2 = R^3$. However, if we had measured $R$ in feet (or furlongs) and $T$ in seconds (or hours), then, Kepler's constant $k$ would be a different number. Kepler's law does not change—only the units we use to measure $T$ and $R$. Newton's constant, $G$, performs the same function here. We measure the masses by weighing them—convenient units might be pounds or grams. Convenient units for the distance could be feet or centimeters. Similarly, force is measurable in various units. $G$ makes the units come out properly.

Newton answered one question immediately. Why do we not feel the force of gravity between two small objects?

> I answer that since the gravitation towards these ( small) bodies is to the gravitation towards the whole earth as these bodies are to the whole earth, the gravitation towards them must be far less than to fall under the observation of our senses

In other words, the earth's mass is so huge compared to the objects around us that we cannot feel a table or chair's gravity compared to the earth's.

## AN IMPORTANT MATHEMATICAL THEOREM

Newton's law of gravitation would only be correct if using it would show that the earth's mass acted on falling bodies and on the moon as if all its mass were concentrated at its center. He had made this assumption in 1666, when Newton had guessed that gravity held the moon in orbit. It was not an obvious conclusion. Consider the falling ball, sketched in Fig. 6-4. Let us imagine we can split the earth up into many pieces, each of the same mass. The pieces of the earth nearest the ball exert the strongest individual gravitational attraction on the ball, because the force decreases as $1/R^2$. The piece labelled number one in Fig. 6-4 is a good example. It exerts a downward force on the ball—towards the center of the earth. The piece labelled two also exerts a downward force on the ball, but the force is very much smaller since instead of being, say, 10 feet away, it is the diameter of the earth 8000 miles away. What is the effect of all the other pieces of mass distributed along the line joining the two extreme pieces, 1 and 2? It was not even obvious that balls should fall downward. If the ball felt only the force of attraction of a piece like the one labelled number

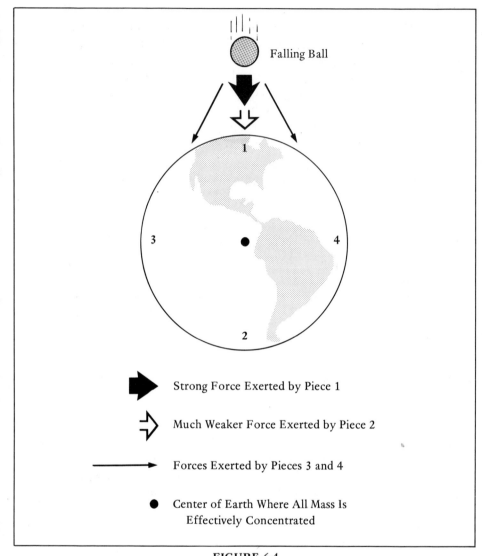

**FIGURE 6-4**
*The Earth Acts As Though All Its Mass Were Concentrated at Its Center.*
Each piece of the earth exerts a force of attraction on the falling ball directed along the line joining the ball and that piece. The "sideways" forces from pieces like 3 and 4 cancel, while the "downwards" forces add. Pieces like 1 and 2 have downwards forces that add. Adding up all the gravitational forces from all the pieces, Newton proved his famous theorem. To do this, he had to assume that the earth was a perfect sphere. Later, he showed the earth was not exactly a sphere.

# An Important Mathematical Theorem

three, it would be attracted, not towards the center of the earth, but towards piece 3. On the other hand, a piece opposite to 3, labelled 4, also pulls on the ball. If you add up the forces from 3 and 4 on the ball, you can see that they both tend to pull the ball "downwards," but that their "sideways" pulls are in the opposite direction, and cancel. But, when you add up all the contributions from all the pieces, will the "sideways" components of force still exactly cancel?

Newton proved the following remarkable theorem. If the earth were a perfect sphere, its gravitational force would act as though all its mass were concentrated at its center. The total "sideways" forces from all pieces like 3 and 4 would exactly cancel, and the forces downwards would add up. (This would not be true for example, if a greater mass were concentrated at point 4.) Thus, the total force is directed towards the center of the earth. Then, adding up the stronger downward forces from the nearer pieces and the weaker forces of the more numerous, more distant pieces, Newton found that the earth acted as though all its mass were concentrated exactly half way between the nearest piece, number 1, and the farthest piece, number 4: in other words, at the center of the earth. Newton's theorem would not be true if the earth were not a sphere, or if his law of universal gravitation had a different form. For example, if $F = GM_1M_2^2/R^2$ or if $F = GM_1M_2/R$ had been the form of the law, his theorem would not be true.

Just as Newton had guessed in 1666, the earth's gravitational force on an object at its surface is given by:

$$F = \frac{GM_e M}{R_e^2}$$

where $M_e$ is the mass of the earth, $R_e$ is the radius of the earth, and $M$ is the mass of the object. The acceleration of a freely falling object, $g$, is just the force, $F$, divided by its mass, $M$, according to Newton's second law.

$$\frac{F}{M} = g = \frac{GM_e M}{R_e^2 M} = \frac{GM_e}{R_e^2}$$

This acceleration does not depend upon mass $M$: all falling objects accelerate at the same rate.

The above equation does not permit us to measure the mass of the earth, $M_e$, or Newton's constant of gravitation, $G$, separately, but only the combination $GM_e$. We can rearrange this equation to put all the unknowns on the left-hand side and all the measurable quantities on the right-hand side:

$$GM_e = gR_e^2$$

To measure $G$, a separate experiment is required, taking two *known* masses, $M_1$ and $M_2$, at a known distance, $R$, and then measuring the extremely weak gravitational force, $F$, between them. Then $G$ could be calculated; with $G$, the mass of the earth would be known. But measuring $G$ was much too difficult for even Newton; this had to wait a century for Henry Cavendish's experiment, which we discuss in a chapter Appendix.

### INERTIAL AND GRAVITATIONAL MASS

Two different ways of speaking about mass appear in Newton's laws. The first comes from Newton's second law where mass is a quantity that measures an object's inertia—its ability to resist changes in its state of motion. Let us call this "kind" of mass inertial mass. The second comes from Newton's law of universal gravitation where the mass measures an object's gravitational force on another object. Let us call this kind of mass gravitational mass.

There is no particular reason to expect that the inertial mass and the gravitational mass of a given object should be equal. In fact, we may assume they are not equal, and see what happens. We return to the example of a falling ball. Suppose it has inertial mass, $M$, and gravitational mass, $N$, and that the earth has an inertial mass, $M_e$, and a gravitational mass, $N_e$. Then, the mutual gravitational force between the earth and ball would be:

$$F = G \frac{NN_e}{R_e^2}$$

where, as before, $R_e$ is the earth's radius.

The ball's acceleration, $a$, would be given by this force divided by the ball's inertial mass, $M$:

$$a = \frac{GN_e}{R_e^2} \cdot \frac{N}{M}$$

Suppose the ratio of gravitational mass to inertial mass, $N/M$, were different for different bodies. Then different bodies would have different accelerations under gravity. But, this would violate Galileo's remarkable discovery that all bodies have the same acceleration, $g$. Thus, $N/M$ must be the same for all bodies. We can take $N/M = 1$, since making it any other number would only change the number $G$. Therefore, the gravitational and inertial masses are equal.

The equality of inertial and gravitational mass was to Newton a remarkable accident of nature. Nothing required them to be equal. Therefore, their equality was something that had to be established experimentally. He performed many experiments verifying that they are equal to at least one part in a thousand, a great improvement on Galileo's accuracy. Several centuries later, Albert Einstein made the equality of inertial and gravitational mass the central hypothesis of the General Theory of Relativity. Einstein's theory motivated many more experiments; no one has ever detected a difference between inertial and gravitational mass, to an accuracy of at least one part in a billion.

### PRECESSION OF THE EQUINOXES

The earth, Newton calculated, could not be exactly a sphere. Since the parts of the earth near the equator are moving faster than those near the poles, the earth's rotation pulled out the earth's equator somewhat, creating a very small bulge at the equator. A diameter through the equator is 27 miles longer than one through the poles. This meant that the acceleration of gravity was not the same everywhere over the earth, as Galileo had thought, but would be slightly different at the equator than at the poles. Newton recounted that in 1677 his friend Halley, "arriving at the island of St. Helena, found his pendulum clock to go slower there than at London." Perhaps this difference was due to a slightly different pull of gravity at St. Helena than at London.

Though he could not measure the earth's bulge directly, he provided a spectacular proof that it was there. If the earth bulged, then Newton calculated that the moon's and sun's gravity would pull slightly differently at the equator than at the poles, and this would cause the spinning earth to wobble slowly like a top. This meant that the earth's rotation axis would point to different positions on the celestial sphere at different times. Since the rotation axis points to the celestial poles, the poles would appear to wander slowly through the sky. Newton showed that the poles would very slowly trace out circles on the celestial sphere. In a mathematical *tour de force*, Newton calculated the rate of wobbling and found that he could account for the precession of the equinoxes and even compute its 26,000-year period—a puzzle to all astronomers since the time of the Greeks.

## PROOF OF KEPLER'S LAWS

Newton could show that Kepler's laws of planetary motion were exact consequences of his law of universal gravitation, provided one assumption was made: that the sun's mass was so large that the mutual force between the sun and any planet was much larger than the mutual forces between any two planets. In other words, in calculating Saturn's orbit Newton would take account of the strong gravitational force between Saturn and the sun and neglect the much weaker force between, say, Jupiter and Saturn. Moreover, he knew that to a very good approximation, the sun's and planets' masses acted as though they were all concentrated at their centers.

Before proving Kepler's first law, Newton proved Kepler's second law: that a line drawn from the planet to the sun sweeps out equal areas in equal times. He proved that the equal areas law is even more general than Kepler had imagined: it is a consequence only of the fact that the force points towards the sun and not on the way it diminishes in strength with increasing distance from the sun. Newton's proof was surprisingly simple. Instead of considering a planet's continuously changing speed and direction, he broke the orbit into zig-zag pieces as in Fig. 6-5.

The planet moves in a straight line, is struck a hammer blow that accelerates it toward the sun, and then moves inertially at a different speed and direction until it is struck again, and so on. If we imagine the time interval between hammer blows getting smaller and smaller, any real continuously changing motion can be approximated as closely as you please.

In Fig. 6-5, the zig-zag path that approximates the true curved path is labelled $ABCD$. $S$ is the center of force (the sun). We only require that the force is directed towards the sun at each hammer blow, but we do not need to specify its size or how it changes with distance from the sun. We choose the time intervals for segments $AB$, $BC$, $CD$, etc., to be the same. The essential point is that, provided the force is always directed towards the sun, the speed *perpendicular* to a line drawn from the sun to the planet (e.g., $AE$ or $CF$) is unchanged during each small time interval between hammer blows. Therefore, it is convenient to study the motion parallel to $SB$ and perpendicular to $SB$ separately in analyzing the total motion from $A$ to $B$ to $C$.

The time interval between hammer blows is chosen to be the same. Kepler's equal areas law will be proven, therefore, if we can show that the area of the triangle $ABS$ equals the area of the triangle $CBS$

# Proof of Kepler's Laws

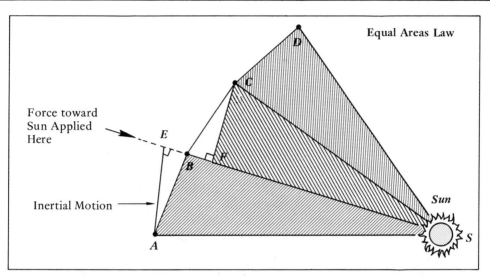

**FIGURE 6-5**

*Equal Areas Law. Any force towards the sun accelerates a planet toward the sun. Assume this force acts intermittently so that the planet moves in an inertial straight line from A to B, at which point it is accelerated toward the sun, thus changing its direction of motion. Subsequently, it moves from B to C inertially. The path ABCD approximates the true orbit if the time interval between the action of the force is very small. The net effect of the inertial motion plus the acceleration towards the sun leads to the true orbit, ABCD. If the same time passes in going from A to B and from B to C, and the triangles ABS and BCS have equal areas, Kepler's second law is proved.*

The area of any triangle is ½ its base times its height. $ABS$ and $CBS$ both have the same base, $SB$. the height of $CBS$ is the line $CF$ drawn perpendicular to $SB$. The height of $ABS$ is the line $AE$ also drawn perpendicular to the common base, $SB$. (We had to extend the line $SB$ to $E$.) The area of triangle $ABS$ is ½ $(AE)(BS)$, and the area of triangle $CBS$ is ½ $(CF)(BS)$. The two will have the same area if $CF = AE$.

The lines $CF$ and $AE$ are both *perpendicular* to $SB$, a line drawn from the planet to the sun. Therefore, the speed along $CF$ and that along $AE$ are unchanged by a force toward the center. Since the time intervals are chosen to be the same, the length $AE$ equals the length $CF$. Thus, $ABS$ and $CBS$ have equal areas. In a similar way, we can prove that the area of $SBC$ equals that of $SCD$. Since nothing depends upon the size of the time intervals, we can choose them to be arbitrarily small. Thus, in equal times, the areas swept out are equal.

Kepler's equal areas law is true for any force that points along the line joining the planet and the sun. It would be true if the force decreased as $1/R$ from the sun, or even increased as $R^2$. However, the planets have elliptical orbits only if the force decreases as $1/R^2$. The mathematical proof of this fact had eluded Newton's great contemporaries, Halley, Hooke, and Huygens. In fact, Newton proved more than this: all possible orbits following from the law of universal gravitation were conic sections: they could be hyperbolas, parabolas, or ellipses. (Recall that a circle is a special case of an ellipse). Hyperbolic and parabolic orbits are open: a planet would come in from very large distances, swing around the sun once, and then fly away. The only possible closed orbit was an ellipse. Newton did not show why the planets have the particular elliptical orbits they do, only that if at the beginning a planet were started off with a particular speed at a particular distance from the sun its orbit would be a particular ellipse.

Thus, Newton first used Kepler's third law to guess at the law of universal gravitation; and then proved that all three of Kepler's laws were a consequence of universal gravitation and the three laws of motion.

Newton could derive Kepler's three laws exactly, by keeping only each planet's gravitational interaction with the sun and by neglecting the gravitational interactions of the planets with each other. In other words, if the sun and Saturn were the only objects in the solar system, Saturn would obey Kepler's laws exactly. But Saturn is not the only planet. For example, Jupiter's gravitational pull on Saturn, while much smaller than the sun's, still should accelerate Saturn. Jupiter's gravity thus perturbs (changes slightly) Saturn's orbit, and by Newton's third law Saturn's gravity perturbs Jupiter's orbit. Using telescopic observations that were accurate to *seconds* of arc (as compared with Brahe's four minutes of arc), Edmond Halley had found that Jupiter was lagging behind the place predicted for it by Kepler's laws and, conversely, that Saturn was being accelerated. Even more puzzling, Halley, upon delving into ancient observations, found that earlier Jupiter had been accelerated and Saturn slowed down. Newton estimated that the interaction between Jupiter and Saturn could account for Halley's observations. Moreover, their interaction seemed to destroy their orbits in several thousand years—comparable with the biblical age of the universe. Would the solar system survive much longer? Perhaps God might reset the solar system and prevent it from collapsing or flying apart.

About a century after Newton, two great French mathematicians and physicists, Laplace and Lagrange, attacked the problem of the stability of the solar system. They began with Jupiter and Saturn. Since they are by far the biggest planets, their gravitational interactions with the other smaller planets can be neglected as an approximation; and only their interactions with the sun and each other need to be retained. The orbital perturbations are very difficult to calculate. Imagine that Jupiter and Saturn are each in their basic elliptical orbits. Their mutual gravitational force, which depends on the square of the distance between them, will change with time, depending upon where Jupiter and Saturn are. They showed that Jupiter and Saturn's mutual interactions do not cause them to drop into the sun or float out of the solar system, but merely cause their orbits to wobble slightly with a 929 year period, accounting for Halley's results. They subsequently extended their work to the whole solar system and found that all such perturbations were self-correcting. The solar system is stable. Laplace was moved to write in his *Mechanique Céleste*:

> The irregularities of the two planets appeared formerly to be inexplicable by the law of universal gravitation; they now form one of its most striking proofs. Such has been the fate of this brilliant discovery, that each difficulty which has arisen has become for it a new subject of triumph—a circumstance which is the surest characteristic of the true system of nature.

Kepler believed in an ultimately simple harmony of the spheres. More accurate observations would make nature's harmonies more, not less, evident. Ironically, the inaccuracy remaining in Brahe's data permitted Kepler to find that the planetary orbits were simple ellipses with the sun at one focus. This gave Newton a simplified situation to deal with first. Having proven Kepler's three laws and recognized them as approximations, Newton then knew what to do next: calculate the small planetary perturbations. Kepler might not have found his three laws using Halley's more accurate observations. Would Kepler have been so triumphant had he known the orbits were not exactly, but only approximately, ellipses?

## THE TIDES, COMETS, AND NEW PLANETS

According to Newton's law of universal gravitation, the moon exerts a gravitational force on all objects on earth. How can we detect the moon's force? The tides are one answer. The moon pulls on the oceans. His theory predicted two tides a day about thirteen hours apart. On the high seas, these would come regularly. However, they are unobservable on the high seas, since there is no reference point. In the *Principia*, Newton sifted through the tide records for various ports around England and in Europe, and discussed how the configurations of their channels would delay the tides. When all this was included, he could find no inconsistency with universal gravitation.

How far does the gravitational force extend from any gravitating body? Could anyone believe that it actually decreases as $1/R^2$ all the way out to infinite distance? Newton knew that the earth's gravity extends beyond the mountain tops at least as far as the moon, and the sun's at least as far as the orbit of Saturn. Newton demonstrated that the comets were also in orbit around the sun, with very elliptical or perhaps open hyperbolic orbits. He showed that Halley's comet was in a very eccentric elliptical orbit, passing by the sun briefly moving very rapidly, then and spending most of its time in "deep freeze" moving slowly on the distant part of its orbit far beyond Saturn: universal gravitation holds well beyond Saturn.

More evidence that gravity extends beyond the orbit of Saturn came with the discovery of a seventh planet, Uranus, by the English astronomer, William Herschel, during a systematic mapping of the sky in 1781. One "star" refused to remain in place. On closer examination, it had a visible disk, so was clearly a planet. Uranus was soon shown to obey Kepler's three laws. It is about 19 astronomical units from the sun, and its period is, therefore, about 84 years. Kepler's "explanation" of the number of planets, using the five regular solids, was completely destroyed. It is surprising that Uranus had never been noticed before, since it is barely bright enough to be seen on a clear night without a telescope, especially during opposition. Actually, Uranus had been occasionally placed on star maps, only to be crossed off later as a mistake when it could no longer be found at the recorded position.

After a few years, tiny deviations began to appear in Uranus' orbit. By the 1840s, Uranus was twenty-one seconds off the position predicted by Newton's laws, even taking into account the effects of the other six known planets. Did Newton's law of universal gravitation fail to work at such large distances? Two astronomers, J. J. Leverrier in Paris and John Couch Adams in Cambridge, independently and almost simultaneously made a bold assumption: Newton's laws hold exactly; the discrepancy in Uranus' orbit is due to yet another undiscovered planet. Assuming the existence of this eighth planet, they set themselves a difficult mathematical problem: Where would it have to be and how heavy (massive) would it have to be to account for the motion of Uranus? Apparently Adams finished first, and requested the astronomers at Cambridge observatory to search for a new planet near the predicted position. But, the astronomers put off the search for many months, having more important things to do than check out a hare-brained theoretical idea. Eight months later, Leverrier published his calculations. Then he wrote to J. G. Galle, the head of the new observatory in Berlin, who was engaged in making a new complete set of star maps, showing very faint stars. Galle pointed his telescope in the suggested direction, and before morning had discovered the eighth planet. There followed a feud between the English and the French over the right to name the new planet. Adams suggested the name Neptune. Leverrier suggested the name Leverrier. Today, the eighth planet is known as Neptune.

Leverrier also discovered that Mercury did not exactly obey Newton's laws. Putting in the effect of all the known planets, he could not account for Mercury's observed position to within the known accuracy of the measurements. Using the same mathematical techniques that had been so successful in the discovery of Neptune, he predicted a new planet *inside* the orbit of Mercury, and named it Vulcan. But, Vulcan was never found. Instead, the few seconds per century by which Mercury insisted on deviating from Newton's laws eventually led to the downfall of Newton's ideas as the last word on gravity. Einstein's theory of gravity, known as the General Theory of Relativity, was proposed in 1916, and was generally accepted partly because it explained the orbit of Mercury exactly. Newton's description of gravitational motion breaks down not a great distance, but surprisingly, close to a very massive body like the sun. We will learn more about this modern development in Chapter 9.

Incredibly, as the years passed, it became apparent that Neptune's motion also could not be explained exactly. Using methods similar to those of Adams and Leverrier, the American astronomer, Percival Lowell, calculated the position of a ninth planet, which could explain Neptune's anomalies. The planet, so faint that only the largest telescopes can see it, was discovered after Lowell's death, at the Lowell Observatory in Flagstaff, Arizona. Far away from the sun's light, the new planet was named Pluto, after the god of the dark regions of the underworld and, incidentally, to immortalize Percival Lowell in the first two letters of its name.

Neptune, like Uranus, Saturn, and Jupiter, is a giant planet. Pluto, on the other hand, is smaller than the earth. Its effect on Neptune's orbit is extremely small. Its discovery was a triumph of firm belief in Newton's laws. Here are the average distance from the sun ($R$ in A.U.) and periods ($T$ in years) of the three new planets:

|         | R    | T   |
|---------|------|-----|
| Uranus  | 19.2 | 84  |
| Neptune | 30.1 | 145 |
| Pluto   | 39.5 | 248 |

Today, we have much evidence that the law of gravity extends even beyond the solar system. Many stars, when examined through a telescope, appear double. Occasionally (as with the next-to-last star in the handle of the Big Dipper), this is just an accident; two stars very distant from each other, just happen to be on the same line of sight. But, more often, the two stars orbit around each other. As we observe them through the years, we find that their paths are ellipses, and that they obey the equal areas law. They are a gravitational system.

One of the most striking things to be seen in a powerful telescope is a star cluster. These are spherical conglomerations of stars, dense near the center, more sparse near the edges, and often trillions of miles across. The distribution of the stars in a cluster—the way the density of stars increases toward the center—agrees exactly with calculations based on Newton's inverse-square law. All the stars that we can see and all the faint stars that make up the Milky Way are part of a giant system of stars, the Galaxy (Greek for milky way). Today, we know that the galaxy is rotating about its center, at something like 200,000,000 years for one rotation, just as the planets rotate about the sun. The Galaxy rotates at just the speed necessary to keep it from either collapsing or flying apart. Furthermore, the Galaxy has "moons." Two smaller aggregations of stars, called the Magellanic Clouds, so near the south celestial pole that they cannot be seen in most northern latitudes, are rotating around it, obeying Newton's inexorable laws. The larger of the two satellites, the Greater Magellanic Cloud, is 1,000,000,000,000,000,000 miles away and contains millions of stars.

## THE MYSTERY OF GRAVITATION

Newton never did say why gravity is universal, *why* all bodies attract each other. To him, it was sufficient to know that they do. It was not clear *how* the gravitational force is transmitted. Was it really there even in totally empty space, as Newton seemed to be saying, or were there perhaps some unseen "gravitational particles" that transmit the force

*New knowledge comes from study of phenomena at its understanding*
*What is less understood + more ultimate than thought?*

between two gravitating bodies? Again, Newton did not say. He maintained only that with his laws of universal gravitation and motion, he could explain the motions of all the bodies in the solar system, and of bodies falling on the earth. In other words, he had reduced many smaller problems to one big one: the mystery of why there is gravity at all. In a famous passage from the *Principia,* Newton gave his opinion:

> Hitherto we have explained the phenomena of the heavens and of our sea by the power of gravity, but have not yet assigned the cause of this power. This is certain, that it must proceed from a cause that penetrates to the very centres of the sun and planets, without suffering the least diminution of its force; that operates not according to the quantity of the surfaces of the particles upon which it acts (as mechanical causes used to do), but according to the quantity of the solid matter which they contain, and propagates its virtue on all sides the immense distances, decreasing always as the inverse square of the distances. Gravitation towards the sun is made up out of the gravitations towards the several particles of which the body of the sun is composed; and in receding from the sun decreases accurately as the inverse square of the distances as far as the orbit of Saturn, as evidently appears from the quiescence of the aphelion of the planets; nay, and even to the remotest aphelion of the comets, if those aphelions are also quiescent. But hitherto I have not been able to discover the cause of those properties of gravity from phenomena, and I frame no hypotheses; for whatever is not deduced from the phenomena is to be called an hypothesis; and hypotheses, whether metaphysical or physical, whether of occult qualities or mechanical, have no place in experimental philosophy. In this philosophy particular propositions are inferred from the phenomena, and afterwards rendered general by induction. Thus it was that the impenetrability, the mobility, and the impulsive force of bodies, and the laws of motion and of gravitation, were discovered. And to us it is enough that gravity does really exist, and act according to the laws which we have explained and abundantly serves to account for all the motions of the celestial bodies, and of our sea.

People had previously viewed nature with some hypotheses about it already in mind. The hypothesis that circular rotations are natural to the heavens is a good example. People were unused to the rigor of Newton's scientific method: that he would develop his hypotheses only from observation. He would guess the laws of physics, and then check his guess by proving the laws explain everything to which they are to apply. Newton's method does not penetrate to "ultimate" reality: it merely strives to put the observed phenomena of nature into a logically consistent order.

Newton was fundamentally pessimistic about the ultimate significance of science. It could tell us how God had designed the universe, but it did not reveal God's moral commandments or His intentions for man. These were revealed only in the Bible.

## PHILOSOPHICAL IMPACT OF NEWTONIAN PHYSICS

In the two centuries since Copernicus, religious thought had developed without cosmology. Yet, no religious view could be complete without some basic statement about the nature of the universe. If Newton himself doubted the ability of science to penetrate God's revelations, his contemporaries and followers did not. With the physics of heaven and earth unified and accessible to human reason, a new vision of the universe emerged. God again had a place, and was no longer at the outermost crystalline sphere. Rather, God had created the universe and ordered its behavior. In the beginning, God created all matter and placed it in a certain configuration, and invented certain inviolate rules for how this matter should behave—"laws of physics." God then let the universe work out its destiny. People had a divine element in them that allowed them to perceive God's laws of nature. Newton's ability to describe nature with mathematical precision had in itself seemed almost divine.

The religious significance of Newton's scientific work was perceived quickly after the publication of the *Principia Mathematica* in 1687. In 1692, a young minister and theologian, Richard Bentley, seized upon the *Principia's* revelations of mathematical order in the universe, and proclaimed in a celebrated series of lectures that the rational design of the universe was evidence of a Divine Providence. Bentley had taken the precaution of checking his understanding of Newton's science with Newton himself. Their correspondence is one of the best sources for Newton's cosmological thought.

In his first letter to Bentley, in December, 1692, Newton outlined several reasons why he believed the universe to be the creation of an intelligent being. For example, he had shown that the planet's orbits result from a dynamic balance of their inertial motions and the central attractive force of the sun. However, if the planets had been at rest in the beginning, they simply would have fallen into the sun. Who gave them their initial velocities tangential to their orbits, if not God? The planets' orbits are nearly circular with orbital planes, all making very small angles to the plane of the earth's orbit (the ecliptic plane), while the comets have highly elliptical closed orbits and even open orbits about the sun; and their orbital planes can make large angles to the ecliptic plane. The planets' orbits are orderly, the comets chaotic. Is the solar system not Divinely ordered? Newton continued this train of thought in 1706, in his *Opticks*.[4]

> What is there in places almost empty of Matter, and whence is it that the Sun and Planets gravitate towards one another, without dense Matter between them? Whence is it that Nature doth nothing in vain; and whence arises all

that Order and Beauty which we see in the World? To what end are Comets, and whence is it that Planets move all one and the same way in Orbs concentrick, while Comets move all manner of ways in Orbs very excentrick; and what hinders the fix'd stars from falling upon one another? . . . does it not appear from Phaenomena that there is a Being, incorporeal, living, intelligent, omnipresent, who in infinite Space . . . sees the things themselves intimately, and thoroughly perceives them, and comprehends them wholly . . . . And though every true Step made in this Philosophy [i.e., science] brings us not immediately to the knowledge of the first Cause, yet it brings us nearer to it, and on that account is to be highly valued.

## ON ABSOLUTE SPACE AND TIME

Space had been absolute in Aristotelian physics. All motion should be referred to the center of the universe. Vertical motions towards the center of the universe were natural for heavy bodies; horizontal motions were unnatural. The only meaningful speeds were those calculated relative to the center of the universe. Galileo's principle of inertia had overturned the notion of absolute space. To Galileo, the only meaningful speeds were those measured by an observer, and calculated relative to that observer. Furthermore, any two observers in straight-line motion relative to one another would make measurements and derive from them the same laws of physics. There is no way to detect absolute motion, and no way to detect motion relative to the center of the universe, if such a center exists. Newton, of course, incorporated the Galilean principle of relativeity into his mechanics.

In Newton's physics, not motion itself but accelerations and forces were important. Could acceleration be given an absolute meaning? Newton thought so, and he performed one experiment that seemed to prove that the idea of absolute acceleration had meaning. He set an upright bucket of water rotating. Because it is moving in a circle, the water is accelerating. When the bucket is rotating, the surface of the water is no longer flat, but higher at the edges than at the center. Newton proved that the shape of the surface is always a parabola. His solution indicated that water in a bucket rotating all alone in otherwise empty space would always have a parabolic surface. An observer rotating with the water would still see a curved surface. This meant to him that space itself was absolute, that the accelerated motion of the water had an absolute meaning. The rotating observer, though at rest relative to the water, could prove he was rotating because the surface of the water was not flat.

Time was also absolute. It flowed regularly and every observer measured the same time in his experiments. Galileo offered no principle of relativity for time. Newton's conceptions of space and time agreed with his deepest religious convictions. He once characterized the universe as "God's Sensorium." By this, it seems he meant that space was the receptacle for all God's creations, that their motions and transformations constituted part of the activity of His mind, and that the universe evolved in time, a time whose flow was determined by the pace of His thoughts.

### NEWTON'S LATER LIFE

After his struggle with the *Principia*, Newton had a nervous breakdown. He begged his friends, particularly Charles Montagu, Lord Halifax, to help him leave the ascetic atmosphere of Cambridge for the civilized attractions and diversions of London. After several attempts, Montagu finally got him appointed Warden of the Mint. This was not an easy job, and Newton devoted considerable energy to it. He put England on the gold standard, which survived in economics longer than did his laws of motion in physics. He introduced an important innovation: the milling on the edges of coins. He supervised the replacement of English coins with new, milled, coins; the old coins had been of questionable value because many people shaved the edges, sold the shavings for their value as precious metals, and then passed off the coins at face value. Milling could help you tell when a coin had been "shaved." Newton attended to his complex administrative job with great energy and attention to detail. He pursued counterfeiters as relentlessly as he had pursued his laws of motion. In 1699, Newton was promoted to Master of the Mint, a job he attended to faithfully the rest of his life.

Newton by no means lost his intellectual powers despite his age and preoccupation with Royal business. At the turn of the eighteenth century, a famous Swiss family of mathematicians, the Bernouillis, posed a number of problems of great difficulty for the mathematicians of the new century to solve. Upon receiving the list, Newton sat down one evening after work, and solved one of the problems before going to bed. He gave the completed solution to his secretary at the Royal Society with the instruction that it should be mailed anonymously to the Bernouillis. They immediately perceived its author, however, proclaiming, "the lion is known by his claws."

In 1703, he was elected president of the Royal Society. Newton, who shunned contact with his scientific colleagues in his early life, ruled English science through the Royal Society with an iron hand at the end of his life. In 1705, he was knighted by Queen Anne. He devoted himself to minor scientific work, to the publication of his work of decades ago, the *Opticks,* and second and third editions of the *Principia.* While Newton was always gracious to young scientists who came to him with questions, he seemed little interested in science, other than to prolong bitter feuds about priority with Leibnitz and with Flamsteed, the astronomer. He was preoccupied with early church history and his own theological works. The old man, by now a legend in his time, lived to the end of a life of dignity and ease. Through shrewd investments, he accumulated a fortune of 32,000 pounds at his death. When he died in 1727, he was buried with full honors in Westminister Abbey. The Lord High Chancellor, two dukes, and three earls were his pall bearers. The French philosopher, Voltaire, then in temporary exile in England, attended Newton's funeral. Struck by the event, he devoted the next few years of his life to mechanical experiments, and to advocating Newtonian thought on the Continent.

Newton was the one man who was in equal measure a creative mathematician and a creative physicist. He was one of the few physicists equally adept at theory and at experiment. His invention of the reflecting telescope, much less the celestial mechanics, would have ensured him a prominent place in the history of astronomy. Newton's contemporaries remarked upon his extraordinary intuition. He seemed to know things that even he could not prove. He speculated, in his *Opticks,* that electricity held atoms together and that matter and light interact electrically, things that were not known for sure until the twentieth century. By far the largest portion of his life was spent in analysis of ancient biblical texts, in which God's spiritual truth was revealed. He sought the most exact enunciation of the Bible's statements, the most precise translation, the most exact dating of its passages. Characteristically, Newton supplemented his careful literary and historical analysis of many different ancient manuscripts with one especially his: when an ancient astronomical event was mentioned, Newton used his laws of motion and Kepler's laws to calculate how long ago it had happened, searching for corroboration in physics and astronomy for the Bible's revelations. As for science itself, it was a definite but limited help in his quest. By studying science, we finite humans can gain some understanding of God's plan for the physical universe, but this unfortunately sheds little light on God's spiritual revelations. Despite this basic reservation, Newton's scientific interests were phenomenally broad. There exist many thousands of pages of unpublished chemical studies: it seems that Newton was interested in alchemy (i.e., the conversion of lead into gold). As for the exact sciences of mathematics, physics, and astronomy, Newton acted as though his great talents were a burden; as though he were annoyed at the fierce concentration demanded of him by the game of finding mathematical solutions to the puzzles of physics, solutions that would lead him only slightly closer to understanding God's revelation. Newton's scientific achievements become all the more striking in view of how little time he spent on them. His life was divided between science, theology, and administration.

Never was there a more reluctant scientific genius. Were it not for continuous pestering by the active scientific community around him, Newton would have kept his thoughts to himself, written and endlessly rewritten in manuscripts shown to no one. Edmond Halley had not only to pay for and arrange the publication of the *Principia Mathematica*, but he had to cajole, persuade, and flatter Newton to keep him at the task. A revised edition of the book would never have appeared without the efforts of the young physicist, Roger Cotes. Newton delayed the publication of his *Opticks* for thirty years until the death of his antagonist, Robert Hooke—ostensibly to avoid public controversy. When he was Master of the Mint and President of the Royal Society, he used to protest that business left him little time for science. However Newton might have viewed himself, to those who followed him this was the person who, for good or ill, ushered in the age of science and reason. Here are some of their comments:

*Nearer the gods no mortal may approach.*
Edmond Halley

*Nature and Nature's Laws lay hid in Night,*
*God said, "Let Newton be," and all was light.*
Alexander Pope

*There could be but one Newton, since there is but one universe to discover.*
Louis de Lagrange
(18th century French astronomer)

*Newton was not the first of the Age of Reason. He was the last of the magicians, the last of the Babylonians and Sumerians, the last great mind which looked out on the biblical and intellectual world with the same eyes as those who began to build our intellectual inheritance.*
John Maynard Keynes

*Newton himself was better aware of the weaknesses inherent in his intellectual efforts than the generations which followed him. This fact has always aroused my admiration.*
Albert Einstein

## And finally:

*I do not know what I may appear to the world, but to myself I seem to have been only like a boy, playing on the seashore, and diverting myself, in now and then finding a smoother pebble or prettier shell than ordinary, whilst the great ocean of truth lay all undiscovered before me.*

Isaac Newton

## Summary

At this point many readers are having trouble with the logic behind Newton's thought. Each little piece is relatively simple, but one is not used to the very extended sequence of logical steps in which each piece is fitted together. This elaborate fitting together of little pieces is what modern science is all about, and it was Isaac Newton who taught us how to do it. From his example, we hope the reader will learn how to marshall evidence, how to draw conclusions from each piece of evidence, and how to put these conclusions together in a coherent way. It is not easy, especially the first time. It wasn't easy for Newton, and it isn't easy for a contemporary scientist. What is important is not the frantic acquisition of new facts, but the careful pondering about the relations among the various pieces of information.

Let us summarize the various steps involved in arriving at the law of universal gravitation, given Newton's three laws of motion as a framework.

1. Newton's cannonball thought-experiment showed that Galileo's laws of mechanics ought to apply to the moon, and by implication, to other heavenly bodies.
2. Because of Galileo's principle of inertia, the nearly circular (elliptical) planetary orbits need explaining. They are not natural.
3. A central force is required, since the acceleration of an object moving in a circle is toward the center.
4. If a planet has a circular orbit of radius R and its speed around the orbit is V, then, the central force must produce an acceleration $a = V^2/R$ in order to keep the planet in its circular orbit. This statement is true for planets with any arbitrary speed, V, and radius of orbit, R.
5. A specific relationship exists between the observed planetary speeds and the radii of their orbits, given by Kepler's third law. Using this, we can compute the actual accelerations, $a = V^2/R$ for each planet, which the sun's force must be exerting to keep each planet in its orbit. These actual accelerations vary as $1/R^2$. Therefore, the sun's force falls off as $1/R^2$ from the sun.

6. Both Jupiter and Saturn have moon systems that obey Kepler's third law. Jupiter and Saturn also must exert a central force to keep their moons in orbit because their moons are observed to obey Kepler's third law; Jupiter's and Saturn's central forces also fall off as the square of the distance from Jupiter and Saturn.
7. The moon has an elliptical orbit (Kepler's first law), and obeys the equal area law (Kepler's second law). It requires a central force from the earth.
8. From the moon's observed orbital speed, V, and radius, R, we can calculate the acceleration $a = V^2/R$ required to keep the moon in its orbit; it is 1/3600 times the earth's gravitational acceleration at its surface, but:

$$\frac{1}{3600} = \left(\frac{Radius\ of\ Earth}{Radius\ of\ Moon's\ Orbit}\right)^2$$

Therefore,

(a) the earth's gravity obeys a $1/R^2$ law also and
(b) the force keeping the moon in its orbit is gravity—the same one that pulls terrestrial objects to earth.

9. Jupiter's, Saturn's, and the sun's forces must, therefore, be gravity.
10. According to Newton's third law, since the earth exerts a force on small terrestrial bodies, there must be an equal and opposite force exerted by the small body on the earth. Therefore, all small bodies have gravitation, too.
11. Thus, the law of gravitation is universal: All objects in the universe attract each other with a gravitational force, proportional to both their masses.
12. The form of this law is:

$$F = G\frac{M_1 M_2}{R^2}$$

where G is a constant of proportionality, $M_1$ and $M_2$ are the masses of any two bodies, and R is the distance between them. The force is directed along the line joining the two masses.

13. Newton then went on to prove that if the law of gravity is as in (12) above, the planets will:

(a) have elliptical orbits (Kepler's first law),
(b) obey the equal area law (Kepler's second law), and
(c) obey the $T^2 = kR^3$ law (Keper's third law).

# Appendix Chapter 6

## More About the Moon

The moon always fascinated Newton. Earlier in this chapter, we showed that the moon's period, $T_m$, came out roughly right using the inverse square law for accelerations. It is instructive to do the calculation more carefully. Kepler's third law should hold exactly if for $R$ one uses the average distance from the center. The average radius of the earth, $R_e$, as Newton predicted (it is not exactly a sphere), is 3956 miles. The accurate value of $g$ is 32.1740 ft/sec/sec or 78,980 miles/hour/hour. Therefore, ignoring air resistance, the period of an imaginary satellite just skimming the earth's surface (call it $T_0$) would be:

$$T_0 = 2\pi\sqrt{R_e/g} = \sqrt{3956/78980} = 1.4062 \text{ hours} = 84.372 \text{ minutes}$$

Now, the moon's average distance is $R_m = 239{,}000$ miles, Kepler's third law requires that:

$$\frac{T_m^2}{R_m^3} = \frac{T_0^3}{R_e^3}$$

or,

$$T_m^2 = R_m^3 \frac{T_0^2}{R_e^3}$$

Putting in the numbers, we get:

$$T_m^2 = (239000)^3 \frac{(1.4062)^2}{(3956)^3} = 436{,}033$$

or,

$$T_m = \sqrt{436033} = 660.33 \text{ hours}$$

The answer is in hours, since we have used miles and hours consistently here as our units. The observed sidereal period of the moon (the time it takes to go once around the zodiac) is closer to 656 hours.

This time we have been careful with the arithmetic so the four hour discrepancy is not due to sloppy claculations. Newton considered it a substantial triumph of his theory to be able to account for these four hours. The mathematical details taxed even Newton's skill, but he showed that the four missing hours could be explained. The important points were: first, that the moon pulls on the earth also, and this effect is not entirely negligible; and second, the sun, as well as the earth, attracts the moon, causing small but calculable deviations from the idealized case of an object moving under the influence of a single, very massive attracting center.

## Weighing the Earth, Sun, and Planets

How do we know how much the sun, the planets, or even the earth weigh? We will show next how it is possible to infer, using the law of universal gravitation, something that is impossible to measure directly. (This process of inference of new facts, which is possible only when you have a solid theory to back you up, is one of the most powerful techniques at our disposal.)

In the last chapter, we saw that a satellite, with speed $v$ in an orbit of radius $R$, has an acceleration $v^2/R$. According to the Law of Universal Gravitation, this acceleration is $GM/R^2$, where $M$ is the mass of the attracting object at the center; therefore:

$$\frac{V^2}{R} = \frac{GM}{R^2}$$

or,

$$GM = V^2 R$$

Now, for the planets, Jupiter's satellites, or the moon, $V$ and $R$ are measurable. Therefore, the combination, $GM$, is known for the sun, and for any planet with moons.

Evidently, it is necessary to know $G$ in order to discover these masses from observing the orbits and periods of the satellites. But, the *ratios* of the masses can be calculated without knowing $G$. For example, let the subscripts $s$ and $j$ stand for the sun and Jupiter; then:

$$GM_s = V_s^2 R_s$$

where $V_s$ and $R_s$ are the speeds and radii of the orbits of the planets. The combination, $V_s^2 R_s$, is the same for any planet; this is just Kepler's third law. Similarly, $GM_j = V_j^2 R_j$, where $V_j$ and $R_j$ are speed and orbital radius of any of Jupiter's moons. Thus, we know $GM_s$ and $GM_j$, but:

$$\frac{GM_s}{GM_j} = \frac{M_s}{M_j}$$

Newton's constant, $G$, cancels out in the ratio. Thus, from astronomical observations alone, it is possible to deduce $M_s/M_j$. The answer is that the sun is about 1000 times heavier than Jupiter. In this way, the relative masses of all the planets with moon systems are known.

How much does the earth or the sun weight in pounds or tons? Astronomy provides no answer, since it cannot tell us the value of Newton's constant, the number $G$. Only one way can determine $G$. According to Newton's law of gravity, two objects of masses $M_1$ and $M_2$ attract each other. The force on the second object is:

$$F = \frac{GM_1M_2}{R^2}$$

And, its acceleration is:

$$a_1 = \frac{F}{M_1} = \frac{GM_1}{R^2}$$

We must know the mass, $M_1$, of the object that is doing the attracting. In other words, we must take two objects, weigh them, and then observe the gravitational attraction between them. Henry Cavendish (1731-1810), who did the experiment in 1781, was an eccentric recluse who set out to measure, experimentally, the most sensitive and subtle prediction of Newton's theory of gravitation: That a force of attraction exists between all bodies. As often happens, Cavendish's experiment was made possible because of the development of a new precision instrument, the torsion balance. A sketch of the balance as shown in Fig. 6-6. A fine wire was hung from the ceiling; at the bottom was a cross bar with small lead balls

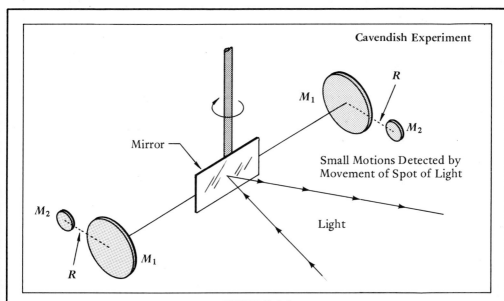

**FIGURE 6-6**
*Weighing the Earth. Two masses, $M_1$, were balanced on a long bar suspended on a fine wire. Two other masses, $M_2$, were brought near the torsion pendulum. The small gravitational forces between $M_1$ and $M_2$ were measured by measuring the acceleration of the torsion pendulum. Light was reflected from a mirror attached to the pendulum in order to measure small twists of the pendulum.*

attached at either end. A mirror is placed in the middle so that the experimenter could shine a light on it and watch the reflection on a wall. If the balance rotated, the light spot would move. Being careful not to disturb the equilibrium, Cavendish placed two very large, heavy lead balls next to the small lead balls at the ends of the balance, one in front, and one behind, as in Fig. 6-6. The large lead balls attracted the small lead balls, according to Newton's law of universal gravitation. One small ball was pulled forward, the other backward; the net effect was to give the torsion balance a twist. The point of light reflected from the mirrors moved slowly around the room.

By careful measurements, Cavendish already knew how much force it took to twist the balance a certain distance. Therefore, measuring the twist he knew the size of the force exerted by the large lead balls. He also knew the distance from the big balls' centers to the little balls' centers and the masses (weights) of the balls. He could, therefore, calculate $G$:

$$G = \frac{R^2 F}{M_1 M_2}$$

The force that Cavendish measured is fantastically small. Even though on an astronomical scale gravitation is the only important force, it is extremely weak compared to the everyday forces we know (e.g., muscular forces and chemical forces). For example, for two, two-and-one-half-pound lead balls whose centers are four inches apart, the force is equal to the weight of a drop of water with about 4/1000 inch radius. No wonder we do not feel this force in daily life, and need carefully built laboratory apparatus to observe it. Cavendish eventually used six-ton lead balls to measure $G$ precisely.

Cavendish did not call his experiment "verifying the law of gravity" or even "measuring Newton's constant;" he called it "weighing the earth." Why? Remember, the acceleration of any object near the earth's surface is $G M_e / R_e^2$, where $M_e$ is the mass of the earth and $R_e$ is its radius (4000 miles). This acceleration is measured to be $g$, 32 ft/sec/sec; therefore:

$$g = \frac{GM_e}{r^2}$$

Solve for $M_e$:

$$M_e = \frac{r^2 g}{G}$$

Cavendish's result was

$M_e = 13 \times 10^{24}$ pounds = 13,000,000,000,000,000,000,000,000 pounds. By comparing the acceleration of the earth around the sun with $g$, we can find the mass of the sun. It is 333,000 times as heavy as the earth. Jupiter turns out to be the heaviest planet, weighing over 300 times as much as the earth.

# QUESTIONS

1. A rock is placed at Saturn's distance from the sun, then, at the earth's distance from the sun. What is the ratio of the gravitational force between the sun and the rock in the two cases?

2. The moon's radius is about 1000 miles, or about ¼ the earth's radius. Suppose the material of which the moon is made is the same density as the material of which the earth is made. How much would a man who weighs 100 pounds on the earth's surface weigh if he stands on the surface of the moon? (That is, the gravitational force of the earth is 100 pounds when he stands on the earth. What is the moon's gravitational force on him if he stands on the moon? Remember, the volume of a sphere whose radius is $R$ is $\frac{4}{3}\pi R^3$.)

3. In fact, objects on the moon's surface weigh about 1/6 what they weigh on the earth's surface. Compare this with your answer to problem 2; what is the origin of the discrepancy?

4. (a) The radius of the earth's orbit around the sun is 93,000,000 miles. What is the earth's speed along its orbit? Get your answer in feet per second. Remember, 1 mile is 5280 feet. (b) Next calculate the acceleration of the earth toward the sun. (c) Compare this with the acceleration, $g$, of an object due to the earth's attraction at the surface of the earth, and use this fact to compute the ratio of the mass of the sun to the mass of the earth.

5. A cloud releases a raindrop at an altitude of 6400 feet. Assuming the drop falls in vacuum, calculate:
   (a) the time in seconds it takes the raindrop to fall,,
   (b) its speed (in feet/sec) when it hits the ground, and
   (c) its speed in miles/hour when it hits the ground.
   Remember that 1 mile equals 5280 feet and that there are 3600 seconds in an hour.

6. An astronaut performing a spacewalk broke the line tethering him to his spaceship. The line snapped back out of reach, leaving him floating motionless with respect to his spaceship. Thoughts of Newton's third law ran through his panicked mind. He fumbled around for the pair of pliers he had been working with, and threw them in the direction opposite to the spaceship. This immediately helped. Why?

   His satisfaction with his cleverness did not last long. He realized his emergency oxygen supply would only last 15 more minutes. He wondered whether he had thrown the pliers hard enough. He knew that the tether line, which had been stretched tight when it broke, was thirty feet long. He had weighed in at 150 pounds, and he guessed that the pliers had weighed 1/10 of a pound. He figured that, encumbered by his spacesuit, he could only throw the pliers at about 50 feet/sec (only 30 miles/hour). As he gradually floated back towards the spaceship, he had time to calculate whether he would make it alive. Did he make it back in time?

7. In problems 2 and 3, we said that a man standing on the moon weighs 1/6 what he does on earth. We also said that the moon's radius is ¼ the earth's

241

radius. Compute the acceleration the moon's gravity imparts to objects on earth. (The earth-moon distance is 60 earth radii, which is the same as 240 lunar radii.)

8. Recently, astronomers have discovered neutron stars—stars made up entirely of neutrons. They are about a million times more massive than the earth, and all this mass is compressed into a sphere of radius 8 miles, compared to a radius of about 4000 miles for the earth. Neutron stars exert just about the strongest gravitational forces found in nature. Compute the ratio of the acceleration of gravity at the surface of the earth and at the surface of a neutron star.

9. As we argued previously in this chapter, the earth's gravitational force results from all the pieces of the earth added up. According to Newton's law of universal gravitation, each piece of the earth also exerts a gravitational force on every other piece of the earth. Why does the earth just not pick itself up and move due to these gravitational forces?

10. Calculate the moon's orbital period, assuming it accelerates at $g$ instead of $1/3600$ of $g$.

### REFERENCES

1. More, L. T., *Isaac Newton*, Dover Publications, Inc., New York, 1934, 1962, p. 298.

2. Newton, Sir Isaac, *Principia, Vol. II, The System of the World*, Motte's translation, edited by F. Cajori, University of California Press, Berkeley and Los Angeles, 1966, p. 397.

3. Newton, Sir Isaac, *Ibid.*, p. 546.

4. Newton, Sir Isaac, *Opticks*, Dover Publications, Inc., New York, 1952, p. 369 ff.

$$\frac{10^6}{c^2} \quad \frac{1}{4000^2}$$

**HUYGENS**

*(Photographed by Barry Donahue and by permission of the Houghton Library, Harvard University)*

CHAPTER

# 7

# LIGHT

Our knowledge of the external world comes to us through our senses, principally our vision. The first science, astronomy, relied solely on vision. Those who studied astronomy had to ask, "How do we see?" Therefore, in our study of light we shall meet many friends again—Aristotle and Ptolemy, Kepler, Galileo, and Newton.

What is light, and how is it related to vision? Is it matter, or is it a disturbance in some intervening medium? Are there simple, quantitative laws governing light, like Newton's laws for the motion of matter; and if so, what are they? These are some of the questions this chapter is concerned with.

First, we shall summarize what the Greeks thought about these questions. As usual, they asked many good questions. Then, we shall describe the simplest facts about light, much as we summarized the results of naked eye observations of the heavens in Chapter 1. The remainder of the chapter describes the gradual accumulation of new knowledge about light, and the formulation of models to explain what was discovered, up to the triumph of the wave model for light in the early nineteenth century.

### ANCIENT MODELS OF LIGHT

Little is known about how many ancient civilizations thought about light. The Hebrew Bible gives the subject startling prominence. "Before He had even created the sun and the moon, God said, 'Let there be light,' and there was light." The ancient Greeks first formulated clear questions, and disputed the answers. Just as they proposed and argued almost every conceivable answer to the questions, "What does the universe look like?" and "What is everything made of?", they also asked, "What is light?", "How does it move?", and "How do we see?".

We see with our eyes. How? Is seeing like touching, something the eye does to the things seen; or rather, does something move between the things that are seen and the observing eye? How do we see things? Do pieces of them, or some image of them, move from seen things to our eyes? Do pieces of stars come all the way to our eyes? Why can we see things when the sun is up, or when there is a fire nearby, but not in the dark? Is light something objective, "out there"? These were questions the Greeks asked.

Euclid, who lived in Alexandria about 300 B.C., wrote the *Elements of Geometry*, a classic textbook that rigorously explored the logical structure of mathematics. It was widely used in schools, unmodified, well into the twentieth century. He wrote another book, *Optics*—from which the subject derives its name. His ideas dominated much of the thinking about light until the Renaissance. He proposed that the sense of vision is similar to the sense of touch. The eyes emit a stream of particles—rays of light; these rays go out and "apprehend" the thing seen. Evidently, the power of the eyes to do this depends on the presence of the sun, or some burning fire. This idea is usually attributed originally to the Pythagoreans. Here is Plato struggling to explain it in his *Timaeus:*[1]

> For they (the Gods) caused the pure fire within us, which is akin to that of day, to flow through the eyes in a smooth and dense stream; and they compressed the whole substance, and especially the centre, of the eyes, so that they occluded all other fire that was coarser and allowed only this pure kind of fire to filter through. So whenever the stream of vision is surrounded by mid-day light, it flows out like unto like, and coalescing therewith it forms one kindred substance along the path of the eye's vision, wheresoever the fire which streams from within collides with an obstructing object without. And this substance, having all become similar in its properties because of its similar nature, distributes the motions of every object it touches, or whereby it is touched, throughout all the body even unto the soul, and brings about that sensation which we now term "seeing." But when the kindred fire vanishes into night, the inner fire is cut off; for when it issues forth into what is dissimilar it becomes altered in itself and is quenched, seeing that it is no longer of like nature with the adjoining air, since that air is devoid of fire. Wherefore it leaves off seeing...

The emission model of vision seems odd to us today, and full of obvious pitfalls, such as the one that bothered Plato in the passage above, namely, why can we not see in the dark.

The atomists had a particle theory of light. All objects, they said, emit a hail of tiny particles in all directions, some of which hit the eye. However, the fact that we perceive an accurate image of the object, that our visual sense gives us a good idea of its nature, was not easily explainable by simple atomism. Consequently, the model was refined; the image itself, quivering on the surface of the object, detached itself, flew through the air, and hit the eye. Since objects do not actually move toward us when we see them, it is obvious that they must emit something. The image carries the object's shape and color to our mind. Here is a summary* of this idea by the Roman poet Lucretius:[2]

> Wherefore more and more you must needs confess that with wondrous swiftness there are sent off from things the bodies which strike the eyes and awake our vision. And from certain things scents stream off unceasingly; just as cold streams off from rivers, heat from the sun, spray from the waves of the sea, which gnaws away walls all around the shores. Nor do diverse voices cease to fly abroad through the air. Again, often moisture of a salt flavour comes into our mouth, when we walk by the sea, and on the other hand, when we watch wormwood being diluted and mixed, a bitter taste touches it. So surely from all things each several thing is carried off in a stream, and is sent abroad to every quarter on all sides, nor is there any delay or respite granted in this flux, since we have sensation unceasingly, and we are suffered always to descry and smell all things, and to hear them sound.

Notice this model fits naturally into a belief that matter is composed of atoms bouncing around with nothing inbetween them. It could also fit well into a picture of the universe in which the planets move through a vast, empty space. This connection between ideas about cosmology and the nature of matter on the one hand, and about the nature of light on the other, was to persist for many centuries.

Aristotle, of course, would have none of this, especially the concept of empty space for the particles of light to fly through. Furthermore, what could they be made of? Surely, not one of the four terrestrial elements, since light can come to us from heavenly bodies.

He objected to the emission model also. If rays of light are emitted from the eyes, how is it that when we open our eyes we see things immediately? You might reply that light travels very fast; but we see

---

*Lucretius' book, *De Rerum Natura*, was an attempt to summarize in Latin verse the scientific ideas of the atomists, and is the source of much of our knowledge about them.

even the distant stars instantaneously, and the stars are very far away in anyone's cosmology. Perhaps the light waves travel with infinite speed; but we already know Aristotle's abhorrence of this idea.

As a solution, he proposed that light was some kind of disturbance in the medium that pervades all space:[3]

> To say, as the Ancients did, that colours are emissions and that this is how we see, is absurd. First of all they should have proved that all our perceptions are due to the sense of touch. . . . Once for all, it is preferable to accept that perception arises from a movement, produced by the body we perceive, in the interposed medium, rather than to consider it to be due to a direct contact or to an emission.

Thus, the idea that light is a wave propagating in a medium was stated in primitive form.

In summary, then, the discussion of vision was quite complicated. The sun's light, or light from fire or candles, was seen as a precondition for vision, but the idea that we see light from such a source reflected from the objects around us into our eyes apparently was not widely held. It was simpler either to assume the image itself is carried through space or to assume that our eyes actively probe the universe. Nearly everyone agreed that light, whatever it is, must travel incredibly fast, if not infinitely fast. It was also agreed that color was a property of the object seen, not a necessary property of light, the carrier of vision.

Thus, the Greeks formulated some clear issues about light, but failed to resolve them. Is light emitted from the eye to the thing seen, or from the thing to the eye? How fast does it travel? Is it a stream of particles, or a sort of wave in the intervening medium? With hindsight, we can attribute their failure to answer these questions satisfactorily to an insistence on mixing up the physiological and psychological questions about the act of seeing with the question of the propagation of light.

## THE GEOMETRY OF LIGHT

The path of a ray of light connecting something seen to an observer's eye has a few simple properties that were known to all the thinkers of ancient Greece, whatever their interpretation of these facts. First of all, light travels in straight lines. This fact was important to those who tried to make models to explain astronomical observations. The straight lines that rays of light follow were also physical examples of the straight lines of geometry. That is why Euclid, a geometer, was interested in light. His *Optics* lays out the geometrical properties of light. Many of these properties can be understood by starting with the idea that the path of a light ray is a straight line, but only people used to applying mathematics to the study of natural phenomena would have thought of doing it.

In Fig. 7-1, let $S$ be the sun, and $AC$ an upright wall; $BC$ is the ground. Then, from inside the shaded triangle, the sun is hidden. You can measure carefully and determine that $AB$ is a straight line. As an application, consider two sticks of equal length. The one farther away appears smaller. In Fig. 7-2, if the eye is at $E$, and $AB$ and $CD$ are the same height, the angle $AEB$ is larger than angle $CED$. Similarly, if as in Fig. 7-3, one object is twice as far away as another but is also twice as big, they will both appear the same size. This is a theorem about similar triangles.

Why does light travel in straight lines? Aristotle suggested an analogy with falling bodies that fall straight down because they are trying to get to their natural place, the center of the universe. Perhaps light attempts to get to its destination as quickly as possible; as any student of geometry knows, a straight line is the shortest distance between two points.

When light interacts with matter, it no longer travels in straight lines only. Its path can be bent in two different ways, which we call reflection and refraction.

Reflection is the bouncing of light off a smoothed, polished surface, like a mirror or the surface of calm water. In Fig. 7-4, the ray of light between the object, $O$, and the eye, $E$, is bent at the mirror, $M$. To the eye, the object appears to be at $N$ instead of $O$. The quantitative law of reflection was first formulated by Euclid: The angles $OMA$ and $EMB$ are always equal. Light bounces off a mirror at precisely the same angle at which it strikes it. This simple, precise rule enabled Euclid to explain how mirrors form images.

The Greeks noticed that light is also bent—*refracted*—when passing from one transparent medium to another. For example, when light passes through an air–water surface, it is bent as in Fig. 7-5.

You probably have noticed refraction when looking at a straight object sticking up out of a glass of water. Another interesting example is that of a swimmer under water. In Fig. 7-5, suppose $S$ is the sun and $E$ is the eye of an underwater swimmer. To the swimmer, the sun appears to be at $T$ rather than at $S$ because the light rays are refracted at the surface.

Is there a simple law for the angle of refraction, as there is for the angle of reflection? Here are the facts: If light falls vertically on the surface, it is not bent but continues straight down. If the light strikes

Light

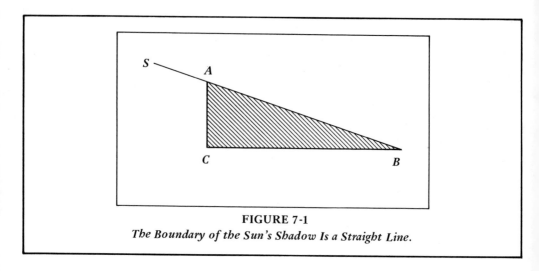

**FIGURE 7-1**
*The Boundary of the Sun's Shadow Is a Straight Line.*

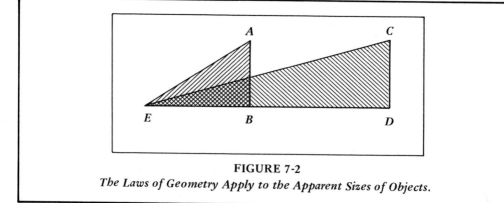

**FIGURE 7-2**
*The Laws of Geometry Apply to the Apparent Sizes of Objects.*

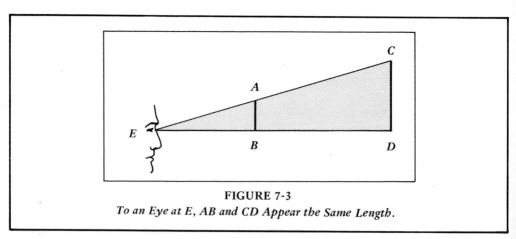

**FIGURE 7-3**
*To an Eye at E, AB and CD Appear the Same Length.*

# The Geometry of Light

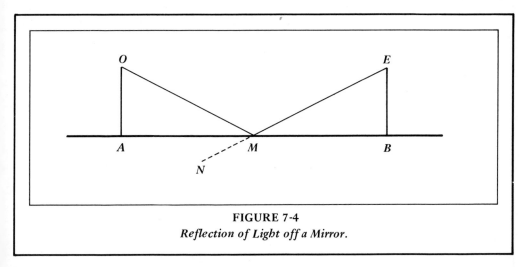

**FIGURE 7-4**
*Reflection of Light off a Mirror.*

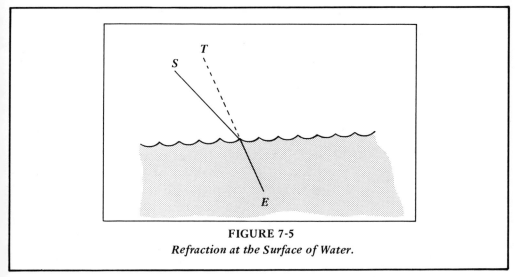

**FIGURE 7-5**
*Refraction at the Surface of Water.*

the surface at an angle, it is bent so that the ray in the water makes a smaller angle with the perpendicular to the surface than the ray in air.

The angle that a ray of light makes in air with an imaginary line perpendicular to the surface is called the "angle of incidence." The angle in the water is the "angle of refraction," as illustrated in Fig. 7-6. In Fig. 7-6, the line *MN* is perpendicular to the surface, and we have drawn a particular ray of light that is refracted at *P*. *SPM* is the angle of incidence, and *EPN* is the angle of refraction. The "angle of refraction" in the water or glass is always smaller than the angle of

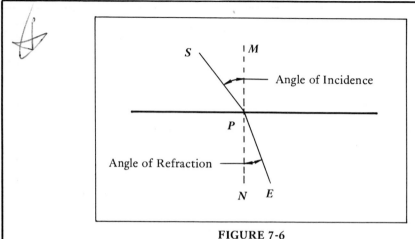

**FIGURE 7-6**
*Definition of Angles of Incidence and Refraction.*

incidence. The terminology seems to apply only to light emitted in the air and entering the water, but it also follows the same path in reverse, moving from water to air.

Is there a quantitative rule relating the angles of incidence and refraction? When a light ray strikes the surface of water or glass at a certain angle of incidence, it always is bent at a particular angle of refraction. As the angle of incidence increases, so does the angle of refraction, but the angle of refraction remains smaller than the angle of incidence. Up to about 40°, a reasonably good rule is that the angle of refraction is three-fourths the angle of incidence for water. (The fraction is closer to two-thirds for common kinds of glass.) For example, for water, when *SPM* in Fig. 7-6 is 20°, *SPN* is 15°. Many thought that this was the general rule, but careful observers knew that this was not so and that it becomes a poor rule for larger angles. The astronomer and geographer Claudius Ptolemy described a careful experiment for measuring the angles of incidence and refraction, and wrote down a table of values. Ptolemy's figures are reproduced in Table 7.1 To these we have added a column showing the value according to the "three-fourths" rule, and another with accurate modern values for these angles. His numbers for the angles of refraction are accurate to within half a degree up to a 70° angle of incidence, above which they become very inaccurate.* Ptolemy made such careful and

---

*There is something suspicious about Ptolemy's table. It is easy to dismiss the last two entries and to be impressed by the accuracy of the others. It is instructive to try to measure the refraction of water at various angles yourself, using only materials available in the second century A.D. (No lenses!). You will do well to get within a degree of the accepted answer.

But, remember that Ptolemy was a master of measuring angles accurately. Would he really have made an experimental error of ½°, the diameter of the moon? In fact, although he

accurate measurements that his star charts and maps of the world were used for centuries. His optical tables survived equally as long.

TABLE 7.1. Angles of Refraction in Water according to: the Three-Fourths Rule, Ptolemy, and Modern Measurements, Which Are Accurate to the Nearest ½ Degree.

| Angle of Incidence | Angle of Refraction | | |
|---|---|---|---|
| | ¾ Rule | Ptolemy | Modern |
| 0° | 0° | 0° | 0° |
| 10° | 7½° | 8° | 7½° |
| 20° | 15° | 15½° | 14½° |
| 30° | 22½° | 22½° | 22° |
| 40° | 30° | 29° | 29° |
| 50° | 37½° | 35° | 35° |
| 60° | 45° | 40½° | 40½° |
| 70° | 52½° | 45½° | 45° |
| 80° | 60° | 50° | 47½° |
| 90° | 67½° | 54° | 48½° |

## ALHAZEN

Although the scientific tradition of antiquity died in Europe after the fall of Rome, it flourished in the great Islamic centers of learning, which stretched from India through the Middle East and Northern Africa to Spain. The Arabs made some basic advances in mathematics—"algebra" is an Arabic word and our decimal number system comes from them; however, in physics and cosmology they were content to expand on Aristotle and Ptolemy.

One of the most influential Arab scholars was Abu Ali Mohammed Ibn Al Hasan Ibn Al Haytham (965-1039). He lived in the new capital of Cairo, not far from Ptolemy's Alexandria. Some of his works were translated into Latin in the twelfth century, ascribed to "Alhazen," as he came to be known. They influenced later Europeans, primarily because they refuted the idea that light is emitted from the eye and established the view that light is external and objective, emitted by a source and eventually perceived by the eye.

does not say so, all his numbers can be calculated from a very simple formula. Let $i$ be the angle of incidence and $r$ the angle of refraction. Then, in modern notation, you can calculate his table from the simple formula:

$$r = \frac{33}{40}i - \frac{1}{400}i^2$$

The suspicion, of course, is that Ptolemy had a preconceived idea that the correct rule should be of the form above—no powers of $i$ higher than $i^2$—and found the numbers 33/40 and 1/400 that best fit a few measurements at small angles. If you make a graph of Ptolemy's rule, plotting against $r$, you get a parabola.

253

Alhazen began as an astronomer, and at one point tried to invent a mechanical model for Ptolemy's heavens, with real linkages for all the epicycles. He ran into great difficulties, and began to wonder whether perhaps things are not really as we see them, but that we are fooled by the way light propagates, just as the underwater swimmer sees the sun in the wrong place.

Alhazen's principal experimental tool was the pinhole camera, or *camera obscura* (Latin for dark room). The *camera* was a dark chamber with a hole in one wall, and a screen opposite the hole. The *camera* was used by astronomers especially to study eclipses of the sun, the sun being too bright to look at directly. The image is upside down and inverted. Since light travels in straight lines, a ray from the top of the object crosses a ray from the bottom at the little opening, and arrives on the screen at the bottom of the picture. In fact, all the rays that reach the screen cross at the small hole. Alhazen considered this simple fact a proof that the eye has nothing to do with light, and that Euclid's emission theory was wrong. If the light from a bright object fell on the hole, an image of it appeared on the screen, as in Fig. 7-7.

Even more convincing was the fact that the *camera* would show an inverted picture of something that was not, a source of light, like the sun or a candle; it could form a picture of a brightly lit landscape or of the moon. Clearly, the formation of these inverted images was independent of the source of light.

Alhazen performed many careful experiments with his *camera*.* From these he reached the following conclusions: Glowing objects emit rays of light in all directions. These rays are emitted from *each* point on the object and rays travel in straight lines. If the something is your eye, you see the glowing object. If a light ray strikes a surface of water or glass, it is refracted according to the regular laws of Ptolemy. If it strikes a smooth, polished surface, it is reflected according to the simple law of reflection. Finally, if it strikes a rough surface, like the surface of the moon and much of the earth, or the screen in Alhazen's *camera obscura* (or a piece of white paper), *each* point reflects the light in *all* directions. Eventually, if it strikes the eye, it is seen.

For Alhazen, the domain of optics stopped when light reaches the eye; the mechanism of vision was the province of physiology or

---

*For example, what happens if the screen is placed very close to the hole? The image obviously, is in the shape of the hole, not the shape of the object outside. What happens between? How far away does the screen have to be before you see the inverted image of th

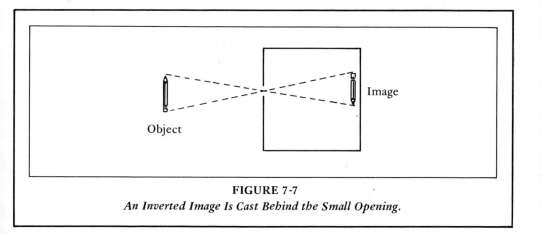

**FIGURE 7-7**
*An Inverted Image Is Cast Behind the Small Opening.*

psychology. Thus, he separated the question of vision from the question of the propagation of light, paving the way for the study of the physics of light.

Alhazen himself made the first step toward a physical explanation of the laws governing light by making the following observation. When a ball bounces off a hard surface, it leaves at the same angle that it strikes, as is well known to ping-pong or billards players. (This is an example of the law of conservation of momentum.) Now, this rule for balls is exactly the same as the law of reflection of light. Apparently, light is not so mysterious after all. It bounces off mirrors just as balls bounce off hard surfaces. However, his main contributions were to make clear the objective nature of light and to understand what happens when light strikes an ordinary, rough, surface. It is reflected in all directions, not just in one direction, as happens on a smooth or polished surface.

## KEPLER

The work of Kepler's European predecessors was relatively insignificant compared to Alhazen's; however, there was a growing belief that light

---

outside rather than the shadow of the hole itself? In general, what effect does the shape and size of the hole have on the image? Alhazen's geometrical theory answered all these questions, and his experiments confirmed the answers.

is physical and objective, and a growing emphasis on the mathematical understanding of its propagation. Also, there was a new interest in anatomy, and an increasing knowledge of the structure of the eye.

Like Alhazen and Ptolemy, Kepler was led to study optics out of a concern for the accuracy of astronomical observations. He asked, "To what extent is one deceived in measurements of the heavenly bodies by the properties of light itself?"

The particular problem that first troubled Kepler was Brahe's attempt to measure the variation in the moon's distance from the earth by measuring the apparent change in the moon's diameter. The measurement is hard because the change in the moon's diameter is small, only a few minutes of arc, close to the limits of Brahe's precision. Brahe tried Alhazen's method to improve the accuracy of his measurements. During a partial eclipse of the sun, he observed the dark side of the moon in a *camera obscura*. By moving the screen far from the hole, the image of the partially eclipsed sun could be made quite large, and the apparent diameter of the moon could be measured quite accurately.

Kepler was disturbed by these measurements, for they did not fit in with his other knowledge about the moon's orbit. Therefore, he examined the functioning of Brahe's *camera*. Brahe had thought that the image on the screen was an accurate picture of the eclipsed sun. Kepler pointed out that it was not, because the hole was not a point but had a certain size. The screen was illuminated at all points, where from anywhere on the sun's surface—on the part unobstructed by the moon—a ray could get through the hole and strike the screen. The dark part, which Brahe had taken to represent accurately the moon, was smaller compared to the bright disk than was really the case. Thus, during an eclipse the moon appeared smaller and, hence, was assumed to be farther than it really was. Kepler showed by carefully tracing the rays of light that he could correct for the size of the hole and find the apparent diameter of the moon correctly; the corrected answer for the moon's distance then agreed with his astronomical calculations of where it should be.

Kepler next tried to understand the optics of the human eye. Where does optics stop and physiology begin? Tradition had it that "seeing" happens when light strikes the front of the eye, probably just the pupil. But, anatomy had made progress during the sixteenth century, especially through the anatomical studies by artists such as Leonardo da Vinci, Michelangelo, Raphael, and others. Kepler thus knew that the eye is a small chamber and that the pupil is a transparent opening. At the back of the chamber is the retina, and the space in between is filled with clear fluid. Most important, the nerves connecting the eye to the brain end at the retina. His conclusion was inescapable. The eye is a *camera,* forming an image on the sensitive retina just as images were formed on Alhazen's screen; there, nerves sensitive to the intensity and color of the light receive the information and report to the brain. Exactly how this works is an interesting question, but not a part of optics. Figure 7-8 is a diagram of the eye.

This picture of the eye has a flaw. The pupil of the eye can be as wide as $1/3$ inch. Kepler had just gone to a lot of trouble to show that the effect of the width of an opening of a *camera obscura* was to blur the image. Nevertheless we see clear, distinct images. How?

Over three-hundred years before, in the later thirteenth century, someone, somewhere, probably in Italy, had made a sensational discovery. An old person whose eyesight had deteriorated could look through a piece of glass shaped like a lentil bean and see as clearly as he had in his youth. This lentil shaped glass—the word became "lens" in English—was soon put into a frame and hung over the nose. Spectacles were gradually improved on a trial and error basis. It is surprising that there is no discussion of eyeglasses in the scholastic texts of the time. However, so long as one believed that vision was due to the direct impact of complete images upon our minds, it was believed that eyeglasses,

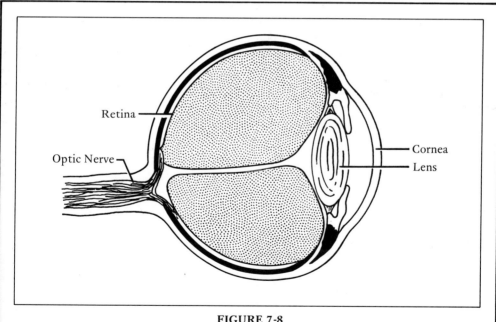

**FIGURE 7-8**
*Schematic Drawing of the Human Eye.*

however magical their rejuvenation of vision, could only interfere with the direct perception of nature. How could one be certain that spectacles did not add or subtract something essential to the incoming image? Sensible Aristotelian principles stated that science should be based upon the *direct* evidence of the senses.

Nonetheless, lensmaking did progress slowly. Artists learned that the image in a *camera obscura* could be improved by putting a lens in the opening. But nobody scientifically investigated lenses until Kepler, in 1609, completing his war on Mars, received a copy of Galileo's *Starry Messenger*. With astonishment he read that the Venetian professor had constructed an instrument with two lenses that magnified and brightened the heavenly bodies that enabled him to make several observations confirming the Copernican theory. But, Galileo's critics claimed that observations through the telescope, not being direct sense-perceptions, were not valid sources of information about the heavens. Kepler decided to investigate the functioning of lenses.

First, Kepler studied the law of refraction. From careful experiments, he concluded that Ptolemy's table, the third column in Table 7.1, was definitely wrong, and that, furthermore, the angle of refraction for the same angle of incidence was not the same for water as for glass. Kepler could find no general rule to explain the correct angles, but made a table of his own, correct up to the accuracy of the measurements he could make.

Kepler went on to study lenses. His explanation of how they work is till the accepted one. Figure 7-9 shows a typical lens. Its two surfaces are sections of spheres. At $S$ is a point source of light,

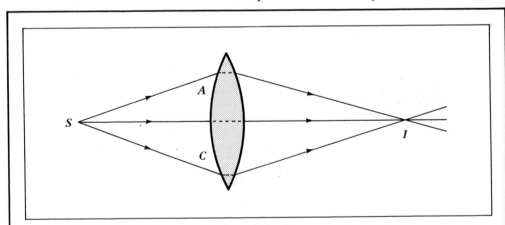

**FIGURE 7-9**
*The Formation of an Image by a Lens.*

emitting rays in all directions. A ray that strikes the lens perpendicularly will not be bent but will proceed straight on. A ray that strikes the lens off center, for example, at $A$, will be refracted as shown. At each of the two surfaces, this light ray is nearer the perpendicular to the surface *inside* the glass than outside. This light ray is bent back and crosses the ray that goes straight through at some point, $I$. One can find the exact location of $I$ by carefully tracing rays, using the table of angles of refraction.

Next consider another ray that strikes the lens at $C$, as far below the center as $A$ is above. It is easy to believe that it is refracted as shown and also crosses the unrefracted perpendicular ray at $I$. What is not obvious, but nevertheless true, is that *any* ray that strikes a good lens is refracted twice and meets with *all* the others at $I$. Thus, an image of the point $S$ is formed at $I$. The best way to prove this is to draw a lens and carefully trace a few rays through, using the law of refraction at the surfaces.

Now, a point off the central axis of the lens will also form an image, but it will not be at $I$. Since the surfaces of the lens are spheres, one ray will get through unbent and again all rays will meet at the same point. If the point is above the central axis of the lens, its image will be below. In this way, as in Fig. 7-10, a lens forms an image of an extended object. If you put a piece of paper or a white screen at the place where the rays of light cross, you will see an inverted image of the object. Try it with any magnifying glass. A more detailed discussion of how lenses form images is in a section of the Appendix.

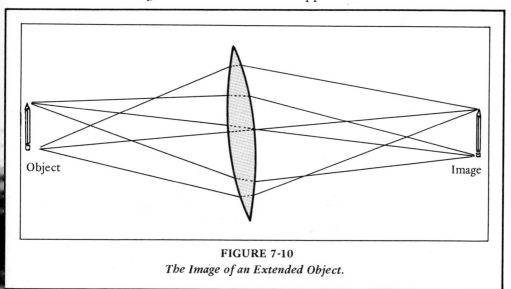

**FIGURE 7-10**
*The Image of an Extended Object.*

From all this, Kepler reached an important conclusion. The *camera obscura* with a lens in the opening, not the simple pinhole camera that Alhazen and Brahe used, is the correct model for the eye. Inside the pupil is a lens, which in a person of good eyesight focuses the light sharply on the retina, where it is sensed by the ends of the optic nerve. Thus, Kepler explained the operation of the eye, of lenses, and of the telescope. He published his theory in the *Dioptrice*, written only a few weeks after Galileo's *Starry Messenger*. Kepler's *Dioptrice* is a classic book on the geometry of reflection and refraction.*

In 1621, a general form of the law of refraction was discovered by a Dutchman, Willebrord Snell. Snell's law is illustrated in Fig. 7-11. *ASM* is the angle of incidence, and *BSN* is the angle of refraction. Table 7.1 shows these angles for water; the table would be somewhat different for each kind of glass. Snell's law can be stated as follows: Choose the points *A* and *B* along the rays so that *AS* and *BS* are the

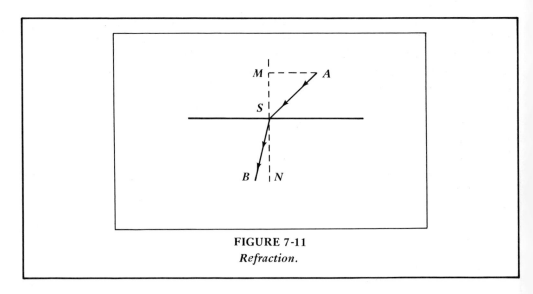

**FIGURE 7-11**
*Refraction.*

---

*One further discovery about light is due to Kepler: The inverse square law for the intensity of light. This law says that an object's apparent brightness diminishes as the square of the distance from the eye. For example, a candle 10 feet away will appear four times as bright as a candle 20 feet away; precisely, this means that four candles twenty feet away will look as bright as one candle ten feet away, or nine candles thirty feet away. This rule is important in estimating the real brightness of distant stars.

Kepler did not just happen on this rule. He pictured light as being emitted in a spherical surface, an idea not unlike Aristotle's. The total intensity of a light depended on its real brightness. But, as one moves farther away from the source, the fraction that enters the eye diminishes. Since the area of a sphere of radius $R$ is $4\pi R^2$, the fraction that enters the eye is $A/4\pi R^2$, where $A$ is the area of the pupil of the eye. Thus, the fraction that enters the eye, the apparent brightness, diminishes inversely as the square of the distance from the source.

same length, and draw *AM* and *BN* parallel to the surface so that *ASM* and *BSN* are right-angled triangles. Snell's law then says that the ratio *AM/BN* is the same for all angles of incidence. This ratio is a property of any substance, and is called the index of refraction. For water, the index of refraction is almost exactly 4/3. Check that the last column in Table 7.1 obeys this rule.*

## DIFFRACTION

For Kepler, as for the Greeks, Optics was a branch of geometry. Light travelled in straight lines, except when it was reflected or refracted. In 1665, a Jesuit professor of mathematics at Bologna, Italy, Francisco Maria Grimaldi, published a challenge to this assumption. His book is remarkable chiefly for its study of diffraction, the name Grimaldi gave to the bending of light rays around sharp obstacles.

In one experiment Grimaldi let sunlight pass through a small hole into a darkened room where he placed a solid object in its path. Its shadow on the wall was wider than it should have been, based upon rectilinear propagation. Peculiar dark and light lines appeared at the shadow boundary, and its edges were colored.

Another of Grimaldi's experiments is sketched in Fig. 7-12. Light is beamed perpendicular to a circular opening, of which *GH* in the figure is a cross section. The diameter, *JK*, of the disk of light seen on a distant screen was larger than the diameter, *NO*, which, according to the rules of geometrical optics, should be the same as *GH*. Thus, in Grimaldi's own words: "Light propagates not only in a straight line, reflected or refracted, but also in a certain other direction, diffracted." Why this might be so, he did not know.

## COLOR

Kepler and most of his predecessors imagined that white sunlight was pure light, and that light added its color when reflected by a colored

---

*The rule is simple to state in the language of trigonometry: Let *i* be the angle of incidence, *ASM*; and let *r* be the angle of refraction, *BSN*. Let *n* be the index of refraction. Then,

$$\frac{\sin i}{\sin r} = n$$

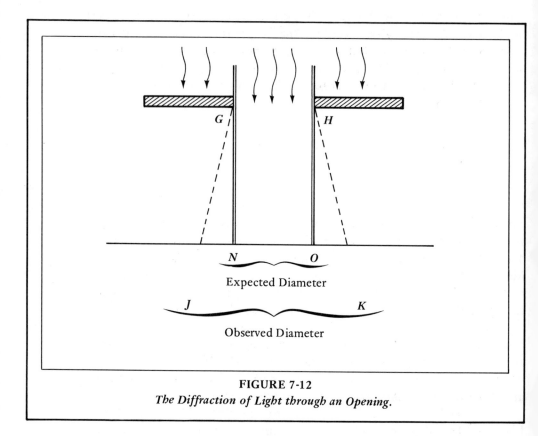

**FIGURE 7-12**
*The Diffraction of Light through an Opening.*

object. But in the 1660s and the 1670s, Isaac Newton performed a brilliant series of experiments that revolutionized the understanding of color.

As more powerful telescopes were constructed, a problem, called "chromatic aberration," had emerged. The images of stars appeared blurred, with colored fringes, and it was feared that it would be impossible to continue to increase the size of telescope lenses. Newton believed that how colors are produced when a ray of light is refracted was not understood. Home at his mother's farm during the plague year of 1666, Newton purchased a set of glass prisms "to investigate the celebrated Phaenomenon of colours." A prism is a length of glass whose cross section is shaped like a triangle, as shown in Fig. 7-13.

First, Newton allowed sunlight, shining through a small hole in the wall of his room, to be refracted by a prism, and observed the light on a screen far away. The light was colored like the rainbow, blue at one edge and red at the other as in Fig. 7-14. This phenomenon, of course, was well known. People generally believed that the

Color

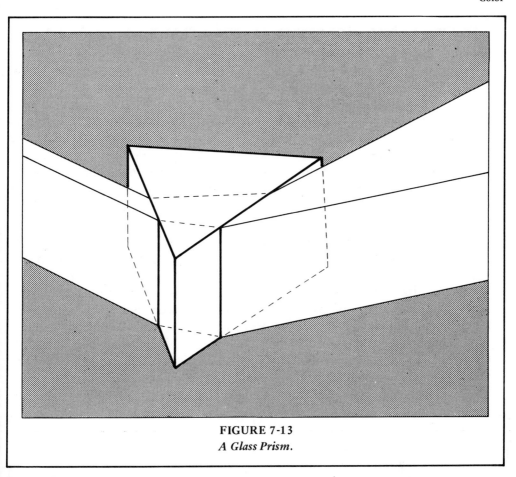

**FIGURE 7-13**
*A Glass Prism.*

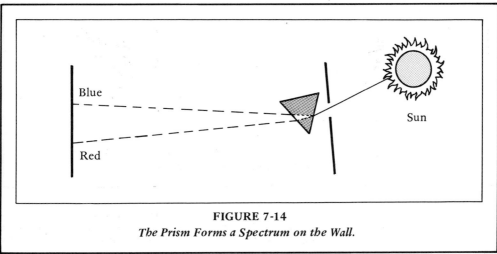

**FIGURE 7-14**
*The Prism Forms a Spectrum on the Wall.*

prism added colors to the pure white sunlight. Newton realized the key point that different colors of light arrived at the screen at different places. Therefore, sunlight must actually be made up of light of all the different colors and that white light is just our name for the mixture. The colored spectrum formed by the prism would then be explained if each color is refracted at a slightly different angle from the others as it enters or leaves the prism; thus, the prism separates out the pure colors.* (Using a prism enabled him to make his measurement in air, rather than under water. Since the prism refracts the light twice, once on entering the prism and once upon leaving it, thereby doubling all the angles over a single refraction, and increasing the separation of the colors.)

Newton next proved that the prism did not add color to white light. He set up two prisms, shown in Fig. 7-15. The sun's light was admitted through a small hole, $F$, then refracted by the prism $ABC$. The light struck a board $DE$, with a gap at $G$, and then a distant board, $de$, where part of the spectrum could be seen. Next Newton made a small hole in $de$ at $g$, in the green part of the spectrum. When a second prism $abc$ was placed behind the gap at $g$, it refracted the green light at the correct angle for green light, but did not add any other colors. Evidently pure green light could not be broken down further by a prism; only green light appeared on the screen at $MN$. By rotating the

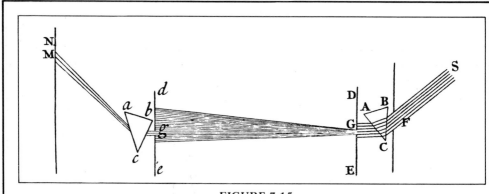

FIGURE 7-15
*Newton's Drawing of His Experiment Showing That a Prism Does Not Add Color to Light. (From His* Opticks.)

---

*The effect is very small indeed. For example, for a common kind of optical glass, "telescope crown," a ray incident at 45° will be refracted so that the angle of refraction is about 27.5°. More precisely, the angle of refraction for red light is 27°20′, and for blue light, 27°37′.

prism *ABC,* Newton showed that the same was true for all the colors of the spectrum.

Finally, Newton performed the decisive, "crucial" experiment. Newton arranged two prisms so that the first prism separated sunlight into its colors. A second prism then recombined the colors into white light, proving that the separate colors were not produced by some unusual action of the first prism. His first prism bent the rays of differently colored light by different amounts. When the second prism bent them all back into the same direction, the light appeared white. In the final form of this experiment reported in the *Opticks*, the second prism was replaced by a lens. Newton's drawing is reproduced in Fig. 7-16. Sunlight enters at *O*, and the different colors are refracted at different angles by the prism *ABCabc*. A lens focuses them at *X*, and the prism *EDGedg* recombines them all into a parallel beam of white light. Testing the light with a third prism at *Y*, Newton found it to be completely indistinguishable from sunlight, which had not been dispersed and recombined.

Newton's experiments showed that sunlight, and white light in general, is composed of rays of every color and that the colors are

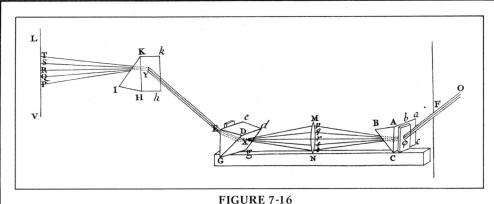

**FIGURE 7-16**
*Newton's Drawing of His "Crucial Experiment", Demonstrating That Colored Light Can Be Recombined into White Light. (From His* Opticks.*)*

---

Thus, the angle between the red and the blue light is only 17′. By using a prism, Newton doubled the difference by refracting the light twice. Furthermore, he put his screen over twenty feet from the prism. At that distance, two lines making an angle of thirty minutes of arc are two inches apart.

properties of the light, not manufactured by the prism. In an unusual burst of self-congratulation, Newton described this as: "the oddest if not the most considerable detection which hath hitherto been made in the operations of nature."

Newton understood how different substances can appear to have different colors. White objects reflect all the light they receive. Under red light, they appear red, and under blue light, they appear blue. Colored things, however, far from adding anything to white light, as had been generally thought, absorb some of the colors in white light. A red object, for example, subtracts the blue and the yellow and the green from white light, reflecting only red. A pure red object, viewed under blue light, appears black. Generally, the colors that make up white light are absorbed in different degrees, so that the light reflected from a colored object is a mixture of some, but not all, of the colors in the spectrum. The eye interprets these combinations to be one of the colors we know, some of which, like purple and brown, are quite different from any "pure" color that occurs when white light is broken up by a prism.

Finally, Newton returned to the chromatic aberration of telescope lenses, which had started him on the study of color. In a refracting telescope (Galileo's kind), different colors are refracted by the lens by different amounts and are, therefore, focused at slightly different places, causing chromatic aberration. For example, a telescope focused for green would be slightly out of focus for blue. Since this effect is an inherent property of refracting lenses, Newton despaired of ever improving refracting telescopes. So, he constructed the first reflecting telescope, using in place of the large lens a mirror, for which no chromatic aberration occurs because the law of reflection is the same for *all* colors. Nearly all of the largest telescopes in use today are of this kind, called reflecting or simply Newtonian telescopes. How one works is explained in the Appendix to this chapter.

### THE SPEED OF LIGHT

The idea that light had a speed, that one did not see things the instant they happened, was as controversial as it was ancient. Aristotle had argued that light must travel at some finite speed, since infinite speeds, like action at a distance, were logically ridiculous. Kepler took a curious but faintly Aristotelian point of view: Light is capable of propagating into illimitable space (i.e., everywhere); moreover, it requires no time to propagate. This is because light, being immaterial, offers no resistance to any moving force and, thus, moves with infinite speed. In this way, reasoning like an Aristotelian, Kepler reached a conclusion that Aristotle himself had ruled impossible.

Galileo, typically, thought the speed of light could not be determined by pure reasoning, and so, resorted to experiments. As usual, his own words provide the best description (from the Dialogue on the Two New Sciences):[4]

> The experiment which I devised was as follows: Let each of two persons take a light contained in a lantern, or other receptacle, such that by the interposition of the hand, the one can shut off or admit the light to the vision of the other. Next let them stand opposite each other at a distance of a few cubits and practice until they acquire such skill in uncovering and occulting their lights that the instant one sees the light of his companion he will uncover his own. After a few trials the response will be so prompt that without sensible error the uncovering of one light is immediately followed by the uncovering of the other, so that as soon as one exposes his light he will instantly see that of the other. Having acquired skill at this short distance let the two experimenters, equipped as before, take up positions separated by a distance of two or three miles and let them perform the same experiment at night, noting carefully whether the exposures and occultations occur in the same manner as at short distances; if they do, we may safely conclude that the propagation of light is instantaneous; but if time is required at a distance of three miles which, considering the going of one light and the coming of the other, really amounts to six, then the delay ought to be easily observable. If the experiment is to be made at still greater distances, say eight or ten miles, telescopes may be employed, each observer adjusting one for himself at the place where he is to make the experiment at night; then although the lights are not large and are therefore invisible to the naked eye at so great a distance, they can readily be covered and uncovered since by aid of the telescopes, once adjusted and fixed, they will become easily visible.
>
> In fact I have tried the experiment only at a short distance, less than a mile, from which I have not been able to ascertain with certainty whether the appearance of the opposite light was instantaneous or not; but if not instantaneous it is extraordinarily rapid—I should call it momentary; and for the present I should compare it to motion which we see in the lightning flash between clouds eight or ten miles distant from us.

Olaus Roemer (1644-1710), a contemporary of Newton's, was a Danish astronomer in the tradition of Tycho Brahe. In the late sixteenth century, Roemer carefully timed the eclipses of Jupiter's moons, trying to verify that Jupiter's four moons moved exactly according to Kepler's laws. Instead, he measured the speed of light.

Like the earth, Jupiter casts a shadow. The earth's shadow is relatively small, and the moon enters it and is eclipsed only once in a while. Since Jupiter's shadow is much larger, its moons are eclipsed almost every revolution. Jupiter's moons move much faster than our moon. Therefore, it usually takes Jupiter's moons only a few minutes to be totally eclipsed. Timing their eclipses is one of the most accurate ways to check on the regularity of the moons' revolutions about Jupiter. For example, one observes two successive eclipses of a moon. Given the exact interval between the two, the third can be predicted. Roemer made a long series of eclipse measurements. As time went on, Roemer found discrepancies in his eclipse predictions that depended in a regular way on the relative position of the earth, Jupiter, and the sun. Precisely, if the original predictions were made by observing eclipses when Jupiter was in opposition (i.e., when the earth was closest to Jupiter), then half-a-year later, when the earth was on the opposite side of its orbit and farthest from Jupiter, the eclipses were 16½ minutes late. Half-a-year later, when Jupiter was in opposition again, the eclipses were on time.

Before Newton, one might have thought that Roemer had discovered a tiny correction to Kepler's laws, just as people used to add small corrections to Ptolemy's or Copernicus' circles. Perhaps some new strange force between the earth and Jupiter's satellites made them go slower when the earth is going away from Jupiter, faster when the earth was approaching Jupiter. After Newton, it was more difficult to make such an explanation. The strange force could not be gravity, since the properties of gravity were known. Besides, Mars did not influence Jupiter's moons, or Jupiter ours.

Roemer made a bold hypothesis: Kepler's laws were, indeed, right for Jupiter's moons, and that the delay was due to the finite speed of light. In other words, that the 16½ minute delay simply represented the extra time it took for the light from Jupiter's moons to cross the earth's orbit.

The speed of light can immediately be computed. It takes 16½ minutes for light to travel the diameter of the earth's orbit. Today we know the radius of the earth's orbit is about 93,000,000 miles, so its diameter is 2 × 93,000,000 = 186,000,000 miles.* Sixteen and one-

---

*Just about the same time, a group of French astronomers, by observing an opposition of Mars simultaneously in France and in South America measured the terrestrial parallax of Mars; i.e., they "triangulated" Mars. Since they knew the distance between France and South America, the astronomers could calculate the distance to Mars in miles. Thus, the scale of the solar system and, therefore, the distance from the earth to the sun was known in miles, rather than in the unknown "Astronomical Unit."

half minutes is almost exactly 1,000 seconds. So the speed of light is:

$$c = \frac{186{,}000{,}000}{1{,}000} = 186{,}000 \text{ miles/second}$$

The letter $c$ (from the Latin *celeritas*, speed) is the traditional modern symbol for the speed of light.

The speed of light ranks with Newton's gravitational constant, $G$, as one of the fundamental numbers of nature; and much effort has been spent in measuring it accurately. The first measurement, Roemer's, was not very accurate because the dimensions of the solar system were not very well known.

Having studied reflection and refraction, diffraction, the speed of light, and the nature of color, we come to the construction of the first comprehensive models to explain the "facts." Let us summarize briefly. What was known about light? The eye, acting like a modern camera, focuses light on the retina, where it is sensed by the nerves. Light, generally, propagates in straight lines. At smooth surfaces, light can be reflected or refracted according to known laws: For reflection, the angles of incidence and reflection are the same; refraction is completely described by Snell's law. Light deviated from straight-line motion (i.e., was diffracted) in passing through a small hole or by a sharp obstacle such as a knife blade. Light was known to travel in empty space at a large but not infinite speed. The speed of light in glass or water was not known. Newton had shown that pure light was of a single color and that different colors were refracted at slightly different angles at surfaces of glass or water. White light was a combination of all the colors.

We might compare this jumble of facts to the knowledge about the motions of the heavenly bodies we outlined in Chapter 1. They were the "data," obtained by careful observation. What was needed next was an explanation, a model, an answer to the question, "What is light?" which would explain everything known about it. Two rival models had been proposed to explain the motions of the sun, moon, stars, and planets, the earth-centered model and the sun-centered model. Similarly, two rival models for light arose in the seventeenth century, the particle model and the wave model. In a sense, the contest between the two models had been going on since ancient times. The Greek atomists believed that light was a stream of particles, while Aristotle had taught that light was a propagation of a disturbance, or vibration, in the aether.

The "particle" model claims that light consists of a stream of tiny material objects—particles, which are presumably too small to be seen individually and not heavy enough to have any observable mechanical effect. The "wave" model states that no matter actually travels from the source of light to the eye but rather that a disturbance propagates; just as when one plucks a string at one end it vibrates at the other, even though no piece of the string travels from one end to the other. Another analogy is with water waves or sound waves in air.

## NEWTON AND THE PARTICLE MODEL

Newton considered a wave model seriously but finally rejected it. He reasoned as follows. Most types of wave motion are vibrations of an identifiable medium. Ocean waves are the vibrations of water near the surface; sound propagates as vibrations of the air. Now, Aristotle's universe, filled with aether, seemed capable of transmitting disturbances seen as light. But, most of Newton's universe was empty. Nothing was out there to vibrate, to transmit light waves from the sun and the stars. Nevertheless, light comes to us from great distances through this emptiness.

Material particles, on the other hand, do not need a medium to travel through. Indeed, Newton's mechanics was formulated to describe objects moving through a vacuum. When his Law of Universal Gravitation is applied to the orbits of the planets, no resisting medium is necessary to account for the observations. A particle theory of light could be consistent with Newtonian mechanics, and was reinforced by the way light travels. Unimpeded, it moves in straight lines just like material particles in the absence of forces. Of course, gravity should act on light particles, but since they must move at a fantastically high speed the deflection of their paths from straight lines by gravity is too small to be observed.

The law of reflection provided an even stronger argument for the particle model of light. Alhazen had already noticed the similarity of the law of reflection for light to the way balls bounce off hard surfaces. Newton understood reflection as follows: When a light particle—or a ball—strikes a smooth surface, the surface exerts a force on it; otherwise, the particle would continue in a straight line. The force is perpendicular to the surface, since on impact the surface bends inward slightly then springs back to its original position. Therefore, no force is exerted on the particle parallel to the surface and so, from Newton's second law, the particle has no acceleration in that direction either, which means that its speed parallel to the surface is the same before and after the collision.

In Fig. 7-17, let $v_x$ and $v_d$ be the speeds parallel and perpendicular to the surface before the collision, respectively. The actual path of the light particles during some time, $t$, is marked with arrows; it is the hypotenuse of a right triangle whose sides of length $v_x t$ and $v_d t$ are shown. Since the speed of light is $c$ (186,000 miles per second), the length of the light path is $ct$. By Pythagoras' theorem, $(v_d t)^2 + (v_x t)^2 = (ct)^2$. After the collision, the speed parallel to the surface is still $v_x$, as we just argued; and $c$ never changes. If we call the speed perpendicular to the surface after the collision $\bar{v}_d$, the right triangle in the left-hand part of Fig. 7-17 has sides $v_x t$, $\bar{v}_d t$, and $ct$. From Pythagoras' theorem, just as before, $\bar{v}_d^2 + v_x^2 = c^2$. So, $\bar{v}_d^2$ is the same as $v_d^2$. Since $\bar{v}_d$ is not equal to $v_d$—the perpendicular speed does change, the solution is $\bar{v}_d = -v_d$. After the collision, the vertical speed is the same size as before but in the opposite direction. The two triangles in Fig. 7-17 have the same size and shape, and the angle of incidence is equal to the angle of refraction.

Newton showed that the particle model could explain the law of refraction (Snell's law) just as easily. He supposed that in a dense medium (water or glass) the speed of light was *faster* than in vacuum. Perhaps by forces similar to gravity, the atoms of the medium accelerated the light particles to a higher speed. Newton again argued that the surface would exert no force on the light particles parallel to the surface. So, inside, the parallel speed is the same as outside. Since the total speed is greater, the perpendicular speed must increase, and the ray is refracted toward the perpendicular.

Quantitatively, in Fig. 7-11 remember that $AS$ and $BS$ are the same length. Let $\bar{v}$ stand for the speed of light inside the medium, and $c$, as always, for its speed outside. The time it takes for light to go from $A$ to $S$ is $AS/c$, and the time it takes from $S$ to $B$ is $BS/\bar{v}$. Then, the parallel speed in air is just:

$$\frac{AM}{AS/c} = c\frac{AM}{AS}$$

while the speed parallel to the surface inside the glass or water is:

$$\frac{BN}{BS/\bar{v}} = \bar{v}\frac{BN}{BS}$$

These must be equal, since there is no horizontal acceleration; and it follows that:

$$\frac{\bar{v}}{c} = \frac{AM}{BN} = index\ of\ refraction$$

which, according to Snell's law, is the same for all angles of incidence!

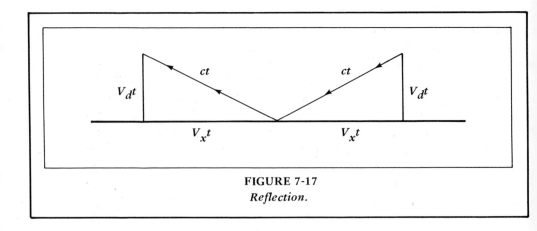

**FIGURE 7-17**
*Reflection.*

Or, to turn the argument around, if $\bar{v}$ is greater than $c$ and if there is no acceleration of the light particles parallel to the surface, Snell's law must hold. Newton's picture of light was very plausible indeed.

Newton also performed many experiments on diffraction. He was not able to develop any quantitative theory, but the phenomenon did not seem to contradict a particle theory of light. When light particles pass very close to matter, they are apparently deflected from their straight-line path by some force, perhaps the same one that speeds up light in glass or water.

Even though light had some properties, like diffraction, which had no quantitative explanation in a particle model, such a model fit well with Newton's mechanistic view of nature, and explained quantitatively the laws of reflection and refraction. After much consideration and vacillation, he finally speculated that light was a hail of tiny particles moving at high speed through empty space. That, somehow, when these particles passed near matter, they were deflected slightly causing the various phenomena of diffraction. And, that their natural speed in water or glass was greater than in air or empty space, explaining Snell's law for refraction. He suggested that light of different colors consisted of particles of slightly different masses. Different mass particles would accelerate differently on entering a medium and would, therefore, be refracted differently. Newton himself was always careful to distinguish between his speculations and his hypotheses, on the one hand, and what he really knew from experiment or mathematical deduction, on the other. But, to those who came after him, his authority was comparable to Aristotle's in the Middle Ages. His speculations, like Aristotle's, became unquestionable dogma, especially in eighteenth century England.

## THE WAVE MODEL

The "wave model" for light was first worked out in detail by the Dutch astronomer and mathematician, Christiaan Huygens (1629-1695). Huygens was almost the only contemporary of whom Newton always wrote with respect. Huygens had been the first to prove that the acceleration of an object in uniform circular motion is given by the formula, $v^2/r$. He proved the law of the pendulum, and invented the first pendulum clock. He built the most powerful telescopes of his time, discovering one of Saturn's moons and seeing Saturn's rings clearly.

Today, Huygens' fame rests principally on a short little book he wrote in 1678. Living in France at the time, he wrote in French, "Traté de la Lumière" or "Treatise on Light," which attacked the particle model on several counts, all "physical." He thought he had an irrefutable objection. If light is a stream of particles, why do two streams not collide when they cross? Why don't we see the collisions? Further, he found the high speed required of the particles hard to believe. It was easier to believe that light is a disturbance, a "wave," in analogy to water waves or sound waves. Water waves travel fast, even though the water simply moves up and down. Huygens performed experiments to determine the speed of sound; in air it was about 1000 feet per second or 600 miles per hour—clearly no air is moving that fast, only the disturbance. For a more familiar analogy, line up a series of pennies and strike the end one with another penny. The penny on the far end will shoot off while the ones in between remain almost motionless, transmitting the disturbance.

Huygens' book begins by analyzing the geometry of wave propagation. He knew that if one drops a stone into a pool of water a circular wave crest spreads out from the center. (In three dimensions, a disturbance creates a spherical wave crest; only for simplicity do we restrict the discussion here to two-dimensional waves.) The wave crest travels outward at some speed characteristic of the medium, here, the water. The water itself does not travel at that speed. The surface at a particular place simply rises and falls as the crest passes by.

A wave causes the surface of the water to rise and fall, just as dropping a stone does. Each wave crest should, therefore, create secondary wavelets propagating outwards in a circular pattern. Huygens proved that adding up all the secondary wavelets recreates the original primary wave. From this, he showed that waves can travel in straight lines, as light does. Figure 7-18, taken from his book, illustrates his primary wave. From this, he showed that waves can travel in straight lines, as light does. Figure 7-18, taken from his book, illustrates his proof.

A disturbance is made at $A$ (a pebble is dropped in water or a

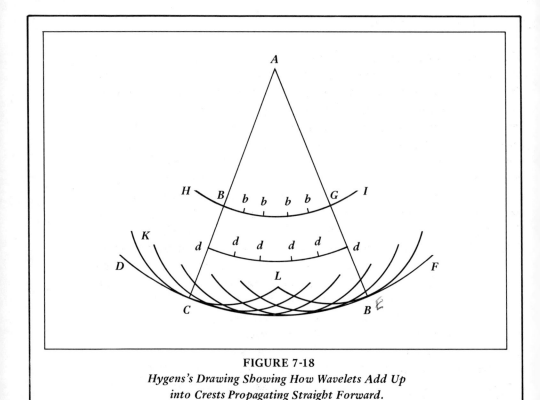

**FIGURE 7-18**
*Hygens's Drawing Showing How Wavelets Add Up into Crests Propagating Straight Forward.*

light flashes). Some time later, the crest is along *HBGI*. Sometime after that, the crest is along *DCEF*. Does that make sense? Why does not each point along *HBGI* emit a circular expanding crest?

Huygens' constructions are a geometrical rule for finding out how a wave crest propagates. In Fig. 7-18, Huygens has drawn sections of circles about *B* and *G*, and four points inbetween marked *b*, representing the crests of the circular waves emitted by these six points at some later time. The crests all lie along the circle *DCEF*. So, at most points, if one is at the crest of some "wavelet" from some point along *HBGI*, one is not at the crest of most of the others. But the crests of all the wavelets pass *DCEF* at the same time; they all add up, and the wave propagates outward. Thus, an observer anywhere sees a crest advancing in a straight line from *A*. Huygen's construction works only if the speed of the wavelets is the same everywhere.

Huygens went on to show that his way of thinking about waves can explain reflection and refraction. Using the same technique of adding up wavelets from the travelling crest, and asking along what line would they all add up together at a later time, he was able to show that when light is reflected the angle of incidence equals the angle of reflection. Also, he showed that Snell's law holds for refraction provided the speed of light propagation in a medium is *less* than it is in vacuum.

Huygens' explanation of reflection is illustrated in Fig. 7–19, taken from his treatise. The wave front, or crest, *AHHHC* in the illustration, strikes a mirror, *AKKKB*. This time we imagine that the light comes from a source so far away that the crest *AHHHC* can be taken as a straight line. When the part of the crest strikes the mirror at *A*, it emits a circular wavelet, as always. Sometime later, the portion of the crest that was at *C* when *A* struck the mirror strikes it at *B*. Where is the wavelet emitted by *A* at that time? It has gone the same distance, *CB*, so is a circle whose center is *A* and whose radius has the length *CB*. The arc *SNR* is a portion of that circle, and *AN* is a radius of

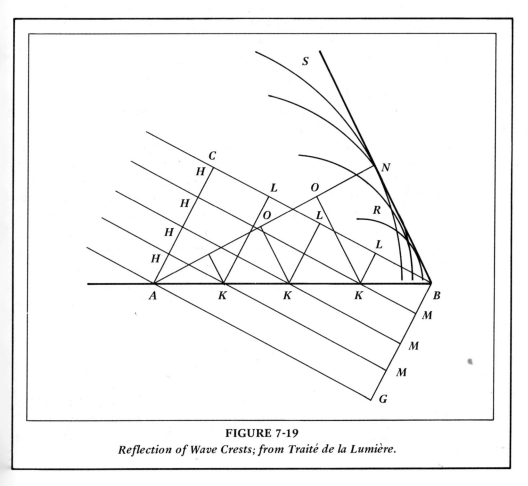

**FIGURE 7-19**
*Reflection of Wave Crests; from Traité de la Lumière.*

the same length as *CB*. In the diagram, *BN* is drawn tangent to the arc *SNR*. Now, when light from one of the points inbetween, *H*, strikes the mirror at *K*, it will emit a circular wavelet, which is shown in the diagram when the portion of the wave that was originally at *C* reaches *B*. Since the speed of light has not changed, these radii are the same as if the mirror had not been there, namely *KM*. Now, it is easy to see that all these circles are tangent to the same line, *BN*. (For the original wavelet, for example, *AN* and *AG* are the same length, *G* is a right angle because *BG* is perpendicular to the wave crest, and *N* is a right angle because *BN* is tangent to the circle; so the triangles *ABG* and *ABN* are congruent.) All the wavelets add up along the common tangent, *BN*, so this is the new wave front. *NA* is the new direction of propagation of the wave. Since the angle between this direction and the mirror, *NAB*, is the same as the angle *GAB*, the angle of reflection is equal to the angle of incidence.

Refraction is explained in a similar way. In Fig. 7-20, wave crest *AHHHC* strikes a transparent surface, *AB*. Suppose that below the surface the speed of light is not $c$, but $c/n$. When the point *A* on the crest strikes the surface, it emits a circular wavelet. When the point on the crest that was at *C* arrives at *B*, the circular crest emitted at *A* has become the arc line *RNS* when the crest arrives at *B*. The new wave front is *BN*, which is tangent to all the wavelets at the same time.

The new wave front, *BN*, is not parallel to the old one. Since the speed of light above the surface is $c$, and the speed of light in the medium is $c/n$ (Huygens assumed this to get the result!), the length *AN* is $1/n$ times the length *CB* ($n$ is the index of refraction, and is larger than 1). But this is just another way of stating Snell's law. Qualitatively, it is obvious from the diagram that the direction the light is moving is closer to the perpendicular to the surface below the surface than it is above; it is bent from the direction parallel to *CB* to the direction parallel to *AN*. (Here is the proof. Suppose, in the diagram, *E* and *F* were points on the perpendicular exactly to the left of *C* and *N*, respectively. Huygens has drawn them slightly off. Then, since $AN/BC = 1/n$:

$$\frac{AN/AB}{BC/AB} = \frac{1}{n}$$

also. The triangle *ABC* is similar to the triangle *AED*, and the triangle *ABN* is similar to the triangle *AFN*. Since the ratios of lengths of corresponding sides of similar triangles are the same, $AN/AB$ is the

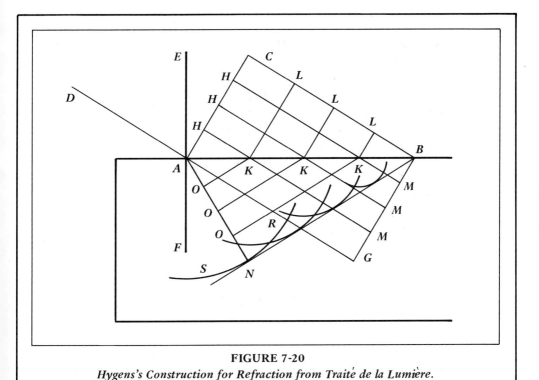

**FIGURE 7-20**
*Hygens's Construction for Refraction from Traité de la Lumière.*

same as $FN/AN$, and $BC/AB$ is the same as $DE/DA$; therefore:

$$\frac{EC/AC}{FN/AN} = \frac{1}{n}$$

which is Snell's law.)

Huygens' wave model was every bit as good as Newton's particle model. Both explained the fact that light moves in straight lines; and both explained reflection and refraction. Neither explained very well either diffraction or the different angles by which different colors were refracted. On both points the particle model had a slight edge. One could imagine that diffraction was caused by the light particles being accelerated by forces as they passed near matter. Newton had speculated that light of different colors consisted of particles of slightly different masses. What could be the explanation of different colors in Huygens' picture?

In fact, as we shall see shortly, Huygens' ideas could provide a natural explanation for Grimaldi's diffraction experiments although ironically Huygens, unlike Newton, seems to have been unaware of

Grimaldi's work. For over a century, nobody worked out the details, and Huygens was generally thought to be wrong. The simplicity of Newton's particles, as well as the authority of his name, made his speculations unchallengeable.

One experiment, if it could have been done, would have provided a crucial, unambiguous test. According to the construction for the refraction of waves in the wave model, if a transparent medium has index of refraction $n$, the speed of light inside the medium must be $c/n$, less than $c$; whereas for Newton's particle model to make sense the speed of light inside the medium has to be $nc$, greater than $c$. If anyone could measure the speed of light in water or glass, he would have a crucial test to distinguish between the two models, comparable to Galileo's observation of the phases of Venus.

## THOMAS YOUNG AND THE REVIVAL OF THE WAVE THEORY

The particle theory of light remained in favor until the first decade of the nineteenth century when an Englishman, Thomas Young (1773–1829), discovered a test, other than the speed of light in a medium, which could distinguish between particles and waves.

Young was born in Somerset, England, into a Quaker banking family. He was one of the most versatile scholars ever to turn his mind to science. He practiced medicine for a while but gave it up, finding that he was more interested in linguistics and scientific problems. In 1801, he became a professor at the Royal Institution in London, where Faraday was to carry out his life work. There, among other studies, he turned his attention to optics and, in particular, to the wave (rather than a particle) description of light.

Newton himself, though he finally adopted the particle description, had at one time seriously considered the proposition that light had a wave nature. At one time Newton had written:

> That fundamental supposition is, that the parts of bodies, when briskly agitated, do excite vibrations in the ether, which are propagated every way from those bodies in straightlines and cause a sensation of light by beating and dashing against the bottom of the eye, something after the manner that vibrations in the air cause a sensation of sound by beating against the organs of hearing.

Young wanted to re-establish this early opinion of Newton's.

Young added a feature to the description of waves that was not explicit in Huygens' original work. A realistic description of a wave is not a single crest, but a series of crests and troughs, regularly spaced, moving in the same direction with the same speed. The spacing between the crests is called the wavelength of the waves. The time interval between the passage of two successive crests by a fixed point is called the period of the wave. Evidently, a particular crest travels the distance of one wavelength during one period, so that the speed of the wave is the wavelength divided by the period. That is, if the wavelength is $\ell$ and the period is $T$, the speed the wave is travelling must be $\ell / T$. One often speaks of the frequency of the wave rather than its period. This is just the number of wave crests that pass by a fixed point per second, which is $1 / T$ if $T$ is the period measured in seconds. Thus, the frequency $f$ is:

$$f = \frac{1}{T}$$

And, the speed, $v$, is:

$$v = \frac{\ell}{T} = \ell f$$

Now, suppose two waves with the same wavelength and frequency arrive at the same place. What is the result? It matters whether the crests of both arrive simultaneously, or not. Suppose we add up two waves whose crests and troughs just match, as in Fig. 7-21. In this case they reinforce each other, and the result is a wave whose crests and troughs are twice as big as each of the original ones. On the other hand, suppose two waves add the other way around—so that the crests of one arrive together with the troughs of the other. Then, the crests and troughs just cancel, as in Fig. 7-22; and the result is no disturbance at all! This effect is called destructive interference; the effect pictured in Fig. 7-21 is called constructive interference. Here is a fundamental difference between waves and particles. Waves can interfere, but particles never can. Two streams of particles arriving at the same place at the same time can never cancel each other.

Now we can understand Huygens' constructions somewhat better. In Fig. 7-18, imagine that as the wave passes the points marked $B$, $G$, and those marked $b$, these points emit a sequence of crests and troughs that spread out at the speed characteristic of the wave. Then, some time later, the crests of all these "wavelets" lie along the arc $DCEF$, and interfere constructively so that is where the large wave is seen. At the same time, anywhere else, some crests and some troughs from different points along $BbbbbG$ arrive together, interfere destructively, and create no large crest.

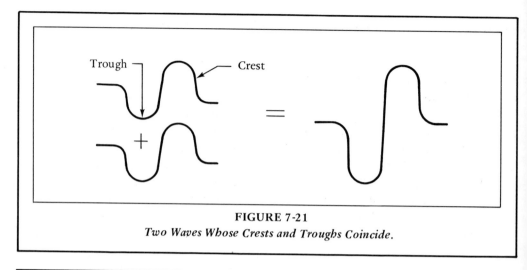

**FIGURE 7-21**
*Two Waves Whose Crests and Troughs Coincide.*

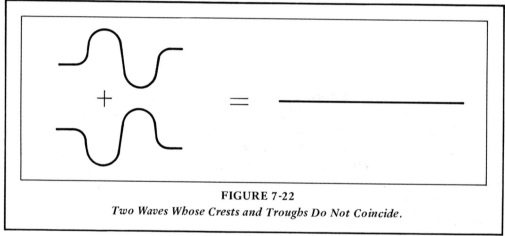

**FIGURE 7-22**
*Two Waves Whose Crests and Troughs Do Not Coincide.*

Diffraction can be understood in a similar way. Imagine a wave crest striking a barrier—a "breakwater" if we are talking about ocean swells—with a small gap $BC$ as in Fig. 7-23. The wave strikes the barrier $NK$ from the left. Following Huygens' constructions, we ought to see a series of spherical waves on the far side of the screen, caused by the "disturbance" at the gap. Water waves and sound waves do indeed exhibit effects like this. One does not have to have an unobstructed view of the source of a sound wave in order to hear it. Figure 7-23 was drawn by Newton to illustrate precisely this property of wave motion, and to argue that since one cannot see light through a hole unless one is looking directly at the source, light could not be a form of wave motion. If light from a source $A$ consisted of a stream of particles, they could be detected to the right of the obstruction only in the region between the lines $PB$ and $QC$. Waves, on the other hand,

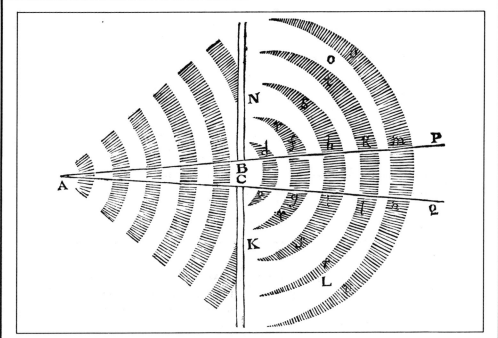

**FIGURE 7-23**
*The Diffraction of Waves through a Small Opening. (From Newton's* Principia.*)*

could be detected elsewhere, for instance at *O* or *L*. Next imagine a barrier with not one, but two, small gaps close together, as in Fig. 7-24. A wave is arriving at the barrier from the left. To the right will be a sequence of circular (in three dimensions, spherical) crests and troughs, spreading out from each gap in a regular succession. In Fig. 7-24, the solid lines represent the troughs. Where the crests cross each other, or where the troughs cross each other, will be a large crest or a large trough: constructive interference. Where a crest crosses a trough, they will interfere destructively, and no "wave" will be there. Therefore, if a screen is held at some distance from the two gaps, one should observe a sequence of bright and dark spots, corresponding respectively, to regions of constructive and destructive interference.

This was the experiment that Thomas Young performed in 1804. He let sunlight fall on a card in which two parallel narrow slits had been cut very closely together, and observed the light that got through on a screen several feet away. Figure 7-25 is a diagram of Young's "two-slit interference" experiment.

What is the intensity of light expected at different places on the screen? If light were a stream of particles, one should expect to see the

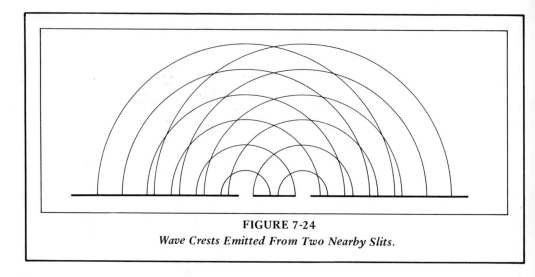

**FIGURE 7-24**
*Wave Crests Emitted From Two Nearby Slits.*

images of the two slits, somewhat widened by diffraction as shown in Fig. 7-26, which graphs the intensity of light against position on the screen. In Fig. 7-26, the points marked $C$ and $D$ are the same points marked $C$ and $D$ in Fig. 7-25.

What Young found was completely different. Instead of two bright spots, he observed a series of light and dark regions, characteristic of the way waves interfere but unlike anything that could be explained in the particle model. Figure 7-27 is a photograph of such an "interference pattern." The matter was settled. Light was definitely some kind of wave.

From his experiment, Young was able to calculate the wavelength of light, the distance between the wavecrests. Figure 7-28 is a schematic drawing of the setup shown in Fig. 7-25. The point $X$ on the screen is the same distance from the slits $A$ and $B$. The crests from $A$ and $B$ strike $X$ simultaneously, since they arrive simultaneously from the distant sun at $A$ and at $B$, and then travel the same distance $AX$ or $BX$ at the same speed $c$. Somewhere above $X$, at $Y$, a wave from $A$ will take longer to arrive at the screen than a wave from $B$, since it has farther to go. A crest from one will arrive together with a trough from the other, leaving a dark spot. Still further, at $Z$, the next crest from $A$ will arrive together with a crest from $B$, forming the next bright spot. If $A$ and $B$ are, in fact, long parallel slits as in Fig. 7-25, the interference pattern will be a sequence of light and dark lines, called interference fringes.

Now we see how Young calculated the wavelength of light; the distance $XZ$ must be just such that $AZ$ is longer than $BZ$ by just one

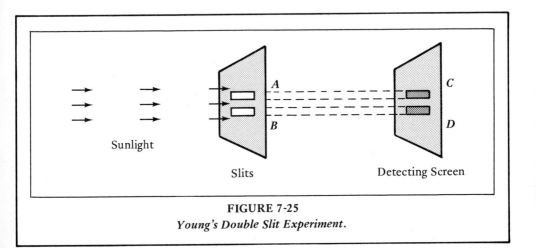

**FIGURE 7-25**
*Young's Double Slit Experiment.*

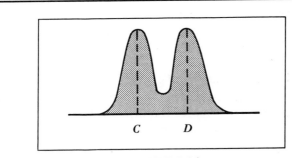

**FIGURE 7-26**
*The Intensity of Light in the Detecting Screen in the Double Slit Experiment.*

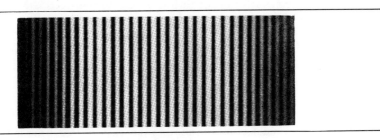

**FIGURE 7-27**
*Photograph of an Interference Pattern.*

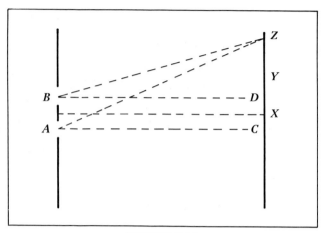

**FIGURE 7-28**
*Schematic Drawing of the Slits and Detecting Screen in Young's Experiment.*

wavelength. This is a simple problem in geometry, and is worked out in the Appendix.

With white light, Young was able to observe only a few interference fringes. He discovered that many more were visible with light of a single color. Furthermore, the distance between the successive bright fringes changed with color; therefore, the wavelength of differently colored light was different. Thus, Newton's experiments on colors meant that light of different colors has different wavelengths and that white light is a mixture of light waves of many different wavelengths. Here is Young's description of his results:[5]

> The middle of the two portions [of his interference pattern] is always light, and the bright stripes on each side are at such distances, that the light coming to them from one of the apertures must have passed through a longer space than that which comes from the other, by an interval which is equal to the breadth of one, two, three, or more, of the supposed undulations, while the intervening dark spaces correspond to a difference of half a supposed undulation, of one and a half, of two and a half or more.
>
> From a comparison of various experiments, it appears that the breadth of the undulations constituting the extreme red light must be supposed to be, in air, about one 36-thousandth of an inch, and those of the extreme violet about one 60-thousandth: the mean of the whole spectrum, with respect to the intensity of light, being about one 45-thousandth.

The wave model explains diffraction through a single slit; i.e., the fact that light falling perpendicularly on a very narrow slit spreads out wider than the width of the slit, in violation of the rule that light travels in straight lines. The extreme case is illustrated by Newton's drawing, Fig. 7-23. There, the width of the slit is only a small fraction of the wavelength, and the light spreads at all angles. More commonly, a slit is large compared to the distance between crests. Why can no light be seen at a large angle in that case? In other words, why is there a shadow? It is because the crests and the troughs from different parts of the slit do not all arrive at the same time. A crest arriving from near the edge of the slit is cancelled by a trough arriving simultaneously from a little farther away, and so on. The crests and troughs arriving from different parts of the slit cancel each other (interfere destructively) almost exactly. Only straight ahead do they all add up, as Huygens' construction shows. In the intermediate case where the width of the slit is a few wavelengths, the cancellation of crests and troughs is not complete; and the beam spreads somewhat. The calculation is outlined in the Appendix.

So great was the prestige of Newton's name that Young was condemned for daring to question Newton's speculation that light was a stream of particles. So severe was the criticism of his wave model and his masterful experiments, especially in England, that he became discouraged with science and turned to other matters, principally linguistics. An accomplished linguist, he mastered Greek, Latin, Hebrew, Chaldean, Syriac, and Samaritan, as well as the modern European languages. In 1799, the Rosetta stone had been unearthed in Egypt by Napoleon's soldiers. This stone bore parallel inscriptions in Greek and ancient Egyptian hieroglyphics, implying the tantalizing possibility of deciphering the ancient Egyptian writing. Young contributed importantly to deciphering the Rosetta stone. Later in life, he became interested in the problems of life insurance, and made pioneering contributions to actuarial science. On his deathbed, he continued to work on a dictionary of the Egyptian language. A man of many and varied accomplishments, Thomas Young is remembered today as the one who proved that light is a wave.

Young's interference experiment was a classic definitive experiment; rather than passively compiling data Young was able to decide between two conflicting hypotheses, through the logic designed into the experiment. Nevertheless, his wave model was not accepted until the details, including all the complicated diffraction effects, were worked out. The wave theory was taken up next by Augustin Fresnel (1788–1827). In a series of masterful memoirs to the French *Académie des Sciences* in 1815, 1816, and 1817, Fresnel formulated the wave theory with much greater mathematical refinement and clarity. Abandoning Huygens' geometrical constructions, Fresnel developed more powerful methods for calculating the pattern of intensity of light behind slits or barriers of any shape. The mathematician Denis Poisson objected to Fresnel's methods because, as Poisson pointed out, they predicted the absurd result that in the middle of the shadow behind a perfectly circular opaque disk there ought to be a bright spot. Fresnel agreed. Francois Arago, a French physicist and astronomer, performed the difficult experiment and observed the bright spot. The surprising spot occurs because the center of the shadow is the same distance from all points along the edge of the circular disk; the wave crests from every point on the edge arrive at the center of a screen behind it simultaneously, interfering constructively. The experiment is difficult because the opaque screen must be exactly circular, within an error less than the wavelength of light.

The work of Fresnel, Arago, and Poisson did for light what Newton's *Principia* had done for mechanics. It set up a mathematical program, which, in principle, could answer *all* questions about the propagation of light waves. The doubters, of course, continued to make up particle explanations of light, just as the motions of the planets can be described in a geocentric model with enough epicycles. But in 1849, another French physicist, A. H. L. Fizeau, succeeded in making a terrestrial measurement of the speed of light with a clever arrangement of mirrors and rapidly rotating toothed wheels. Fizeau's experiments were forerunners of the very accurate measurements by Michelson, which we will describe in Chapter 9. Soon afterward he was able to measure the speed of light in water, and found it to be slower, not faster, than in vacuum. This was the final confirmation of the wave theory of light. For the nineteenth century, light was a wave.

One big unanswered question remained. What kind of wave is light? Surf waves propagate in water, sound waves in air; in what medium does light propagate? The mysterious medium was given the name aether, Aristotle's old name for the material that fills all space. The nineteenth century was faced with the difficult proposition that mechanics required a vacuum and light, a medium

*Summary*

*The Greeks knew that light, unimpeded, travels in straight lines; and knew the law of reflection, that the angle of incidence equals the angle of reflection. Light seemed to be a physical realization of the abstract straight lines of geometry. Refraction was also studied, especially by Ptolemy. A popular theory of vision was that one's eye* emitted something that saw objects; the sense of sight was analogous to the sense of touch. But, Aristotle speculated that light was a disturbance, a vibration, of the aether.

The Arabs, particularly Alhazen, and later the Europeans, particularly Kepler, established that light is emitted from bright objects and eventually is seen by the eye. Kepler also corrected Ptolemy's theory of refraction and, with the more correct law, explained the workings of lenses. Soon after learning of Galileo's telescope, he explained how it worked from first principles. The law of refraction has a simple form, which today is known as Snell's law.

Newton performed a brilliant series of experiments, showing that white light is composed of many different colors, each of which is refracted at a slightly different angle. Olaus Roemer, a Danish Astronomer, patiently timing the eclipses of the moons of Jupiter, discovered that light does have a finite speed—about 186,000 miles per second. Grimaldi, and later Newton, studied diffraction, the slight bending of light that occurs when light passes near straight edges or through very narrow slits.

In the eighteenth century, there existed two rival models for light, Newton's particle model and Huygens' wave model. The particle model seemed to fit in with Newton's mechanical picture of the world, especially with the fact that space was empty. Huygens explained how to calculate the propagation of waves, introducing the idea that little "wavelets" were emitted by wavecrests at all times in all directions, the resulting wave being the sum of all the wavelets. Both models could explain reflection and refraction quantitatively.

Two light waves of exactly the same wavelength can interfere destructively; if the crests of one and the troughs of another arrive simultaneously, they will all cancel, and no signal will result. In 1804, Thomas Young passed light through two nearby, narrow slits and observed an interference pattern, a sequence of light and dark fringes where the light from the two slits interfered, alternatively, constructively and destructively, thereby proving that light has properties characteristic of waves and not particles. The wavelengths are very small, about 1/50,000 of an inch, which is why the experiment is difficult and the wave properties unobservable in ordinary circumstances.

## Appendix Chapter 7

*More on Lenses*

In the text we saw how a simple lens forms an image. A more common application of the simple lens is as a magnifying glass, as in Fig. 7-29.

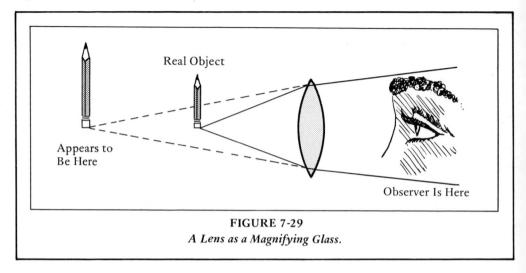

**FIGURE 7-29**
*A Lens as a Magnifying Glass.*

The lens in Fig. 7-29 might be the same as the one in Fig. 7-9, but the object is so close that the rays of light never converge to a focus. They do bend inward somewhat, however, so to an observer to the right of the lens they appear to be coming from a point farther away. In the illustration, the rays are traced for the bottom of the arrow. The same thing happens for the top. You can see that the arrow appears larger than it really is.

A simple telescope has two lenses, as in Fig. 7-30. The large lens is called the objective. It focuses the light from a great distance at a point $F$, called the focus of the lens. A smaller lens is a powerful magnifying glass called the eyepiece, through which the observer looks at the inverted image formed by the objective. Thus, the telescope accomplishes two things: First, it magnifies the field of view so that two points appear farther apart than they really are; and secondly, since the objective is larger than the pupil of the eye, it collects more light from a distant source than does the unaided eye, so objects appear brighter.

### The Reflecting Telescope

Having discovered that chromatic aberration is an inherent property of refraction, Newton designed and constructed a telescope with the big objective lens replaced by a curved mirror. A curved mirror whose shape is that of a small piece of a sphere will focus light just as a lens does; that is, light rays from some point striking different parts of the mirror will all be reflected to the same point, forming an image of the object. As with lenses, this property is not immediately obvious, but if you draw a portion of a circle and see where the rays go, using the law of reflection, you will verify that it is true.

There is a technical problem: If you look at the image with an eyepiece, as with a refracting telescope, it seems that your head will get in the way of the incoming light. Newton's solution is illustrated in his drawing of his telescope, shown in Figure 7-31.

Light from a distant star enters the telescope along parallel lines like $PQ$ and

Appendix Chapter 7

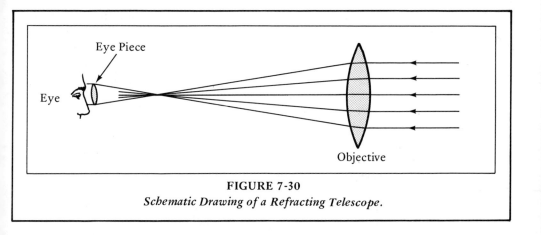

**FIGURE 7-30**
*Schematic Drawing of a Refracting Telescope.*

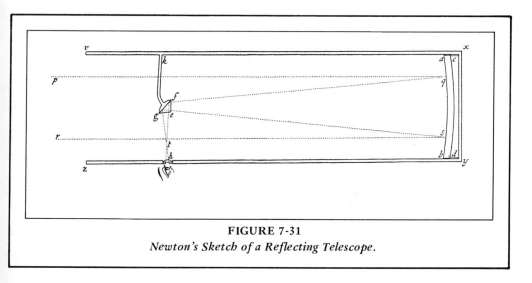

**FIGURE 7-31**
*Newton's Sketch of a Reflecting Telescope.*

*RS*. They are then reflected and, if unimpeded, will come to a focus at some point a little way behind the diagonal mirror. Just in front of the focus, Newton inserted a small mirror, *FG*, which reflected the light at right angles to a focus at *T*. The eyepiece was placed in a small hole in the side of the telescope tube at *H*, where the image was observed just as in a lens telescope.

Unlike a person's head, the diagonal mirror, *FG*, blocked out only a small portion of the incoming light; Newton put in a right-angled prism, *FEG*, to make the reflection at the surface, *FG*, more perfect; it is not an essential part of the design of the telescope.

Newton's aim was to get rid of chromatic aberration; since the law of reflection is the same for all colors, unlike the law of refraction, the big mirror focuses all colors at the same place. (The chromatic aberration introduced by the eyepiece is negligible.)

Today, we know how to make lenses without chromatic aberration by combining two or more pieces of glass with different indices of refraction. These are called achromatic lenses, and most good optical devices have them. Nevertheless, very large lenses are harder to support and to make than very large mirrors, so the largest telescopes are now made following Newton's reflecting design.

### The Wavelength of Light

Refer to Fig. 7-28. Let $d$ stand for the distance, $AB$, between the two slits; and let $L$ be the distance between the card with the slits and the detecting screen. Let $s$ be the spacing between the two bright fringes at $X$ and at $Z$. Since $C$ and $D$ are directly behind $A$ and $B$, respectively, the distance $CD$ is $d$ also. The point $X$ is half way between them, so the distance $CX$ is $d/2$.

Now draw the triangle $BDZ$. It is a right-triangle, with the right angle at $D$. The distance $BD$ is $L$. The distance $DZ$ is smaller than $s$ by the distance $XD$, which is $d/2$: $DZ = s - d/2$. Let $X$ be the distance $BZ$, the hypotenuse of the right triangle. $X$ is the distance the light from $B$ travels to the first bright fringe. Then, Pythagoras' theorem says:

$$L^2 + \left(\frac{s-d}{2}\right)^2 = x^2$$

Now draw another triangle, $ACZ$. The point $C$ is directly behind the slit $A$ so this is a right triangle with the right angle at $C$. The length of the side $AC$ is again $L$, and the length of the side $CZ$ is now $s + d/2$. The hypotenuse, $AZ$, contains exactly one more wavelength than $BZ$ so that a crest from $B$ arrives at $Z$ at the same time the next one from $A$ gets to $Z$. If we let $\ell$ stand for the unknown wavelength, then the distance $AZ$ is $x + \ell$. Pythagoras' rule for the triangle $ACZ$ reads:

$$L^2 + \left(\frac{s+d}{2}\right)^2 = (x + \ell)^2$$

Next we write the two equations together, expanding the squares:

$$L^2 + s^2 - sd + \frac{d^2}{4} = x^2$$

$$L^2 + s^2 + sd + \frac{d^2}{4} = x^2 + 2x\ell + \ell^2$$

Subtract the first equation from the second:

$$2sd = 2x\ell + \ell^2$$

The wavelength $\ell$ is much smaller than $x$, which is nearly $L$, so the term $\ell^2$ is negligible compared to $2x\ell$. We have by now used such arguments several times. The result is that the wavelength is given by:

$$\ell = \frac{sd}{x}$$

The distance $x$, of course, is approximately $L$ so that $\ell = sd/L$ is just as good.

If you are considering doing the double-slit experiment, this formula should be a guide to its feasibility.

## Diffraction through a Slit

In Fig. 7-32, light of wavelength $\ell$ falls perpendicularly on the slit whose width $AB$ is $d$. What is seen on a screen placed at a distance $D$ from the slit, at a distance $s$ above the center of the screen? It is the sum of the crests and troughs from all parts of the slit. At what distance, $s$, do they first cancel exactly? This is the beginning of the shadow. If the light coming from the edge $A$ has to go just one wavelength longer than the light coming from $B$, then light from any point in the half of the slit near $A$ just cancels the light from a corresponding point in the half near $B$. For example, if there is a crest at $A$ and $B$, the trough is half way inbetween and just cancels the pieces of the crest. The geometry is the same as in the previous problem, where the result was $\ell = sd / L$.

The distance, $s$, from the center of the slit to the beginning of the shadow, therefore, is:

$$s = \frac{L\ell}{d}$$

or,

$$\frac{s}{L} = \frac{\ell}{d}$$

The distance to the shadow, divided by the distance to the screen, is the same as the ratio of the wavelength to the width of the slit. (In trigonometric language, this ratio is the sine of the angle the light deviates from the forward direction.) If the width of the slit is sixty wavelengths, the shadow spreads at an angle of one degree.

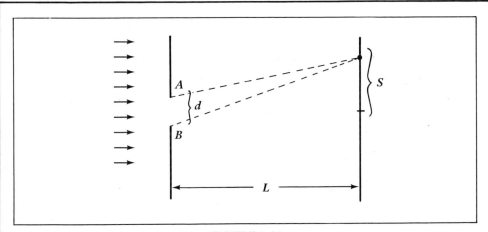

FIGURE 7-32
*Schematic Drawing for Calculating the Distance to the Beginning of the Shadow for Light Diffracted through a Single Slit.*

## QUESTIONS

1. In what sense was Newton's experiment, in which he recombined the colors of the spectrum to form white light, a "crucial," or decisive experiment?

2. According to Young, the wavelengths of red and blue light, respectively, are 1/36,000 inch and 1/60,000 inch. What are the frequencies of red and blue light?

3. The distance between Young's two slits was about one-hundred times the wavelength of visible light. What would happen if the two slits were an inch apart? What would happen if the distance between them were much less than the wavelength of light?

4. A certain kind of glass has an index of refraction exactly 1.5. If a light ray has an angle of incidence exactly $30°$, what is the angle of refraction? (To answer problems 4 and 5, either make a drawing and measure the angles with a protractor or use an elementary knowledge of trigonometry.)

5. A ray of light passes from beneath the surface of water ($n = 4/3$) into the air. Let us still call the angle the light ray makes with the perpendicular in the water the angle of refraction, as in Fig. 7–6. At what angle of refraction will the "angle of incidence" be $90°$, i.e., the light emerges parallel to the surface of the water? What do you think happens if the angle in the water (the "angle of refraction") is even greater than the angle you just calculated?

6. Suppose two slits are 1/16 inch apart, and light passes through them to make an interference pattern on a screen three feet away. What is the distance between the fringes: (a) for red light? (b) for blue light?

7. A single slit is 1/100 inch wide. Red light is passed through it; the image of the slit, broadened by diffraction, is observed on a screen three feet away. What is the width of the image?

8. The law of reflection can be stated another way. Suppose a light ray is emitted at point $O$, reflects off a surface at $M$, and arrives at point $E$, as in Fig. 7–4. The path $OME$ is the shortest one that exists between $O$ and $E$, requiring only that the light be reflected somewhere on the surface $AMB$. Convince yourself, by trying some other paths, that this is, in fact, true. (This fact was discovered by Hero of Alexandria in 150 B.C.!)

9. In Fig. 7–11, the path $ASB$ of a ray refracted at the surface at $S$ is not the shortest path from $A$ to $B$ made up of two straight lines with one angle somewhere on the surface. (The shortest parth is an unbroken straight line.) However, if the speed of light above the surface is $c$ and the speed below is $c/n$, where $n$ is the index of refraction; then, the time for the light to go from $A$ to $B$ is the shortest, provided the angles obey Snell's law of refraction. This general feature was discovered by Pierre de Fermat in the early seventeenth century, and is often useful in analyzing complicated optical systems in which a light ray undergoes several reflections and refractions. Convince yourself that this rule is true by comparing the time for three or four other paths from $A$ to $B$. (The rule can be proved by using elementary differential calculus.)

## REFERENCES

1. Plato, *Timaeus* 45c, trans. by Rev. R. G. Bury-Heinemann, (London, 1929) p. 101; as quoted in Vasco Ronchi, *The Nature of Light,* transl. V. Barocas (Harvard University Press, Cambridge, 1970).

2. Lucretius, *De Rerum Natura,* trans. Cyril Bailey (Oxford University Press, 1947), Book IV, pp. 216-219, as quoted in Ronchi, op. cit.

3. Aristotle, *On the Senses,* Chapter III, as quoted in Ronchi, op. cit.

4. Galileo Galilei, *Dialogues Concerning Two New Sciences,* transl. Henry Crew and Alfonso de Salvio (MacMillan, 1914), p. 43.

5. Quoted by C. C. Gillispie, *The Edge of Objectivity* (Princeton University Press, 1960), p. 418.

**FARADAY**
*(Photographed by Barry Donahue and by permission of the Houghton Library, Harvard University)*

CHAPTER
# 8

# MAGNETISM AND ELECTRICITY

When we began, we said that *physics* attempts to answer two questions: "How do things move?" and "What is everything made of?" We have devoted ourselves almost exclusively to the first question, and will continue to do so. Our reason is that, aside from occasional speculations about atoms and elements, no serious progress in understanding the nature of matter was made until magnetism and electricity were understood. Both magnetism and electricity had been known since ancient times. Until the nineteenth century, both were mainly curiosities; magnetism had a practical application in the compass, but was not well understood. The eighteenth century saw the consolidation of the Newtonian view of nature: matter in motion, obeying his inexorable Laws. The nineteenth century tried to fit all natural phenomena, including magnetism and electricity, into Newton's scheme. In doing so, it unified not only magnetism and electricity but, surprisingly, also light. This unification is the subject of this chapter. The revolution to which it led is the subject of the next.

### MAGNETS

The ancient Greeks were fascinated by lodestones, mysterious rocks that attracted iron but not any other metal.* Lodestones were especially easy to find in Magnesia, a province of ancient Asia Minor. Our

---

*Greek legends recount the magic wrought by the Cabiri, whose lodestones could attract and hold several iron rings without any visible means of attachment. In legend the Cabiri became gnome-like dwarfs who mined their magical metal and fashioned it into various instruments and ornaments; they were immortalized in Grimm's "Snow White and the Seven Dwarfs."

modern word "magnet" comes from the Greek name for lodestones—stones from Magnesia. Not only could lodestones attract single pieces of iron, but iron, once in contact with the lodestone, could attract other pieces of iron. By the Middle Ages, something even more mysterious was known. Take a long thin lodestone—a bar magnet—and labeled one end $N$ for "north" and the other end $S$ for "south." In Fig. 8-1, the left end of the bar magnet is labelled $N$. Now, cut the original bar magnet in half and label the left end of each half $N$ and their right ends $S$ as before. After the bar is cut, the two pieces attract each other. The ends marked $N$ attract the ends marked $S$. Next, rotate one magnet but not the other, so that now two ends marked $S$ are close together. The two magnets push each other away. Next, put two "north" ends close together. Again, the two magnets repel each other. Thus, whether two magnets attract or repel each other depends upon how they are oriented relative to one another.* We will see shortly when we discuss compasses why we chose "North" and "South" to label the two ends of the magnets.

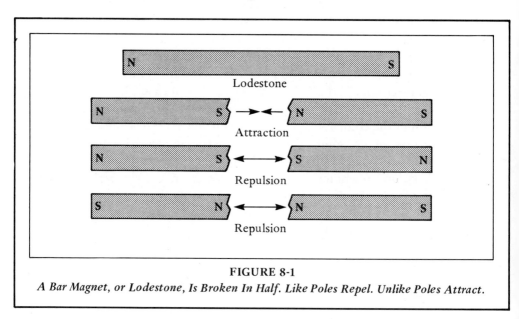

**FIGURE 8-1**
*A Bar Magnet, or Lodestone, Is Broken In Half. Like Poles Repel. Unlike Poles Attract.*

---

*These properties of magnets, at least the fact that like poles repel, were known to Roger Bacon, who lived in England in the thirteenth century. Bacon was an experimental scientist who in many ways was far ahead of his time, and much of what he knew had to be rediscovered later. On magnets, Bacon wrote, "If the iron is touched by the north part of the magnet, it follows that part wherever it goes. . .and if the opposite part of the magnet is brought against the touched part of the iron, it flees from it, as though inimical, as though the lamb from the wolf." A contemporary and friend of Bacon's, Peter Peregrine, discovered that if you broke a magnet in two, a new pair of poles appeared at the fractured ends. Long before, the Chinese had known that opposite poles attract, and like poles repel. To some of them it was a natural manifestation on the opposing principles, Yin and Yang.

Many attempted to explain the magnet's mysterious ability to attract iron. For example, some speculated that the surface of a lodestone might be covered with many microscopic hooks, and that of the iron with little rings, so that when the two were put into contact the hooks would thread the rings, thereby holding lodestone and iron together. However, that did not explain the attraction between lodestone and iron, or the attraction or repulsion between two lodestones, when they were *not* in contact. How could lodestone cause iron to move toward it? Before Galileo's time when inertia began to be understood, only living beings and the celestial spheres were thought to be able to move without being pushed. Motion was a property of life. It was natural to suppose, therefore, that magnets were somehow alive, that they possessed "souls", that they were magical.\* Magnets aroused considerable fear and superstition. Ptolemy, in his *Geography*, describes a magnetic island in the South China Sea that attracted and trapped all ships built with iron nails. In the Middle Ages, some sailors feared that ships put together with iron nails would be pulled apart if they ever sailed over submerged magnetic islands. An obvious precaution was to use wood pegs instead of iron nails. Magnets contradicted Aristotle's teaching that action at a distance was impossible. Nonetheless, few wanted to overthrow his entire unified view of the universe just because of a few strange rocks.

The first compasses were bar magnets fastened to small flat pieces of wood. Once set afloat in a bowl of water, the magnet would rotate so that one of its ends—always the same end—pointed more or less towards the north celestial pole and the other end towards the south.\*\* The more familiar pivoted magnetized iron needle came later. Magnets were used in Europe as reliable navigational compasses beginning about the twelfth century. Because the compass could be counted upon to point roughly north, sailors could tell in what direction their ship was sailing, even during the day or during cloudy weather when it was impossible to calculate their ship's position by observing the stars. During the great age of exploration in the fifteenth and sixteenth centuries, European mariners carried compasses all over the world.

It would have been wonderful for the Aristotelian view of the universe had it turned out that the compass always pointed towards the

---

\*Plato called magnets magical. In one of his *Dialogues*, Plato has Socrates praise his friend Ion for his eloquence, saying[1]: "There is a divinity moving you like that contained in the stone which Euripides calls a magnet. . . .The stone not only attracts iron rings, but also imparts to them a similar power of attracting other rings. . .and all of them derive their power of suspension from the original stone. In like manner, the Muse first of all inspires men herself: and from these persons a chain of other persons is suspended who take the inspiration."

\*\*Ancient Chinese writings tell of a "Yellow Emperor" over 4000 years ago, who successfully used a "south-pointing" instrument at the battle of Tzu-Lu to help his army maneuver in a fog. There are reliable reports of a navigational compass in China in the eleventh century A.D.

north celestial pole, the pivot point around which the Greek heavens rotated. Magnets might then have been pieces of celestial material fallen to earth, revealing their celestial origin by their attraction to the one fixed point in the sky. However, as the earth was explored, it became clear that the compass does not point precisely due north, but a few degrees away from the north celestial pole. Moreover, how far away from the celestial pole the compass pointed depended upon longitude. Christopher Columbus had been fascinated by the fact that his compass pointed east of true north in the Mediterranean, to true north near the Azores Islands, and to west of true north as he sailed further west towards America. Sometimes his compass did not behave so well. As he reported later to the King and Queen of Spain:[2]

> "When I sailed from Spain to the West Indies, I found that as soon as I had passed 100 leagues west of the Azores—the needle of the compass, which hitherto had turned to the northeast, turned a full quarter of the wind to the northwest..."

His crew, knowing their lives depended upon the good behavior of a mysterious piece of metal on the ship's bridge, panicked. Columbus managed to calm them by claiming that it was the north star, and not the compass, which was acting up!

In a practical sense, it seemed promising that the compass did *not* point to true north. It had always been easy to determine a ship's latitude at sea by measuring the elevation of the north star above the horizon. However, without good clocks, it was quite difficult to determine a ship's longitude. What had interested Columbus was the almost systematic variation of the compass' deviation from true north as he sailed westward. This suggested that accurate compass maps would help determine longitude.

Early compasses were balanced so that the needle rotated in a horizontal plane. Thus, one could tell how the compass aligned itself with respect to true north, but not whether it would also rotate up and down as well. In 1576, Robert Norman, a retired sailor and a London compass maker, designed a compass that rotated in a vertical plane. He found that his compass rotated until it made a "dip" angle of 71° 51' with respect to the horizontal at London. In other words, it pointed nearly straight down. This suggested that measurements and charts of the magnetic dip angle might also help determine longitude at sea. Norman also hoped to learn whether the mysterious force on compasses was celestial or terrestrial. By measuring the dip angle at many other locations, he concluded that his compass pointed to a spot inside the earth and not to the heavens.

Since compasses did not point exactly towards the north celestial pole, since there were local anomalies where they did not point northward at all,* and, especially, since they dipped towards the earth, it became clear that whatever influenced compasses was terrestrial. But what was it? In 1600, the same year that Johannes Kepler first encountered Tycho Brahe and his data, the physician to Queen Elizabeth I of England published a large treatise on magnetism. In it, William Gilbert (1544–1603) summarized nearly twenty years of observations and experiments on his hobby—magnets. After disposing of old wives' tales, such as that garlic rubbed on lodestone removes a magnet's virtue, but goat's blood brings it back, Gilbert recounted the experiments with a spherical lodestone which taught him how the earth influences compasses.

Like others before him, Gilbert knew that a lodestone caused an iron compass to point toward it. This had led some to guess that there were large lodestones buried in the earth that caused compasses to move. Gilbert went a step further. He speculated that, except for its surface, the earth was made of the same material as lodestones—it was, in effect, a giant magnet. He conceived and carried out an experimental test of this idea. He fashioned a spherical lodestone, which he called a *terrella*—a "little earth." He then placed a small compass near the surface of his terrella and measured its bearings at many points on his model earth. He ruled circles of longitude and latitude on his terrella, and compared his surface compass readings with mariner's charts for the same longitude and latitudes. Their general similarity told him that the earth was indeed like a magnet. His compass dipped towards the terrella just as Robert Norman's had dipped towards the earth, and his charts of dip angle were similar to those of Norman and others. Gilbert believed that magnetic anomalies were due to irregularities in the earth, and not in the heavens. To bolster his argument, he gouged a small depression in his terrella and discovered an artificial anomaly.

Gilbert had one great advantage over mariners: he could see and map the whole earth. Everywhere his compass needle lined up along great circles, and all the great circles passed through two points, the terrella's north and south geomagnetic poles. Comparison of his great circles with sailors' charts told him that the earth's north geomagnetic pole was displaced by a few degrees from its north geographic pole (see Fig. 8-2). In fact, he predicted where the north geomagnetic pole would be, and moreover, that the dip angle would be exactly 90° at the magnetic poles. Later, on a voyage to Northern Canada, the explorer

---
*Columbus had discovered such an anomaly.

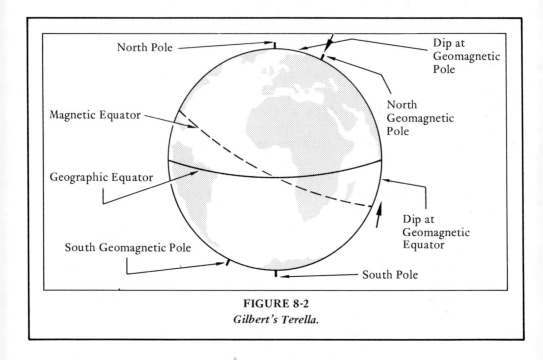

**FIGURE 8-2**
*Gilbert's Terella.*

Henry Hudson found the place where the compass needle points straight down—Gilbert's north geomagnetic pole. Not only did Gilbert's terrella predict compass bearings where no explorer had yet gone, but it also predicted them where no one of his time could ever go: in space. Gilbert found a regular pattern of compass bearings when his compass was removed from the surface of his terrella. Just as his two small magnets, compass and terrella, or indeed any two magnets, interact at a distance, so must also the influence of the giant magnet Earth extend deep into space.

Since the magnet Earth has magnetic poles, ordinary lodestones must also have poles. Gilbert named the end of the lodestone that points toward the earth's north geomagnetic pole, the north pole of the lodestone. The other end he named the south pole. Now it is clear why we labelled the two ends of the magnets in Fig. 8–1 North and South. But, let us not be confused by Gilbert's terminology. He called the north geomagnetic pole north because it is near the north geographical pole. However, magnetically speaking, it is the south pole of the giant magnet Earth.

Compasses rotate until they point north and south because the north end of the compass magnet is attracted to the south end of the magnet Earth, and *vice versa*. Using this idea, we can understand qualitatively why the compass dips. Near the north geomagnetic pole, the north end of the compass magnet is strongly attracted to the relatively nearby south pole of the magnet Earth, while the south end of the compass magnet is far away from the north pole of the magnet Earth and is weakly repelled. The north end of the compass points straight down. Similarly, at the south geomagnetic pole, the south end of the compass magnet points down. Half-way inbetween, the ends of the compass are equidistant from the north and south poles of the earth's magnet, and the compass remains horizontal.

Using Gilbert's terminology, we can describe more precisely how lodestones attract iron. We call permanent magnets those that never lose their magnetism. They are made of the naturally occurring mineral, lodestone, certain iron alloys such as steel, or various chemical compounds containing iron. Pure iron temporarily becomes a magnet when it is brought near a permanent magnet. The magnet "polarizes" the iron. The iron near the magnet's south pole becomes a north pole, and vice versa. Therefore, iron is always attracted to and never repelled from a permanent magnet. For example, suppose a bar of iron is brought near the north pole of a permanent magnet. The end of the iron bar nearest the north pole becomes a south pole, and the opposite end a north pole. The closer south pole of the iron bar is attracted more strongly than the more distant north pole is repelled, and the iron as a whole is attracted to the magnet. If on the other hand, the iron bar is brought near the south pole of a permanent magnet, the iron acquires the opposite polarization, and again is attracted to the magnet. Since iron, once polarized, becomes a magnet, it can polarize and attract other pieces of iron. When pure iron is removed from the permanent magnet polarizing it, it loses its magnetism. But certain mixtures and alloys containing iron can retain their magnetism. These become artificial magnets. The magnetized iron needles in compasses were made in this way. Sailors often kept a lodestone handy to remagnetize their compass when it began to weaken.

Thus, Gilbert did for magnetism what Copernicus had done for astronomy: bring it back down to earth. Gilbert was an ardent Copernican, and one of the first.* Before Copernicus, the cause of the daily rotation of the heavenly bodies had been ascribed to a rotation of the celestial spheres; Copernicus ascribed it to the rotation of the earth. To

---

*A long section of his book is devoted to attacking Copernicus' opponents. Here is a sample of Gilbert's opinion of the idea that the earth is at rest: "Surely that is superstition, a philosophic fable, now believed in only by simpletons and the unlearned; it is beneath derision."

a Copernican, the celestial poles were no longer special points in the sky, and a celestial explanation for the behavior of the compass did not seem natural any longer. It was more natural to say that the compass points to the poles of the earth.

Gilbert suggested that magnetic forces were responsible for the daily spinning of the earth, thereby replacing one of the shakiest of Copernicus' arguments: that it was natural for spherical bodies like the earth to rotate on an axis. Perhaps magnetism also kept the earth from flying apart. Kepler, a contemporary of Gilbert's, was deeply influenced. Gilbert's arguments led him to predict that the sun spins on its axis, radiating a magnetic force that pushes the planets around their orbits. In fact, Kepler tried to show that magnetic forces would move the planets in elliptical orbits. Newton, uncomfortable with the "occult" nature of gravity—that bodies attract each other when there is nothing between them—at least had one example in Gilbert's magnetism to reassure him that action at a distance was possible.

## ELECTRICITY

*Electron* is the Greek word for amber—a pale yellow or brown resin solidified from the remains of fossil plants. If one rubs amber with fur or a wool cloth, it will attract nearby pieces of dust, paper, or other light objects for some time afterwards. To the earliest Greeks, its ability to cause motion suggested that amber, like lodestone, possessed a living spirit. Later, amber puzzled those Aristotelians who disbelieved in action at a distance. This affront to the basic principles of physics was not systematically investigated until two thousand years after the Greeks. Once again, experiments by William Gilbert began the science of electrical (amberlike) attractions.

Would rubbing amber turn it into a magnet? Evidently not, for Gilbert found that whichever end of the compass needle was nearest the rubbed amber was attracted towards it. The compass acted no differently than light pieces of paper in the presence of electrified amber. There was no preferred attraction for north and south poles, no complicated pattern of dip angles. Moreover, the "compass" did not even have to be made of magnetized iron; it could be made of "any sort of metal" and show the same effect. Furthermore, magnets could not attract paper and other light objects, as amber did. How many other substances behaved like amber? Gilbert found that many, but not all, exhibited an amber-like attraction when rubbed. He called those that did attract his compass *electrics* (amberlike). Glass and sulfur are two examples. The other he called non-electrics. By extensive experiment he showed the metals were never electric.

Several branches of physics began with the systematization of primitive observations requiring no special instrumentation: astronomy, with catalogs of naked-eye stellar and planetary observations; optics, with vision, reflection, and refraction; and magnetism, with lodestones and compass charts. However, there are no primitive electrical observations. To make an electrical effect, one must perform an experiment. Even the amber found in nature must be rubbed before it exhibits electrical attraction. Before cataloguing of empirical knowledge could begin, the ancient—and particularly Aristotelian—fear that experiments so changed the state of nature that their results were untrustworthy had to be overcome. Perhaps this is why the science of electricity had to wait for William Gilbert; for he, even ahead of Galileo, relied upon experiment to establish physical law. Since he showed that many substances exhibited electrical attraction, electricity, once a curiosity, became sufficiently general to be worth pursuing. It became scientifically respectable.

The other "electrics" Gilbert had found were not as good as amber. They had to be rubbed harder to produce the same or smaller attraction. Sixty years after the publication of Gilbert's work, in the same year, 1660, that Isaac Newton began his studies at Cambridge, Otto von Guericke, the burgomaster of the German city of Magdeburg, constructed a powerful electrical generator. He poured molten sulfur, one of Gilbert's "electrics," into a glass globe, let it solidify, and then broke the glass. The ball of sulfur, mounted on an axle, could be rotated by means of crank. The rotating ball was electrified by rubbing it with a cloth. In later versions of von Guericke's machine, the ball was rotated even faster by a system of pulleys. Von Guericke's machine was soon copied and improved upon elsewhere.

Electrical generators, like von Guericke's, became popular drawing room entertainments. Audiences were astounded by the spectacular maelstrom of feathers, paper, grains of wheat and whatnot swirling about the spinning sulfur sphere. The air was filled with sparks and the smell of ozone, and those approaching the sulfur globe felt their skin tingle. These electrical machines also permitted serious scientific study to go forward, too. Von Guericke, himself, noted that Gilbert had not been completely right about electrical attraction. Objects were attracted to the electrified sulfur only until they came in contact with it, and then they were immediately repelled!*

Another seventy years passed until the next major discovery about electricity. In 1729, Stephen Gray, a poor retired Englishman, transmitted electrical effects. Gray generated his electricity, not by a machine like von Guericke's, but by rubbing a glass tube into which he had inserted corks at each end to keep the dust out. He was fascinated to note that the corks, which had not been rubbed, also exerted an electrical attraction. Evidently, whatever caused electrical attraction had moved from the glass, where is was generated, to the cork. He wondered how far the electricity could be conducted. He inserted long metal rods into the cork, and found that the ends of the rods exerted a larger attraction than the cork. Finally, he connected the cork to a spool of wire, and succeeded eventually in showing that whatever caused the attraction could be conducted through the wire for a distance of at least 765 feet.

Gray found that if his wire ever touched the ground, there was no effect. Evidently, the electricity he generated just flowed through the wire into the ground and was lost. If he suspended his wire above the ground with metal wires, again, there was no effect. The electricity flowed through the metal wires and again was lost to the ground. But if he suspended it with thin silk threads, he could transmit the electrical effect through the wire. Silk, one of Gilbert's "electrics," evidently did not permit electricity to pass through it. Thus, metals were conductors and "electrics" were non-conductors of electricity.

Gray wondered whether Gilbert's division of materials into those that could be electrified by rubbing ("electrics") and those that could not ("nonelectrics") was the fundamental one. After all, Gilbert had not known that electricity could move. Gilbert's non-electrics were Gray's conductors, and Gilbert's "electrics" were his non-conductors. The real difference between materials was in their ability to conduct electricity. But why had Gilbert been unable to electrify the conducting metals? Gray speculated that the electricity generated by rubbing had found a conducting path to the ground. Gray, therefore, insulated his conductors from the ground by mounting them on non-conductors. When rubbed, they too showed electrical attraction. Thus, it appeared that whatever caused electrical attraction or repulsion was present in *all* matter, that it could be called forth by friction, and that it was free to move in conductors but it remained in place in non-conductors.

---

*About the same time, Newton came close to this discovery. In 1675, he reported the following experiment. Two glass plates were separated by pieces of wood. Between the glass plates, Newton placed little pieces of paper. When he rubbed the glass (one of Gilbert's electrics), he noticed that some of the pieces of paper were repelled to the opposite glass—the one that was not rubbed—rather than being attracted. (The important point for us to note is this experiment was performed with two electrics (glass and paper) rather than with one electric and a metal compass needle, as were Gilbert's experiments.) Other than inducing other members of the Royal Society to repeat this curious experiment, it appears Newton did not pursue this further.

Four years later, Charles DuFay, in Paris, confirmed and greatly extended Stephen Gray's experiments. Moreover, Dufay extended William Gilbert's original results, performing many experiments whose importance Gilbert had missed. For instance, when Dufay rubbed two pieces of amber with fur, the pieces of amber repelled each other—just like the north poles of two magnets. Gilbert had shown that glass rubbed with silk had attracted his compass. When Dufay rubbed two pieces of glass they repelled each other—just like two pieces of rubbed amber. But rubbed amber and rubbed glass attracted each other.

Since electricity moved freely through conductors, such as metals, Dufay began to think of it as a kind of fluid found inside matter. However, two different fluids had to be present to account for both electrical attraction and repulsion. He called these two fluids *resinous* (or amberlike) and *vitreous* (or glasslike). Resinous repelled resinous, and vitreous repelled vitreous, but resinous and vitreous attracted each other. Likes repelled and opposites attracted. These electrical fluids were present in all matter, even conductors. Ordinary matter contained both fluids in equal amounts, so the result was neither attraction nor repulsion. However, when amber was rubbed with fur, the amber gained a slight excess of resinous fluid; and the fur, a slight excess of vitreous. Two pieces of rubbed amber, each with an excess of resinous fluid, would repel each other. Similarly, glass rubbed with silk acquired an excess of vitreous fluid, and would repel another piece of rubbed glass. Rubbed amber and glass would attract each other.

Other gadgets were added to the repertoire of eighteenth century lecturers on electricity. Those who made the biggest spark got the largest audiences. But not until Faraday and Maxwell in the nineteenth century did intellectual giants of the stature of Copernicus and Kepler bother with electricity and magnetism. And, a Galileo or Newton did not reappear until the twentieth century when a gradual accumulation of contradictions and puzzles forced a revolutionary modification of the ideas about space and time that had dominated science since Newton, as we shall see when we study the work of Einstein.

One person stands out in the study of electricity in the eighteenth century. His name may surprise you: Benjamin Franklin (1707-1790). Franklin, better known as one of the leading statesmen of the American Revolution and as an author of homey aphorisms, became so fascinated with electricity that in 1748 he sold his newspaper and publishing business to devote all his time to electrical experiments. Nearly every American child knows Ben Franklin flew a kite into a thunderstorm to demonstrate that lightning flashes were just like electrical sparks (see Appendix). Fewer people know that Franklin also made a very important contribution to the theory of electricity. Dufay's theory had not been very quantitative. For example, it did not specify how much resinous fluid remained on the amber and how much vitreous fluid went to the fur. Franklin concluded that the excess of resinous fluid on the amber exactly equalled the excess of vitreous fluid on the fur. This meant that he could do away with Dufay's two fluids, and replace it with one fluid, which he called electrical charge. All bodies contain a certain natural reservoir of charge; in this state they exert neither attraction nor repulsion. However, when two dissimilar substances are rubbed together, some charge flows from one to the other. The one with an excess of charge becomes positively charged; the one with a deficiency of charge is negatively charged. Positively charged bodies repel each other, as do negatively charged bodies, but positive and negatively charged bodies attract each other.

We still use Franklin's names for positive and negative charge today.* But Franklin did far more than replace Dufay's complicated names for the electrical fluids with simpler ones. His single fluid theory stands for an important quantitative principle—the conservation of charge. In other words, charge never disappears. If glass rubbed with silk acquires a positive charge $+Q$, then the silk loses the same amount so that its charge is $-Q$. Since amber is attracted to glass, we infer that glass is charged negatively when rubbed by fur, which is then charged positively. When amber is rubbed with fur, the fur gains exactly as much positive charge as the amber loses. The total charge is still zero, since $Q + (-Q) = 0$.

We can explain several puzzles using the idea of electric charge, together with the general idea that the nearer two charges are to each other, the stronger the electric force between them. For example, why was the nearer end of William Gilbert's metal compass always attracted and never repelled? Why was the nearer end attracted to both rubbed amber and glass, though in Franklin's terminology, the glass was charged positive and the amber negative? Let us refer to the top of Fig. 8-3, where we show the amber, charged negative, and the metal compass that is grounded, which would happen if Gilbert held the compass in his hand. (The human body is a relatively good conductor.) The

---

*Franklin's choice of positive and negative was completely arbitrary. He could have named the two kinds of electric charge the other way around. From a modern point of view, he made an inconvenient choice. Today, we know that matter is made up of atoms with heavy, positively charged nuclei, surrounded by light, negatively charged electrons. Ordinarily, the electrons are tightly attached to the atoms. Those near the surface of solids can be rubbed off by friction, leaving a deficit on one side and a surplus on the other. Which way the electrons are likely to move depends on the intricate details of the forces that hold them to their atoms. In metals, some electrons are not firmly attached to the atoms but are free to wander. Therefore, it is hard to keep metals charged, unless one is careful to avoid contact with another conductor (including moist air). Electrons move around freely inside a conductor. If there is an excess of electrons, they repel each other and move to the surface of the conductor.

amber attracts a positive charge that flows from the ground to the near tip of the compass. The force of attraction then pulls the near tip of the compass towards the amber. On the other hand, positively charged silk induces a negative charge on the compass tip, again leading to a force of attraction. Even if Gilbert had insulated his compass from ground, he would have obtained the same result, for then charge would have flowed from one end of the compass to the other, having one end charged positive and the other negative, as in the right portion of Fig. 8-3. The end nearest the amber or glass would always have a charge

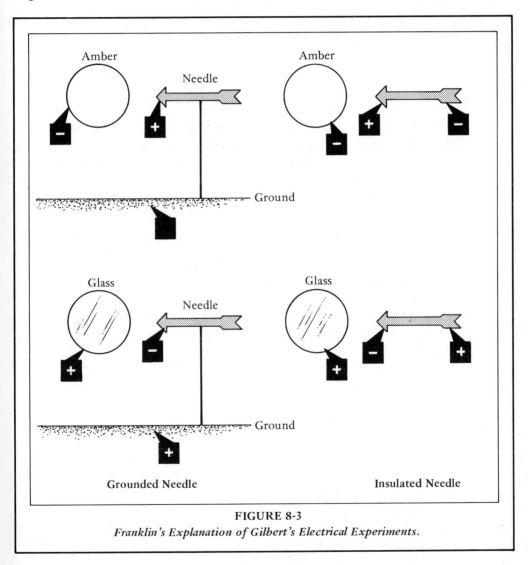

**FIGURE 8-3**
*Franklin's Explanation of Gilbert's Electrical Experiments.*

opposite to that of the amber or glass. In this case, the attractive force would be stronger than the repulsive force, and the compass would point to the charged body just as a magnetic compass points to the pole of a magnet.

Why did von Guericke's sulfur ball first attract and then repel small pieces of paper? The charged ball induced a charge of the opposite sign on the near end of the paper, which was attracted towards the ball. When the paper touched the sulfur ball, some of the charge on the sulfur flowed to the paper. The ball and the paper then had the same sign charge, and the paper was repelled. When the paper hit the ground, it lost its charge to the ground, and it could be attracted again by the sulfur ball.

These simple examples underline the usefulness of the abstract concept of electric charge. The model of charge as a fluid that moves freely through conductors and can be fixed on insulators conveniently summarized the results of many experiments. However, no one in the eighteenth century had ever "seen" electric charge nor had ever isolated any property of charge other than as a source of attractive or repulsive forces. It was not even clear what other properties of real fluids should be ascribed to charge. Real fluids have weight; does charge? Attempts to weigh charge by comparing the weight of amber or other substances before and after they were rubbed proved fruitless. Real fluids move with measurable speed; does charge? People tried to measure the speed with which charge moves through conductors and found it too fast to be detectable. These and other problems with the fluid model of charge remained unresolved. Today, one man, Charles Coulomb (1736–1806), is given most of the credit for recognizing the true significance of charge: it could be made into a quantitative measure of the electrical forces charged bodies exert upon one another. He put electricity into a Newtonian framework.

### COULOMB'S LAW

Coulomb was born in 1736, three years after DuFay proposed the idea of electrical charge, in Montpellier, the son of a minor official in the French government. An early interest in mathematics led him, with his father's encouragement, to apply to the school of military engineering. Coulomb's penchant for applying mathematics to engineering problems failed to impress his severely practical teachers, and he graduated without distinction. As a first lieutenant in the engineering corps, he was immediately posted to Brest where he could have anticipated only a routine life. But then, chance launched him upon a remarkably varied and diverse career. A more experienced engineer, about to depart on a major expedition to rebuild a fortress on the island of Martinique in the West Indies, fell ill and Coulomb took his place at the last moment. Martinique was a dangerous assignment; many engineers had already died there of disease. Coulomb's skill as an engineer soon propelled him to the leadership of the entire project. Coulomb's eight years in Martinique gained him more experience with practical engineering problems than many an older man had.

Coulomb found time to ponder the theory behind many of the problems he encountered in Martinique, and outlined the first of the many papers that would establish his reputation as the leading engineer of his age. All his life he remained interested in how teams of men worked; and later he composed several mathematical papers on the optimum use of labor. His works on friction and on the ability of soil to withstand stresses stood for more than a century. Upon his return to France, Coulomb occupied a number of minor military posts; at each he found something to whet his theoretical appetite, and from each he submitted a stream of mathematical memoirs to the French Academy of Science. Gradually, his interest turned towards basic physics.

Having dealt with very large forces of nature in his engineering life, he began to wonder how the small forces exerted by charges and magnets could be measured in the laboratory so that the link between electricity, magnetism, and Newtonian theory could be forged. To do so, he constructed many different torsion balances, similar in principle to those designed independently by Henry Cavendish to measure the force of gravity.

Coulomb first applied his torsion balance to measuring the earth's magnetism. Small variations in the earth's magnetism that depended upon the time of day had recently been discovered. In 1777, the French Academy offered a prize for the design of a more sensitive compass. Coulomb solved the problem and won the prize by suspending a magnetized iron needle on a long thin thread to make a torsion compass, which easily detected the daily magnetic variation. Subsequent to his memoir on the torsion compass, Coulomb submitted seven more concerned with fundamental aspects of electricity and magnetism over the next fifteen years. He was granted a permanent post in Paris by his military superiors, so he could pursue the busy life of research and public service that was typical of a full member of the French Academy of Sciences.

In 1780, a large scale improved torsion compass had been installed in the basement of the Paris Observatory to make regular magnetic observations. It immediately ran into difficulties because it was *too* sensitive. Whenever anyone touched the instrument's eyepiece to take a reading the compass needle moved. After some thought, Coulomb concluded that the astronomers had frictionally charged themselves, and subsequently, the instrument in walking over the floor, and that the resulting electrical forces deflected the metal compass.

Coulomb, of course, had already been thinking of electrical forces. In 1785, he was ready to present his so-called first Memoir on electricity and magnetism, which described measurements of the electrostatic force law using the torsion balance. His apparatus consisted of two glass tubes in which he enclosed his torsion balance. The tubes isolated his system electrically and from air currents. Inside, a small non-conducting rod was suspended from a silk thread. At one end of the rod was a small pith ball and at the other, a counterweight. Coulomb chose to do his experiments with pith, a material made from the stems of plants, because it was easily charged and was so light that it could easily be deflected by weak forces. Coulomb also inserted a conducting rod with another pith ball at one end into his apparatus. He arranged it so that when the silk thread had no twist, the two pith balls were just in contact. He then charged the end of the conductor protruding outside his apparatus with an electrostatic machine similar in principle to von Guericke's. The charge then flowed to the two pith balls, which he reasoned—since they were in contact and the same size—gained the same amount of charge. Since they both had charges of the same sign, their repulsive force pushed the suspended pith ball away. This ball moved until the silk thread's twist exactly balanced the repulsion. Coulomb recorded the angle through which the string was twisted at the position of force balance, and the distance between the two pith balls. He then twisted the thread, forcing the two balls closer together. How much he twisted was a measure of the force the thread exerted on the suspended pith ball. He measured the distance between the two balls again. Repeating this procedure several times, Coulomb found that the strength of the repulsive force between two similarly charged pith balls decreased as the square of the distance between the centers of the balls. The electrical force had the same dependence upon distance as Newton had found for the force of gravity! Coulomb encountered much more difficulty with the attractive force between two oppositely charged bodies, for the two quickly stuck together. After two more years' work, Coulomb eventually overcame this problem; and in his second Memoir in 1787, he proclaimed a $1/R^2$ force law for both attraction and repulsion. Today, we write Coulomb's law in the form:

$$F = k \frac{q_1 q_2}{R^2}$$

The number $k$ is a universal number like Newton's constant $G$. Its value, of course, depends on the units we choose to measure electric charge in; one common but not always practical choice is to *define* the unit of charge so the $k = 1$. Notice an important property of Coulomb's law. If the charges $q_1$ and $q_2$ have the same sign, the force pushes the charged bodies apart: Like charges repel. But, if $q_1$ and $q_2$ have opposite signs, one object is positively charged and one is negatively charged, the product $q_1 q_2$ is negative. How do we interpret a negative repulsive force? As an attractive force! Opposite charges attract. Both rules are combined in the same law.*

While Coulomb never completely tested that the force was proportional to $q_1 q_2$, he never seems to have doubted it. Earlier he had separately shown that the size of the force is directly proportional to each of the charges. When he charged two identical pith balls in contact, each received half the total charge. If then he took one of the balls and touched another uncharged pith ball, each received half the charge, or one-quarter the original total. The charge was divided again. He then measured the force between the balls with one-half and one-quarter the original charge and found that it was one-half that between the balls with half the original charge each. With this work the meaning of charge became clear. Charge, whatever else it was, was a measurable quantity that determined the size and sign of the electrical *forces* between bodies.

The similarity between Newton's Law of Universal Gravitation and Coulomb's Law is striking. Both laws say that the force is proportional to a product; of charges in this case, and of masses in the gravitational case. More importantly, both laws are inverse-square laws. The force between two objects decreases as the square of the separation between them. The only two fundamental forces we know much about, gravity and electricity, are inverse-square forces.

Evidently, electric (and magnetic) forces are extremely strong, compared to gravitational forces. Simply consider that a small charged ball or rod can pick up an object like another charged ball or a piece of paper against the gravitational attraction of the whole earth pulling

---
*Notice that the masses of the two charged objects do not enter into Coulomb's law. But, if you want to find out how the second object moves under the influence of the electric force of the first, you have to use Newton's second law of motion, so the *effects* of the Coulomb force do depend on the masses of the charged objects. Notice, also, because of the symmetry between the two charges in Coulomb's law, Newton's third law is satisfied.

in the opposite direction! We do not usually encounter electric forces because of the tendency of positive and negative electric charges to cancel exactly. On the large scale of the universe, long-range electric forces do not seem to exist—the positive and negative charges cancel very closely, while gravitational forces all add up and so extend over great distances. Since there is no gravitational analog of negative charge, there can be no cancellation.

In addition to research, Coulomb carried out the many other duties of a busy Academy member. After the French Revolution in 1789, Coulomb retired to his country home. When he died in 1806, the physicist Biot eulogized him as the one who transformed electricity and magnetism from a qualitative science to a quantitative science whose precision began to rival that of Newtonian mechanics.

## MOVING ELECTRICITY

When a conductor is electrified, its charge spreads very rapidly over its surface and then sits still. Up to the end of the eighteenth century, people had succeeded in studying only electricity at rest—static electricity. Yet, buried in the concepts of charge and conductor was the notion that charge can move. We mentioned earlier that people who had tried to measure the speed with which charge moved, by charging one end of a very long conducting wire, failed because the charged moved too fast. Nonetheless, it was natural to ask whether a moving charge produced a force different from one at rest. Before this question could be answered, it was necessary to know how to create steadily moving charges—an electrical current.

The fluid theories of electricity inspired an Italian professor of the University of Pavia, Allessandro Volta, to make the first electric current. Since each substance had its own particular reaction to the generation of electricity by friction, he thought that if he put two different conducting materials together, perhaps electricity might flow from one to the other, like water flowing over a dam. This idea guided him to the invention of the first electrical battery, which he announced in 1800. A sketch of his battery is shown in Fig. 8-4. It consisted of a zinc bar and a copper bar inserted in a pail of salt water. Volta knew that salt water was a conductor. When he touched one of the metal bars to a device that measured electric charge, he found it became very slightly charged. The charge was very weak, much weaker than he could obtain with a frictional generator. However, once the charge was removed, it was immediately replaced at the end of the metal bar. A current had flowed to replace the charge taken away. By repeating the procedure many times, he accumulated a considerable charge, and each time the battery replaced the charge taken away. It was not long

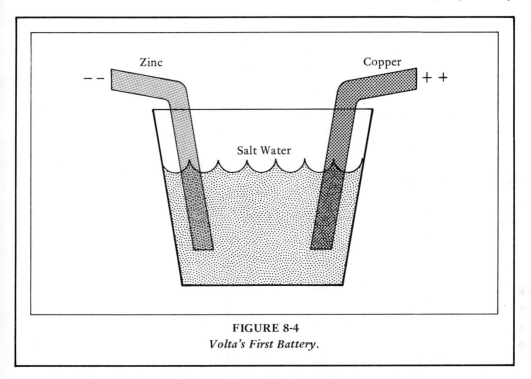

FIGURE 8-4
*Volta's First Battery.*

before he realized that by connecting the two metal bars with a conducting wire, a continuous current could be made to flow through the wire.

Excited by his invention, Volta did not puzzle much longer over how it worked, but devoted the next few years to making bigger and better batteries. He experimented with different metals and different solutions. He found that by connecting them in series, with the copper ball in one such cell connected electrically to the zinc bar in a second cell, the copper in the second cell connected to the zinc one in a third cell, and so on, he could produce more powerful currents. He made his batteries more compact. He found that he could produce a current with a stack of plates of zinc and copper, or zinc and silver, with a layer of cardboard soaked in salt water between every other plate. Soon, Voltaic cells—batteries—became standard equipment in every physics laboratory.* Today, we celebrate Volta's name through the name of the electrical unit, *volt*, a measure of the "electro-motive force" that drives current through wires. Originally, this electro-motive force—the ability to move charge—was measured by the number of standard Voltaic cells connected together, but now it has a more abstract meaning.

---

*These primitive batteries were called Voltaic piles, since Volta made them by piling up a stack of alternating sheets of zinc, copper, and wet cardboard. The limit to the strength of a Voltaic pile was simply how many of these sheets he could stack up before the whole pile fell over.

How Volta's invention really worked did not become clear until his experiment was run backwards.* Within months of the announcement of Volta's discovery, an English chemist, Humphry Davy, using a Voltaic cell of very great power, passed a big electrical current through water. He found that water was broken down into its elements, hydrogen and oxygen, the hydrogen bubbling up at one of the metal bars and the oxygen at the other. This process is "electrolysis." Davy found that the electrical current could break down other substances. He passed it through what was then called "muriatic" acid, and found that hydrogen bubbled up at one metal bar and a new element, which he called "chlorine," at the other. Because of Davy, muriatic acid was renamed "hydrochloric" acid. Encouraged by his success, Davy decomposed other chemicals by electrolysis, thereby discovering sodium and potassium. If Davy was more interested in what electricity could teach him about chemistry, his protegé, Michael Faraday, of whom we will hear much more later, was concerned with what chemistry could teach about electricity. In 1833, he repeated the electrolysis experiments of Davy and others, this time taking very great care to relate the precise amounts of hydrogen and oxygen evolved to the quantity of current flowing. He found that for each unit of electricity flowing, precisely two volumes of hydrogen and one volume of oxygen were evolved. Faraday's experiment made it clearer than ever before that matter was indeed composed of atoms, that electricity was deeply involved in their structure and in regulating their chemical combination. It was obvious, furthermore, that Volta's battery generated its electricity through chemical interactions.

The battery started a new line of research that yielded rich rewards in chemistry and technology; by 1820, however, it seemed that electrical science had played itself out and was headed for a period of quiescence. Then, an accidental discovery by a Danish physicist started a new avalanche of discovery, which this time revolutionized the understanding of electricity and magnetism.

---

*In the late 1700's, an Italian physician and medical researcher, Luigi Galvani, discovered that when a sharply pointed scalpel was brought near the leg nerve of a recently slaughtered frog, the frog's leg twitched violently and a little spark was seen. Even today we say that someone who is startled into a burst of activity has been "galvanized." Galvani imagined that he had discovered some kind of "animal electricity." Volta was disturbed by such an unmodern idea. Did not the whole success of modern, Newtonian physics depend on a clear separation of physical effects from organic phenomena; on a clear separation between, for example, the attractive force of gravity and the desires of the soul or the muscles of the body? Therefore, Volta set about looking for an inorganic, chemical, origin of Galvani's animal electricity. A certain historical irony is inherent. Recently, we have discovered that Galvani was right after all; our nervous system works by electric currents. Electric charges are isolated by biochemical processes and sent as tiny electric currents down the nerves—acting as wires—to signal the muscles how to behave.

Electricity and magnetism had been put into the same category by the ancient Greeks because they were both examples of action at a distance, and William Gilbert had continued this tradition by publishing his research on both together. To the philosophically minded Hans Christian Oersted (1770-1851), this was no mere historical accident. Electricity and magnetism were manifestations of a deeper unity. According to one account, he felt he had guessed the key to their interrelation: Charges at rest produced an electrical force, and charges in motion a magnetic force. Thus, a compass would be attracted to a wire carrying a current. If the law of attraction to a current was anything like Coulomb's law, the force ought to act along a line joining the wire and compass. The compass should point directly at the wire, just as if the wire were a magnetic pole. So sure of the outcome was he that he did not bother to perform this experiment in advance of a major public lecture he gave on the unity of the forces of nature. After some grandiloquent opening remarks, he plunged into the demonstration. He tried various positions and orientations of the compass without finding the expected result. Since his battery was running low, Oersted dismissed his audience, promising to make his apparatus work for the next lecture. On the next try, with his audience filing out, Oersted finally saw the needle move. Triumphantly, he recalled his audience. It was unimpressed by the minute deflection produced by the all but dead battery. Nonetheless, he had found his effect. He had been fooled by his preconceptions; expecting a force of attraction that would point the compass *at* the wire, he found one that aligned the compass needle *perpendicular* to the wire.

When Oersted repeated his experiment, he found that reversing the direction of the current, by reversing the leads to his battery, reversed the direction in which the north pole of his compass pointed. A compass could detect not only the presence of electrical current but also its direction. Simply stated, Oersted discovered that electric currents can attract and repel the ends of a small magnet, and that the force is perpendicular to the direction in which the electric charges move. He circulated his results in a short pamphlet dated July 21, 1820. Several months later, it arrived in France. François Arago read it aloud to the members of the French Academy on September 11. There was an immediate explosion of activity at the Academy, led by Andre Marie Ampère (1775-1836), who read his own first paper on electromagnetism a mere one week after Arago's report. Nearly all the weekly meetings of the Academy for the next four months were devoted to spirited discussion of electromagnetism. After centuries of speculation about the relation between electricity and magnetism, the laws relating the size and direction of electrical currents to the pattern of magnetic forces in the space around the currents were conceived and written down, in the form still used today, in just a few weeks.

Oersted reported that his compass had pointed not toward the wire, but at right angles both to the wire and to an imaginary line from the center of the compass needle to the wire. Ampère immediately generalized this by measuring where the compass pointed all around the wire. He found that if he moved the needle around the wire, it traced out a little circle, with the current-carrying wire at the center. This is sketched in Fig. 8-5. The wire did not act like a magnetic pole at all: the "compass lines" were circular. Ampère went on to show that if the wire was not straight, but bent into a curve, the compass lines near the wire were still circles with the wire at their centers, as shown in Fig. 8-6. Then Ampère bent the wire into a circle. If a straight wire produced circular compass lines, a circular wire ought to produce one straight compass line passing through the center of the circular wire, as sketched in Fig. 8-7. Very near the wire, the compass lines would be circles centered on the wire. At the middle of the loop, the compass line would be a straight line perpendicular to the plane of the circular wire. Along this center line, the north pole of his compass was attracted towards the current loop when it was above the loop, and the south pole was attracted to the loop when it was below. In other words, the current loop acted as if it had its own poles, its south pole above and its north pole below. A circular current behaves like a magnet! Perhaps the earth was not made entirely of lodestone after all; there might be a big electrical current loop in its interior. If we

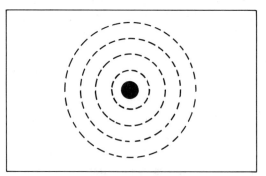

**FIGURE 8-5**
*The Circular Pattern of Magnetic Forces around a Current-Carrying Wire. (This Is an End View. The Central Black Dot Represents a Cross-Section of the Wire.)*

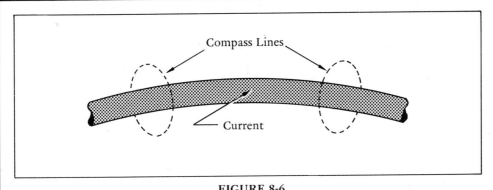

**FIGURE 8-6**
*The Pattern of Magnetic Forces around a Curved Current-Carrying Wire.*

surrounded the current loop of Fig. 8-7 with a sphere to represent the earth or Gilbert's terrella, and then took compass readings outside this sphere, we would find a pattern of "magnetic dip" similar to what Gilbert found. The plane of the current loop would be the geomagnetic equatorial plane of Fig. 8-2.

Ampère also hit upon a clever way to use the same battery to produce an even larger magnetic force. He wound his wire very tightly around a hollow cardboard cylinder so that the same wire made many loops. When a current was passed through this helical wire, which is called a solenoid, a very strong magnetic effect was observed. The compass responded rapidly. A solenoid is sketched in Fig. 8-8. We have drawn very many compass lines close together inside the solenoid to suggest that the magnetic forces are strong there. The pattern of magnetic forces around a solenoid is similar to that around a straight bar magnet.

A permanent magnet could temporarily turn unmagnetized iron into a magnet. Could the magnetism from a current do the same thing? In 1831, a Princeton physicist, Joseph Henry, using this principle, inserted an iron bar inside a solenoid and thereby built an electromagnet that could lift 750 poinds. Only 11 years after Oersted's discovery, magnetism could be created at will. Electromagnets were much more powerful than permanent magnets. By shaping current-carrying wires properly, one could design the pattern of magnetic forces in the space surrounding the wires, as Ampère had demonstrated. By changing the strength of the current with time, one could even change the magnetic forces with time, something that was impossible with permanent magnets. No longer was the study of magnetism limited by what could be dug up out of the ground.

# Magnetism and Electricity

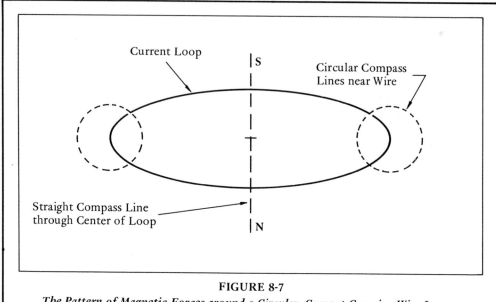

**FIGURE 8-7**
*The Pattern of Magnetic Forces around a Circular, Current-Carrying Wire Loop.*

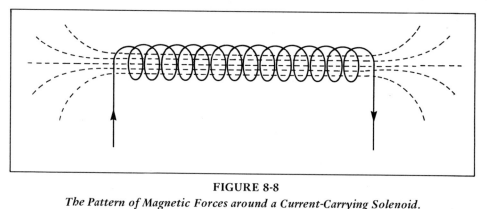

**FIGURE 8-8**
*The Pattern of Magnetic Forces around a Current-Carrying Solenoid.*

According to Newton's third law, if currents exert forces on magnets, a magnet must exert an equal and opposite force on a current. That this was, indeed, the case was shown by Ampère and, independently, by Davy in England. We are close to the discovery of the basic nature of magnetism. If a magnet exerts a force on a current-carrying wire, so does an electromagnet. But an electromagnet is just another current-carrying wire, coiled up. Thus, Ampère realized that two current-carrying wires must exert a force on each other. He placed two wires parallel to each other, arranged so that one was free to move, as in Fig. 8-9. When the current flows in the same direction in both wires, the wires *attract:* when they flow in opposite directions, they *repel.*

What is the nature of the force between two current-carrying wires? It is not the electric force, since the wires are electrically neutral. Since it acts on magnets as well as on currents, it must be the magnetic force. Here one can see the deep connection between magnetism and electricity. For here is a "magnetic" effect without magnets, with only moving electric charges (electric currents). Evidently, moving electric charges have magnetic properties indistinguishable from the properties of a magnet. And magnetic forces, whether they arise from magnets or from electric charges in motion, affect currents just like "permanent" magnets do.

Is a current a magnet? Ampère briefly tried this approach, but found that it led to confusion. For example, the magnetic forces associated with currents could not be associated with north and south poles; he soon realized that it was not useful to imagine a current-carrying wire to be equivalent to some configuration of imaginary magnets. Rather, the magnetic force between two currents should be taken as a fundamental force law, to be added to the list that already contained Newton's law of universal gravitation and Coulomb's law for the force between electric charges.

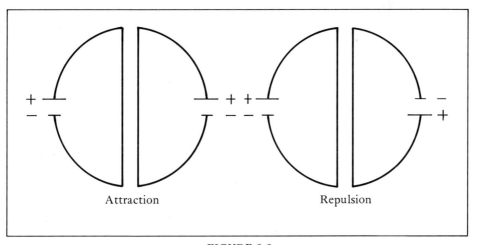

FIGURE 8-9
*The Force between Two Wires Carrying Currents.*

Ampère's law goes on as follows: A numerical value can be assigned to a current, simply the amount of electric charge that passes by a point in the wire per second. The founders of the theory of electric currents have been honored by naming these fundamental units after them. The basic unit of charge is a coulomb. Currents are measured in amperes (amps); one coulomb flowing by per second is a one-ampere current. The letter I is commonly used for current. Ampère's law says that if two wires are parallel, if the current $I_1$ flows through one wire and the current $I_2$ flows through the other, and if the wires are a distance R apart; the force between the wires is given by:

$$F = K \frac{I_1 I_2}{R}$$

The force is perpendicular to two parallel wires, and is *attractive* if the currents have the same sign; i.e., if they are flowing in the same direction. Again, $K$ is a universal constant whose value depends on the units in which current is measured.

If the two wires are not parallel, the situation is a little more complicated. The force depends not only on the sizes of the two currents and the distance between them but also upon the angle the two wires make with each other. Ampère, in his enthusiasm, did not get the general rule quite right. It was correctly stated by Jean Biot and Felix Savart, and bears their names.

The new knowledge presented a great puzzle: What is the relation between the magnetic forces of permanent magnets and the magnetic forces of electric currents? Are there basically *two* kinds of magnetism, one created by currents in a wire and another an inherent property of certain metals like iron? And what about the property of induced magnetization: The fact that if you wrap a wire about a piece of unmagnetized iron and then run a current through the wire, the magnetic field produced will be very much larger than if the iron were not there? Do you need two theories, or possibly three, to account for all these various effects? Why is there not such a thing as magnetic charge?

If the force between currents was to be taken as the fundamental magnetic effect, then permanent magnets and lodestones had to be explained. Ampère suggested that continuous currents circulate inside permanent magnets to create the magnetic forces. In a letter to Ampère, Augustin Fresnel—who had settled beyond a doubt that light was a wave and not a stream of particles—pointed out that if such a current existed in magnets, permanent magnets ought to be hotter than their surroundings, since Voltaic currents heated the wires carrying them. Since Fresnel had found no such effect, he suggested to Ampère that perhaps the molecules of matter itself possessed permanent current loops. So little was known about the molecular constitution of matter, Fresnel argued, that it was conceivable heat would not be generated. Ampère immediately proposed a theory of "electromagnetic molecules," which explained at least qualitatively, all the known phenomena of magnetism in material bodies. Matter, he suggested, is made of molecules, which, whatever their other properties, were microscopic current loops. In ordinary matter, the current loops were oriented randomly so that the net magnetic force averaged to zero, as in Fig. 8-10. In permanent magnets, internal forces within the material itself aligned the molecular currents so that a net magnetic force would be produced, as in Fig. 8-11.

This picture of magnetism explains induced magnetism as follows: When an iron bar is inserted into a coil carrying a current, the magnetic force aligns the molecular current loops, like little compasses. Now, they are all acting together, instead of pointing in random directions. The magnetic forces due to these tiny magnets all add up and

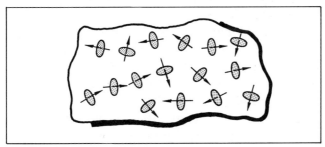

**FIGURE 8-10**
*Randomly Oriented Molecular Currents in Ordinary Matter.*

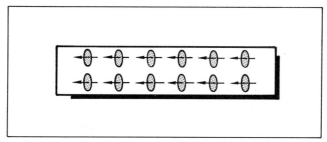

**FIGURE 8-11**
*Molecular Current Loops Aligned in a Permanent Magnet.*

supplement the force of the coil, creating an even more powerful electromagnet (see Fig. 8-12). When the current is turned off, the molecular currents move around, point in random directions again, and no longer have any large scale magnetic effect. But, if the bar is made of certain alloys and compounds of iron, the molecular currents remain lined up, resulting in a permanent magnet. (Lodestone is made of such a substance. It is slowly magnetized over millions of years by the magnetic force of the earth itself.)

It was a beautiful solution. All magnetism was unified into one basic law—Ampère's law for the force between two currents—and the idea of microscopic, molecular current loops. Except for the detailed description of these microscopic currents, it is the explanation accepted today as correct. Why only iron, of all substances, exhibits large magnetic effects of this kind is not completely understood even today. Actually, all substances can be magnetized to a minor extent by very strong magnetic forces; and the metals cobalt and nickel show effects similar to, but not as strong as, iron.

Thus, by the 1820s, much was known about electricity and magnetism. Electric charges exerted electric forces on each other, according to Coulomb's law. Moving charges, a current in a wire, produced and were susceptible to magnetic forces as well. Permanent magnetism resulted from the combined action of small-scale or molecular currents. Ampère developed the mathematical techniques to explain all known effects from Coulomb's law and his own law for the force between currents, as modified by Biot and Savart. However, not all the relationships between electricity and magnetism had been found, and there was no completely comprehensive mathematical theory.

Gilbert, von Guericke, Gray, DuFay, Franklin, Oersted, Volta, Ampère, and many others had each contributed to the rational understanding of electric and magnetic phenomena. Finally, there appeared two men whose achievements begin to compare to those of the men who accomplished the Copernican Revolution: Faraday and Maxwell.

### MICHAEL FARADAY

*Nothing is too wonderful to be true, if it be consistent with the laws of nature, and in such things as these, experiment is the best test of such consistency.*

Michael Faraday (1791-1867)

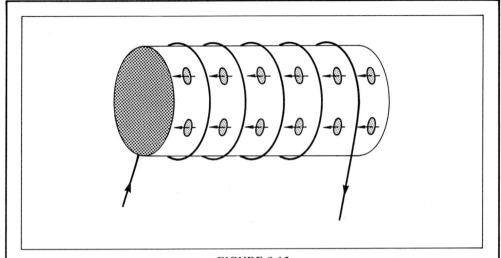

**FIGURE 8-12**
*The Magnetic Force of a Current-Carrying Wire Lines Up the Molecular Currents in a Piece of Iron, Magnetizing It.*

Faraday's life was a classic success story, and deserves some retelling here. His father was a blacksmith in failing health, who could barely provide for his family; at one point, one loaf of bread had to last Michael an entire week. His early education was virtually nonexistent—the bare rudiments of written and spoken English and arithmetic. Thus equipped for life, he was pushed into the world at age 14 to make his way as an apprentice bookbinder. The work was hard and laborious. In his off-hours, the penniless boy did the only thing he could for recreation. "There were plenty of books, and I read them," he said later. At first, his reading was aimless, his mind attracted to whatever came his way; but later articles on electricity in the *Encyclopedia Brittanica* and a popular book on chemistry awakened a curiosity for science that was to consume his time, thought, and personality over the years. Thereupon, he started a relentless program of self-education, more severe and stricter than that of any university, one which was not to be completed for a decade. Goaded by his abysmal ignorance, Faraday forced himself to improve his grammar, to speak and write clearly, and to learn arithmetic, all so that he could further comprehend that mysterious and unattainable science. In 1810, at the age of nineteen, he joined the City Philosophical Society, an informal group of young men similarly bent upon self-improvement. Each week one would lecture on a topic of current scientific interest. Of these young men, Faraday had the furthest to go, but he persevered, taking complete notes and absorbing every lecture systematically. It was not long before he too began giving amateur lectures. Faraday's apprenticeship was completed when he was twenty; this meant he would have to find permanent employment as a bookbinder, an anguishing prospect. In his desperation, he wrote to Sir Joseph Banks, president of the Royal Society, for a job, any job at all, in science, only to receive no answer. But then, fortune smiled. The director of the Royal Institution for Science, Humphry Davy, had hurt his eyes in an explosion in the laboratory and needed a secretary for a few days. Faraday, whose penmanship was by then excellent, got the job. After this brief encounter, Faraday sent Davy detailed notes he had taken of Davy's public lectures on chemistry, with an urgent plea for a permanent job, no matter how menial. At first, Davy could do nothing, but Faraday's chance came a few months later. The boy who washed the bottles at the Royal Institution was fired for fighting. Davy remembered Faraday, and in March 1813, Faraday was rescued. He had his job in science, as a laboratory assistant at the Royal Institution. He would remain there forty-nine years.

The Royal Institution had been founded by an American, Benjamin Thompson, to disseminate useful knowledge to the middle and lower classes.* Its lectures on practical topics were given at night to accommodate the working man. In the aristocratic universities, Faraday would never have prospered, but here was a practicing scientific laboratory and school designed expressly for people like him.

Davy noticed Faraday's extraordinary manipulative skill immediately, and assigned him chemical analyses of ever-increasing complexity. Davy had broadened the Royal Institution's appeal (and financial base) by adding a Friday afternoon lecture series, aimed at the idle young ladies of the aristocracy. Faraday set up the laboratory demonstrations for these lectures. There, nothing escaped him. His diary records complete discussions of the subject matter, of the optimum use of the demonstration equipment, and even analyses of the

---

*Thompson himself was a colorful fellow who made a major scientific discovery. Born in Woburn, Massachusetts, in 1753, he evacuated Boston with the British in 1776. In London, he dabbled in politics, studied the properties of gunpowder, and was elected a fellow of the Royal Society. Knighted by George III, he moved on to Bavaria, where he was named Count Rumford, and for several years was Bavarian minister of war and of the police. He studied the heat generated by the process of boring gun barrels, and discovered a basic relation between the amount of force used in the boring, the time elapsed, and the rise in temperature of the metal of the gun. In this inauspicious way began the incorporation of the study of heat and temperature into the system of Newtonian mechanics. In 1791, Thompson returned to England where he founded the Royal Institution and appointed Sir Humphry Davy its first director.

lecturer's style. Faraday's next chance came when Davy married and took his honeymoon on the continent to visit the leading scientists of the day. He invited Faraday along as his chemical assistant. Davy was lionized by Ampère and Laplace, Biot and Arago, and scientists in Italy and Germany. Faraday quietly listened in on every conversation. And on the long journeys between cities, there was Davy, his facile mind endlessly speculating aloud to the captive and captivated Faraday. The grand tour of the continent was Faraday's university.

Upon his return to England in 1815, Faraday was prepared to be a scientist. Faraday pursued self-advancement as he had self-education. First, he made himself invaluable in the laboratory. Soon, Davy could do nothing in the laboratory without Faraday. Eventually, he began publishing papers under his own name. His first work, following Davy, was in chemistry where he has his own major achievements. He liquified chlorine, and discovered benzene. Trained analytical chemists were rare, and Faraday's skills made him a valuable consultant to industry. The Royal Institution, always financially pressed, lent Faraday out for money.

Faraday's interest in electricity was awakened by Davy's insistence that electrical forces were behind chemical combinations. Furthermore, Oersted's success in converting electrical current into magnetism immediately prompted others to look for the reciprocal effect: the conversion of magnetism into an electrical current. In 1822, Faraday entered into his laboratory notebook the goal, "convert magnetism into electricity," indicating that he too had joined the race. In 1824 and 1825, he undertook an extensive series of unsuccessful experiments in which he moved a permanent magnet through a helical solenoid to see if a current would be generated without benefit of a battery. Characteristically, and correctly, Faraday assumed his difficulty was the familiar one of experimental technique: the effect might have been very weak and detectable only using a better magnet.

In designing his experiment, Faraday was guided by his conception of "lines of force." Heretofore, we have discussed how natural it was to visualize the pattern of magnetic forces surrounding a magnet or a current by joining together with an imaginary line the directions in which a small compass pointed. We have called these "compass lines." Henceforth, we will call them, as Faraday did, "lines of force." The lines of force of a magnet can be seen easily if one sprinkles small iron filings on a paper and holds a magnet underneath the paper. The magnet induces magnetism in the iron filings, which then behave like little compasses pointing out the direction of the lines of force. Faraday extended this idea in a profound way. Not only could lines of force describe the *direction* in which a compass would point, but they could also define the *strength* of the magnetic force. It was only necessary to imagine that where the magnetic force was strong many lines of force were crowded together; and where it was weak the lines of force were spread far apart.

Faraday, almost completely innocent of higher mathematics, could not calculate magnetic forces, as Ampère could. Yet, in his daily work he was faced with designing complex experiments. He had to know where the magnetic force went. For this, he imagined lines of force surrounding all his apparatus; through long experience he developed the uncanny ability to predict how his experiments would behave without doing a single calculation. Ampère, magnetism was a set of equations; to Faraday it was a mental image of lines.

Let us apply Faraday's conception to a solenoidal current. Suppose initially no iron is inside the solenoid. The strength of the magnetic force is characterized by the number of lines of force that thread the ends of the solenoid. When an iron bar is inserted inside, the magnetic force is much more powerful; more lines of force thread the ends of the solenoid. The number of lines of force should be proportional to the strength of the forces in the two cases. In general, the density of the lines of force—the number crossing a given area—is proportional to the strength of the magnetic force.

Faraday's aim was to show that very powerful magnetism—many lines of force crowded together—could induce an electrical current in a wire. He pictured iron as concentrating lines of force. Thus, by bending iron bars into different shapes, he could control where the lines of force went. In 1831, he bent an iron bar into a ring, as shown in Fig. 8-13. He then wound a wire around one side of the iron ring (side A), creating a small solenoid there. When the wire was connected to a battery, current flowed through the solenoid, and magnetic lines of force were created inside the iron ring. Since the iron concentrated the lines of force, most of them would remain in the ring; and very few would escape into the space outside. He then wound another wire around the opposite side of the iron ring (side B) and twisted the ends together. This wire was in no way connected to the battery. He hoped to show that when very many lines of force passed through the solenoid at side B, a current would flow in the secondary circuit. He would detect this current as Oersted had done, with a small compass, placed near the wire of the secondary circuit and far from the iron ring. Here in his own words is what happened:[3]

> I have had an iron ring made 7/8ths of an inch thick and six inches in external diameter. Wound many coils of copper (wire) around. . . . will call

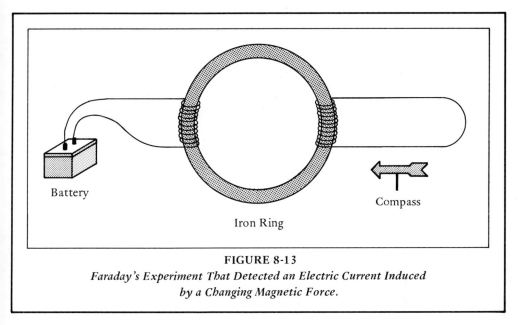

**FIGURE 8-13**
*Faraday's Experiment That Detected an Electric Current Induced by a Changing Magnetic Force.*

this side of the ring A. On the other side, but separated by an interval, was wound wire amounting to about sixty feet in length, the direction being as with the former coils. This side call B. Charged a battery . . . , and connected the extremities [of coil on B] by a copper wire passing to a distance and just over a magnetic needle . . . . then connected ends of A side with battery: immediately a sensible effect on needle.

It oscillated and settled at last in original position. On breaking connection of A side with battery, again a disturbance of needle.

It is a tribute to Faraday's acute powers of observation that he noticed that the little compass was affected, not when the number of lines of force inside the iron ring was large and constant, but that it jiggled ever so slightly only when he connected and disconnected the battery. And it is even more important that he made something of it. For he concluded that only when the number of lines of force was changing with time would a current be induced in circuit B. When he connected the battery, the number of lines of force increased rapidly, and the compass jumped one way; when he disconnected it, the number of lines of force diminished rapidly, and the compass jumped in the opposite direction. The direction of the current in circuit B, therefore, depended on whether the number of lines of force increased or decreased with time.

Later, Faraday showed that the iron ring was not necessary at all. His most striking experiment is sketched in Fig. 8-14. He eliminated the iron core and physically separated the secondary circuit from the

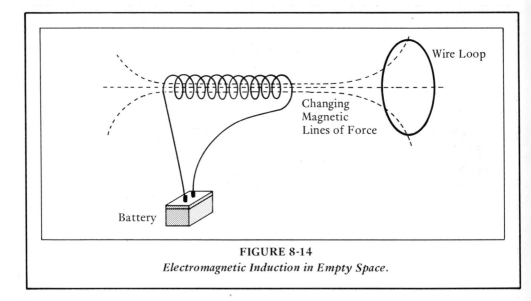

FIGURE 8-14
*Electromagnetic Induction in Empty Space.*

first. Faraday was able to show that any *changing* magnetic force can drive an electric current. For example, you can take a strong permanent magnet, wave it around near a coil of wire, and induce a measurable electric current.

In order to describe this effect quantitatively, Faraday found his concept of lines of force very useful. Recall that he pictured any source of magnetism as being surrounded by continuous lines of magnetic force, whose direction at any point was the one a little compass needle would take at that point. When the number of lines of force passing through a wire loop changes, an electric force is produced and current flows in the wire.

Faraday had such faith in the concept of lines of force that he waved a wire in the air and detected a small, but measurable, current. Why? The current is produced by the magnetism of the earth!

Faraday's will and determination transformed his greatest liabilities into his greatest strengths.* His very lack of formal education led to an indiscriminate storage of facts in his mind; for he was ignorant of academic authority to tell him what was important and what was not. This habitual cast of mind made him an attentive observer of

---

*Here is an enraptured description of Faraday as a lecturer, by one Lady Holland:[3]

First, as he stood at the lecture table, with his voltaic batteries, his electro-magnetic-helix, his large electrical machine, his glass retorts, and all his experimental apparatus about him,—the whole of it being in such perfect order that he could without fail lay his hand upon the right thing at the right moment, and that, if his assistant by any chance made a blunder, he could, without a sign of discomposure, set it right. His instruments were never in his way, and his manipulation never interfered with his discourse. He was completely master of the situation; he had his audience at his command,

those small occurrences in the laboratory that now and again signal an important breakthrough. Since he had to develop the knowledge himself, he became completely self-reliant in the laboratory. Faraday, in 1857, described his life-long mode of working:[4]

> I have never had any student or pupil under me to aid me with assistance; but have always prepared and made my experiments with my own hands, working and thinking at the same time. I do not think I could work in company, or think aloud, or explain my thoughts at the time. Sometimes I and my assistant have been in the laboratory for hours and days together, he preparing some lecture apparatus or cleaning up, and scarcely a word has passed between us.

Faraday learned his first science in a chemistry laboratory. As he increased his chemical knowledge through experimentation, he developed an accurate sense of the complexity of matter and the specific differences between materials, which made him habitually suspicious of casual generalizations—such as the idea that there were two universal electrical fluids unrelated to the specific materials in which they flow, or Ampère's facile electrodynamic molecules, however elegant they might be mathematically. His lifelong incomprehension of higher mathematics made him natively suspicious of it and totally dependent upon experimentation:[5]

> I was never able to make a fact my own without seeing it, and the descriptions of the best works altogether failed to convey to my mind, such knowledge of things as to allow myself to form a judgment upon them. It was so with *new* things. If Grove, or Wheatstone, or Gassiot, or any other told me a new fact and wanted my opinion, either of its value, or the cause, or the evidence it could give in any subject, I could never say anything until I had seen the fact.

The rigorous cycle or experiment, trial and error and proof and counterproof, was a straightjacket upon his theoretical imagination, preventing the transcendent perception of harmonious relationships that moved Kepler, Newton, and Maxwell to the highest creative syntheses.

---

as he had himself and all his belongings; he had nothing to fret him, and he could give his eloquence full sway. It was an irresistible eloquence, which compelled attention and insisted upon sympathy. It waked the young from their visions and the old from their dreams. There was a gleaming in his eyes which no painter could copy and with no poet could describe. Their radiance seemed to send a strange light into the very heart of his congregation; and when he spoke, it was felt that the stir of his voice and the fervour of his words could belong only to the owner of those kindling eyes. His thought was rapid, and made itself a way in new phrases, if it found none ready made,—as the mountaineer cuts steps in the most hazardous ascent with his own axe. His enthusiasm sometimes carried him to the point of ecstasy when he expatiated on the beauty of nature, and when he lifted the veil from her deep mysteries. His body then took motion from his mind; his hair streamed out from his head, his hands were full of nervous action, his light lithe body seemed to quiver with its eager life. His audience took fire with him, and every face was flushed.

Nonetheless, towards the end of his life, as the strain of solitary mental concentration became ever greater and his mental breakdowns ever more frequent, Faraday struggled against his very strengths to generalize upon the insights drawn from a lifetime of experimentation. Nowhere is this clearer than with his lines of force. They were originally a means of getting around his mathematical ignorance, a way of estimating the patterns of magnetic force surrounding the apparatus he created in his laboratory, patterns that he could not calculate. He learned by experimentation how the lines of force would be bent, concentrated, or rarefied in passing through matter. Following his chemical instinct, he originally thought lines of force were the sum total of all the intermolecular forces within matter, and so he treated them as a convenient conceptual tool, without necessary reality. But as the years passed, his experiments led him to the faith that lines of force existed without matter as a kind of strain or tension imposed upon otherwise empty space.

The 1850s found Faraday questioning the basic article of faith of Newtonian mechanics, action at a distance. The problem of gravitation, even, seemed to him ill-posed. Consider a gravitating body, and then introduce a new gravitating body in its neighborhood; Newtonian mechanics states that interaction begins instantaneously; yet how can this happen without something passing between them? The whole problem would be solved, Faraday argued, if you could assume that the lines of the gravitational force from the first body were *already present* in space when the second body was introduced. Einstein's General Theory of Relativity has its roots there.

The ideas of Faraday's old age were treated with gentle derision by his contemporaries. They regarded lines of force as Faraday originally had—as a convenient crutch for a mathematically illiterate man to lean upon. Only a young Scottish aristocrat seemed taken with them. James Clerk Maxwell's excitement is apparent in a letter he wrote to Faraday in 1857.[6]

> Now as far as I know you are the first person in whom the idea of bodies acting at a distance by throwing the surrounding medium into a state of constraint has arisen, as a principle to be actually believed in. We have had streams of hooks and eyes flying around magnets, and even pictures of them so beset, but nothing is clearer than your descriptions of all sources of force keeping up a state of energy in all that surrounds them. . . . You seem to see the lines of force curving round obstacles and driving plumb at conductors and swerving towards certain directions in crystals, and carrying with them everywhere the same amount of attractive power spread wider or denser as the lines widen or contract.

> You have also seen that the great mystery . . . how like bodies attract (by gravitation). . . Here your lines of force can "weave a web across the sky" and lead the stars in their courses without necessarily immediate connection with the objects of their attraction.

Faraday's lines of force would be transmuted by Maxwell's quicksilvery imagination into the greatest theoretical conception of the nineteenth century: the field.

## JAMES CLERK MAXWELL

Maxwell was born in 1831, the year of Faraday's discovery of magnetic induction, into a Scottish aristocratic family. He spent his early years at the family house, "Glenlair," Often in later life, he would return frequently to Glenlair for rest and communion with his beloved Scottish countryside. Many of his greatest papers were written there.

It is said that he was exceptionally inquisitive as a young boy. He showed early signs of intellectual precocity and facility with abstract mathematics. His first paper, on a mathematical topic, was published in the *Proceedings of the Royal Society of Edinburgh* when he was fourteen. He enrolled in Edinburgh University when he was sixteen, where he remained until he was nineteen. Then he went on to Trinity College, Cambridge, then, as now, the most distinguished school for all English scientists. By the time he was twenty-four, he was the professor of physics at the University of Aberdeen in Scotland.

Already his mathematical researches in electricity and magnetism had attracted the attention and warm admiration of the great Faraday. Could it be that Faraday recognized the man who would impart permanent intellectual significance to a lifetime of arduous experimental work, just as Tycho Brahe saw in Kepler the man who would make sense of his observations?

In 1860, Maxwell became professor at King's College, London, where finally he met Faraday and continued his researches into electricity and magnetism. In 1865, he retired to his estate to pursue physics in more solitude than afforded by the bustle and noise of London. In 1871, he became the first Cavendish professor of physics at the new Cavendish Laboratory at Cambridge University. Maxwell died of cancer in 1879, at the early age of 48.

Like Newton, Maxwell made fundamental advances in many branches of mathematical physics. At the age of fifteen, Maxwell submitted his first paper to the Royal Society of London, on some mathematical properties of certain curves. He first achieved fame by deducing the nature of the rings of the planet Saturn. By cleverly combining the laws of gravity with what was known about the non-gravitational properties of materials, he demonstrated that Saturn's rings could not be either solid or fluid, but must consist of a huge number of tiny grains of matter.

Maxwell laid the foundation for the science we call today statistical mechanics. If a gas (like air) is really made up of a huge number of tiny atoms, bouncing around and colliding with one another, what are its large-scale properties? For example, how does it heat up as it is compressed? Why is it that we can produce heat from mechanical or chemical processes; e.g., rubbing two things together, starting a fire, or heating a wire with a battery, but cannot reverse the process and extract useful energy by cooling down the ocean? A large part of the mathematical language for discussing these questions was invented by Maxwell.

Let us describe Maxwell's contribution to the theory of electricity and magnetism. To make Faraday's lines of force meaningful, Maxwell had to find mathematical quantities that described the forces in the space surrounding charges, currents, and magnets. He called them electric and magnetic fields. The fields were a mathematical version of Faraday's lines of force. We begin our discussion of fields by describing Faraday's lines of force for a single, small electrical charge. A charged ball with positive charge $+Q$ is drawn in Fig. 8-15a. Lines of force are

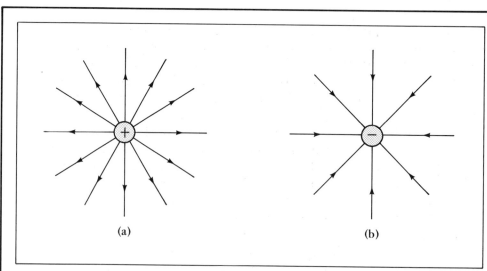

FIGURE 8-15
*The Lines of Force near (a) a Positive Charge and (b) a Negative Charge.*

defined by the direction of the electric force on a small test charge placed nearby. If the test charge is positive (+$q$), it will experience a force pointed along the line joining +$Q$ and +$q$ directed away from the pith ball. The pith ball's lines of force are, therefore, directed radially outward, as shown. According to Faraday's rule, the strength of the electrical force is given by the number of lines of force crossing a unit area. In Fig. 8-15, the lines of force spread out at $R$, the distance between the two charges, increases; and so the force should diminish as $R$ increases. We can even show that Faraday's picture leads to Colomb's law (see Appendix). Faraday's lines of force are equivalent to Coulomb's law, but are useful in more complicated situations than one small charge.

Maxwell's electric field is a mathematical description of electric lines of force. It is a quantity called $E$, which has a value at each point in space and which measures the strength and direction of the electrical force on a test charge at each point. $E$ does not depend on the nature of the test charge, which we imagine to be so small that we can neglect its own electrical forces. Thus, Maxwell defined $E$ to be the force $F$ divided by the test charge $q$, or $E = F/q$. For the pith ball of Fig. 8-15a, according to Coulomb's law, the force $F$ at a distance $R$ is:

$$F = \frac{Qq}{R^2}$$

So, the electric field $E = F/q$ has a magnitude:

$$E = \frac{Q}{R^2}$$

its direction is the direction of the force, radially outward. Faraday's lines of force, shown in Fig. 8-15, can be thought of as a "picture" of the field of the charges.

Suppose the charge on the pith ball were negative, $-Q$. The force on the positive test charge, +$q$, would now be attractive; and the lines of force would point towards the pith ball, as in Fig. 8-15b. The force would be negative:

$$F = \frac{-Qq}{R^2}$$

and the electric field would have the opposite sign, too:

$$E = \frac{-Q}{R^2}$$

All this is straightforward. But now we can compute more complex electric fields. Suppose there were two positive charges, each +$Q$. What is the electric field now? We simply add up the forces on a small test charge $+q$ from both charges. If $+q$ is near either of the charges $+Q$, it will be strongly repelled from one and hardly feel the force of the other. Therefore, near either of the charges, the lines of force will still be directed radially outward. Halfway between the two charges, a test charge $+q$ would feel the force of both charges equally. There is no net force, and the density of lines of force must be zero. The rest of the pattern of lines of force is sketched in Fig. 8-16.

We sketch the lines of force for a positive charge $+Q$ and a negative charge $-Q$ in Fig. 8-17. A test charge $+q$ is always repelled by $+Q$ and attracted to $-Q$. Near $+Q$, the lines of force are radially outward; near $-Q$, radially inward. Note the similarity of the electric lines of force in Fig. 8-17 to the magnetic lines of force surrounding a bar magnet. The charge $+Q$ acts like a south electric pole, and $-Q$ like a north electric pole, as far as the lines of force are concerned. It is obvious that a pole is just a point where lines of force enter or leave. We say that an electric pole (a charge) is a "source" for the electric field—it creates it.

Magnetic fields also exist. There is an important difference between the sources of electric and magnetic fields. Whereas the electric field can have a single pole—a point charge as in Fig. 8-15, the magnetic field cannot. The most elementary source for the magnetic field has two poles, North and South. We call this a magnetic dipole. As Ampère

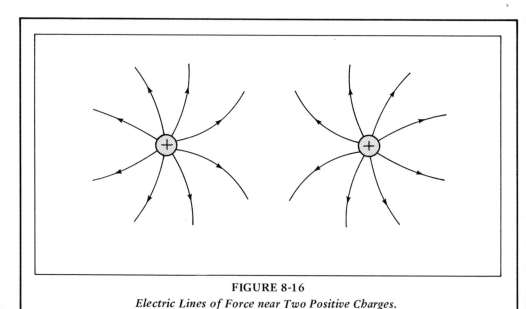

FIGURE 8-16
*Electric Lines of Force near Two Positive Charges.*

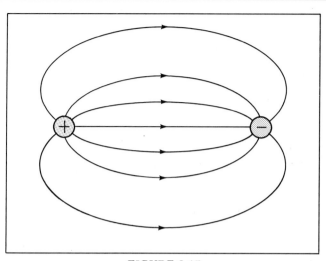

**FIGURE 8-17**
*Electric Lines of Force near One Positive Charge and One Negative Charge.*

showed, a dipole is equivalent to a current loop. If single magnetic poles existed, we could define the magnetic field in the same way as Maxwell did the electric field. We could place a small test pole near a bar magnet, measure its acceleration, compute the strength and direction of the force, and then divide by the "magnetic charge" to get the magnetic field. But since magnetic poles do not exist singly, we do the next best thing: we insert a test dipole (a small compass or current loop) and measure the force on it at every point in space. Computing the magnetic field in this way is a bit more involved than for the electric field, but the principle is identical. The "compass lines" of Figs. 8-5 through 8-8 can now be thought of as magnetic field lines.

There is another completely equivalent way of defining the magnetic field. From the work of Ampère and of Biot and Savart, we know that two currents exert a force on each other. In the language of Maxwell, we would say that the first current is a source for a magnetic field in the space around it. We then could measure the force on a small test current inserted into the magnetic field. What is a test current? A moving test charge. Thus, the magnetic field could also be defined by measuring the force on a moving charge.

Once we know the size and direction of the magnetic or electric field in space, we need not be concerned about what the source of that field is. For example, enclose a bar magnet and a current loop inside two boxes. If the two make identical magnetic fields, a compass would feel the same forces outside the two boxes. There is no way to tell, by studying the magnetic properties of the space outside the boxes, whether the magnetic field is caused by a current or a permanent magnet.

Thus, Maxwell imagined that at every point of space there was an electric and magnetic field. The strength of the field was a number that described the strength of the electric or magnetic force on an object placed at each point. The fields at every point of space also had a direction, which was the direction of the electric or magnetic force on a charge or magnet. The fields, of course, change with time as their sources change. Like Faraday's lines of force, the fields are always there, whether or not something is there for them to exert a force on.

We emphasize that the strength of an electric field $E$ at any point is the electrical force on a test charge $q$ due to all the other charges in the world, divided by the test charge $q$. This number $E$ does not depend on $q$ because in accordance with Coulomb's law the strength of the force on the test charge from every other charge is proportional to $q$, so that $q$ cancels out when the net force is divided by $q$ to get $E$.

An analogy with the more familiar force of gravity is useful. According to Newton's Law of Universal Gravitation, the force a mass $M$ exerts on a small "test mass" at a distance $d$ away is $F = Mm/d^2$. We can define a "gravitational field" at the point where the test mass is by dividing the force by $m$: *gravitational field* $= F/m$. Notice that the field is independent of the size of the test mass, $m$. In this case, it is just the acceleration of any object at that point, which we know is independent of an object's mass. Indeed, the idea of a gravitational field can be just as useful as the idea of electric and magnetic fields.

Now, let us sum up in the language of fields what was known in Maxwell's time about electricity and magnetism. Forces that accelerate electric charges can be described by electric fields. A free charged particle would be constantly accelerated in a constant electric field. When an electric field is applied along a wire, the electrons move at constant speed, since wire resists the electrons' motion. (As in the case of raindrops falling through the air, we are here in Aristotle's limit of motion, which Newton showed meant constant speed under constant force.) Electric charges produce electric fields, according to Coulomb's inverse-square law.

A moving charge is subject to magnetic forces, as well as to electric forces, according to the laws of Oersted and Ampère. A free moving charge is simply deflected from a straight-line path by a magnet. If charges move in a wire (an electric current), the wire is deflected.

Magnetic fields are created by moving charges. The simplest example is an electric current that is surrounded by a magnetic field. The microscopic currents in atoms each create a magnetic field. In most matter, these fields cancel out, since the microscopic current loops are randomly oriented. In a permanent magnet, these atomic currents are all lined up, and a large-scale magnetic field results. A freely moving beam of electric charges (as in an electron tube) also has a magnetic field around it.

Finally, Faraday discovered magnetic *induction,* i.e., that changing magnetic fields produce electric fields. What Faraday actually observed was that changing magnetic fields produced electric currents. When a magnet is moved through a wire or when a nearby current is turned on and off, a current is observed in the wire. Since currents can be caused only by electric fields, the conclusion is that changing magnetic fields "induce" electric fields.

Such, in outline, was the knowledge of electricity and magnetism at the time of the birth of Maxwell. Maxwell's great achievement was to synthesize all this knowledge about electricity and magnetism into a simple set of laws, known today as Maxwell's equations. Instead of Faraday's imprecise ideas about lines of force, Maxwell imagined that at any instant each point of space was characterized by a few numbers: the electric field strength, and its direction; the magnetic field strength, and its direction; and the density of electric charge and the speed and direction in which it moves. The relation among the charges and currents, on the one hand, and the electric and magnetic fields, on the other hand, are Maxwell's equations. Included in Maxwell's equations, of course, were all the discoveries that had gone before. In many ways Maxwell did for electricity and magnetism what Newton had done for mechanics and gravity. His laws systematized in a neat and elegant way all known effects, but also predicted new ones that could be checked by experiment. And, like Newton's law, they were a system of mathematical statements that claimed to explain *all* electrical and magnetic phenomena.

His equations displayed many similarities between the properties of the electric field and the magnetic field. Time-changing magnetic fields produce electric fields, as Faraday had shown. What about the reverse process? Could time-changing *electric* fields produce magnetic fields? If so, there would be a complete symmetry between the two fields.

Maxwell postulated this as an additional law: Changing electric fields produce magnetic fields. Notice the novel origin of this rule. It was not suggested by any known experiments, but by a possible symmetry of the field description. It was required for the beauty or elegance of a set of *mathematical equations.* (Of course, it did not contradict any known experiments.) We will not write down Maxwell's equations here. For the first time in our history of physics, a fundamental discovery is too advanced mathematically to describe in detail at the mathematical level of this book.

Now we come to Maxwell's great synthesis, his discovery of the possibility of electromagnetic waves. Consider the last two laws of electric and magnetic fields we have learned: Changing magnetic fields produce electric fields, and changing electric fields produce magnetic fields. Neither of these statements says anything about the *source* of the fields, namely, the electric charges. Is is possible that a system of continuously changing electric and magnetic fields can sustain itself in the absence of charges?

Simply by studying his equations, Maxwell realized that the answer is "yes." For example, a current produces a magnetic field around it. An alternating current (i.e., one in which the direction of the motion of the charge changes continuously back and forth) produces an alternating magnetic field, one in which the direction of the magnetic field changes back and forth in step with the alternating current. The changing magnetic field produces a changing electric field, in accordance with Faraday's law of magnetic induction; the electric field is also continually changing in direction, pointing first one way, then the other, in step with the changing magnetic field. But the changing electric field produces a changing magnetic field, on top of the one that was already there, and so on and so on. The whole thing amounts to an electromagnetic disturbance that continues reproducing itself. Maxwell called it an electromagnetic wave, for like a water wave, both the electric field and the magnetic field have swells and troughs; furthermore, they move outward in every direction until something stops them.

How fast do electromagnetic waves travel? Or, if someone starts an electromagnetic wave by setting up an alternating current in a wire, how long does it take before an observer a great distance away can detect the system of vibrating electric and magnetic fields set up by the alternating current? Maxwell found that he could solve this problem from his equations. He calculated the speed at which electromagnetic waves propagate, and found it to be enormously fast, though not infinite; the speed was about 186,000 miles per second.

No one was present when he made this great discovery. No one had ever made an electromagnetic wave, let alone measured its speed; yet Maxwell not only predicted the existence of such waves but also predicted that they would travel at *exactly* the measured speed of light. The implication was obvious: Light itself is an electromagnetic wave! Light is no more than a system of electric and magnetic fields, vibrating at just the frequency (number of vibrations per second) that our eyes are tuned to detect. In this way Maxwell unified three apparently separate phenomena: electricity, magnetism, and light.

Instantaneous action at a distance does not occur for the electromagnetic field. For example, suppose one rapidly creates an electric charge. This will create an electric field. Will the electric field appear at a distance from the charge at the same instant as the charge was created? No. In building up the charge, a time-changing electric field was created which propagates as an electromagnetic wave at the speed of light, $c$, across space. The field is first felt only when the electromagnetic wave arrives. Electromagnetic interactions propagate, not with infinite speed, but with the speed of light. This is to be contrasted with Newton's law of gravitation. It remained for Einstein's general theory of relativity to show that information about changing gravitational fields also propagates at the speed of light.

In 1860, Maxwell had left Aberdeen for King's College, London, where he stayed for only five years. In 1865, he retired to his estate in Scotland to devote all his time to research on electricity and magnetism, and the theory of gases. In 1871, he was lured back into university life. He became the first professor at the new Cavendish Laboratory of experimental physics at the University of Cambridge, charged with developing a fully equipped modern laboratory. Maxwell, the theoretician, forsook theory to devote the rest of his life to this exacting administrative task. Soon, the Cavendish became the leading center for experimental physics in the world. But Maxwell did not live to see the things he started bear fruit. Cancer took him in 1879, at age 48, unfulfilled; his laboratory unfinished, his theory of electromagnetic waves untested experimentally.

Had Maxwell lived but nine more years, he would have seen his theory verified. As Thomas Young had shown, light of all colors has extremely short wavelengths, a few millionths of an inch, with blue light having a shorter wavelength than red. According to Maxwell's theory, such short wavelength waves have extremely high frequencies. The electric and magnetic fields oscillate back and forth some quadrillion times per second. No one knew how to make electric fields vary so rapidly, though apparently atoms did it all the time. However, nothing in Maxwell's theory restricted the possible wavelength of electromagnetic waves. Very much lower frequency waves were possible. The real test of Maxwell's theory was to produce a low frequency electromagnetic wave with standard electric apparatus and then show it obeyed the known laws of optics. On Friday, the thirteenth of November, 1888, Heinrich Hertz detected electromagnetic waves emitted by sparks. He proved that short wavelengths are refracted more than long ones in passing through matter, just as Newton had shown in his prism experiment. Electromagnetic waves could be diffracted and showed interference. There was no difference, except wavelengths and frequency, between the invisible electromagnetic waves made by the spark and the light seen by our eyes.

Hertz's experiment provides the basis for today's vast radio and television communications industry. In 1894, a young Italian, Guglielmo Marconi, only twenty years of age, read of Hertz's work and got the idea of using electromagnetic waves for communication.[7]

> It seemed to me that, if the radiation could be increased, developed, and controlled, it would be possible to signal across space for considerable distances. My chief trouble was that the idea was so elementary, so simple in its logic, that it seemed difficult for me to believe that no one else had thought to put it into practice. I argued, there must be more mature scientists who had followed the same line of thought and arrived at almost similar conclusions. From the first the idea was so real to me that I did not realize that to others the theory might appear quite fantastic.

Today we are bathed in electromagnetic waves that are transmitted by radio and TV stations. The oscillating electric and magnetic fields in these waves are very weak so we do not feel them. (They hardly heat us up.) Our radios and TV's are sensitive receivers, designed to be tuned to a given frequency so they can pick up signals of one frequency out of the many all around us and amplify them (increase their strength). In AM radio, when the announcer speaks, his voice is made to modulate the *strength* of the radio wave, which travels at the speed of light. This modulated wave is picked up by an antenna (a wire) in your radio, and the tiny currents driven in the antenna are amplified. This amplified signal then drives a small electromagnet in your loudspeaker. Your loudspeaker pushes the air back and forth, generating a sound wave that propagates through the air to our ears. We do not "hear" the radio waves directly.

Even more important, the atoms in every substance bounce and jostle about. As a result, electromagnetic waves are produced. Everything in the universe radiates electromagnetic waves—planets, the sun, rocks, the trees, and you, all at a frequency of atomic vibratory motion.

Heat produces electromagnetic radiation: the hotter things are, the faster the internal vibrations, and the higher the frequency of radiation.

At night when the sun goes down, the heated ground reradiates electromagnetic waves to the sky, cooling off the ground. If there is water vapor in the atmosphere, in clouds or in droplets, this radiation is reflected back to earth. On the other hand, dry desert climates have cold nights because this radiation escapes to space, cooling off the ground. This is why in Southern California dry Santa Ana days, however hot, are followed by cool nights. In the Eastern United States in the winter, cloudless nights can be very cold due to radiational cooling.

Nearly all our information about the universe beyond the earth is carried by electromagnetic waves of one form or another. The stars give off light and radio waves. The dust in space gives off radiation that is detectable. We would be quite alone without electromagnetic waves.

The only factor distinguishing various kinds of electromagnetic waves is their frequency. All have the same speed. Thus, we can classify waves according to their frequency. The whole range of frequencies we call the electromagnetic spectrum. Because the wavelength of an electromagnetic wave is the speed $c$ divided by the frequency, these waves can be classified by their wavelength as well as their frequency.

Frequency is measured in cycles per second, the number of vibrations the wave makes each second. Waves in the "kilocycle" and "megacycle" range, from about a thousand cycles per second up to ten million, are used for radio and television. These frequencies correspond to wavelengths from a few feet to many hundreds of feet. Radio waves diffract easily around and through small objects, which is why a radio receiver works inside a building, and need not be on a direct line of sight with the transmitter. Frequencies between ten million and a trillion cycles per second are called microwaves because their wavelengths are of laboratory size, from a fraction of an inch to a few feet. They are used in radar and other recent electronic devices.

Waves of somewhat higher frequencies are called infrared and are created naturally by molecular vibrations. Visible light is electromagnetic radiation between about 4,000,000,000,000,000 and 8,000,000,000,000,000 cycles per second. Much higher frequency waves, called x-rays and gamma rays, occur naturally in atomic and nuclear vibrations, and in cosmic rays.

Note the relatively narrow range of frequencies that we can see with our own eyes. But now, the reason is relatively clear. Just as our radio must "tune in" to certain frequencies, so also are our eyes tuned to a narrow portion of the electromagnetic spectrum. With the logic of evolution, our eyes developed over the eons so that they are sensitive over precisely the range of frequencies at which the sun puts out the most light.

Our senses are quite limited. Our life would be considerably richer if we had eyes that were tunable to other portions of the electromagnetic spectrum. Our laboratory instruments, in effect, do this by measuring things we cannot see. The fact that there are so many things we do not sense directly leads to the complete defeat of the Aristotelian program of basing scientific thought solely upon the immediate evidence of the senses.

## FOUNDATIONS OF ELECTRICAL CIVILIZATION

The twin discoveries, that currents produce magnetic fields and that changing magnetic fields produce electric fields, which, in turn, can be used to drive currents through wires, are cornerstones of our modern electrical civilization. Using currents to produce magnetic fields meant that one could design magnetic fields and did not have to depend upon finding a permanent magnet. Moreover, by changing the current one could change the magnetic field at will, bringing the field under much greater control. Combine these facts with the remarkable property of certain metals, notably copper, to conduct electrical currents over very long distances, and one has, for instance, the telegraph. In 1831, Joseph Henry, the Princeton physicist, had constructed a small electromagnet that activated the magnetized clapper of a bell, illustrated in Fig. 8-18. He then connected this electromagnet, through a mile-long wire to a battery in his classroom, which he then connected and disconnected, thereby ringing the bell at his home. This was a demonstration for

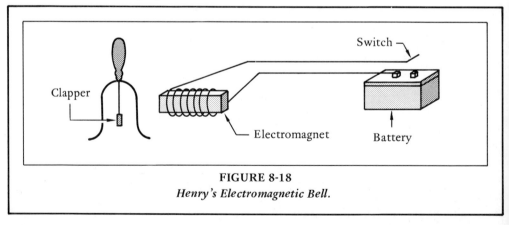

**FIGURE 8-18**
*Henry's Electromagnetic Bell.*

Henry's undergraduate classes in physics; however, Henry fully knew the implications of his toy, and was careful to point out its significance for communication. Henry freely disclosed his ideas to Samuel F. B. Morse in America and Wheatstone in England, both of whom within five years took out patents on the telegraph. Both subsequently became rich. Not only did Henry never profit financially, but neither Morse or Wheatstone credited him for his contribution.

Faraday's discovery of electromagnetic induction led directly to the large-scale production of electrical power. Previously, electricity could be generated only frictionally, or chemically in batteries. Both of these techniques only produce small quantities of electricity. With electromagnetic induction, electricity can be produced by mechanical work. All that is needed is to rotate a magnet, which produces a changing pattern of magnetic lines of force in space. This produces an electric field. This electric field can then drive a current in a wire. If the wire is a good conductor, the current can be carried over very long distances. The electrical power is limited only by the mechanical power available.

The reverse of this procedure is the simple motor. In other words, the electrical power can be reconverted back into mechanical power. Recall Oersted's discovery that current carrying wires (or equivalently, permanent magnets) experience a force upon them. An electrical current can be used to produce magnetic fields on two pole pieces, which are activated alternately; i.e., the direction the current flows through the electromagnet is switched back and forth, thereby constantly changing the direction of the magnetic field at the armature, as illustrated in Fig. 8-19. Activating them alternately creates a magnetic field, which continually reverses its direction. This field then exerts a force on an armature, which may either be a permanent magnet or an electromagnet, to set it rotating. The rotational motion then can

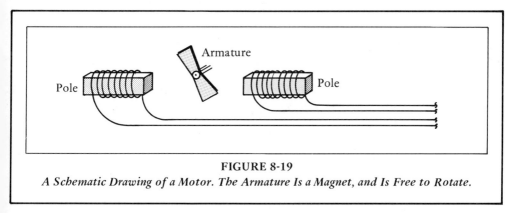

**FIGURE 8-19**
*A Schematic Drawing of a Motor. The Armature Is a Magnet, and Is Free to Rotate.*

drive a machine. The switching, of course, can be done by the motor itself.

We often hear it said that the twentieth century is a century of technology. If so, then our technology is mainly the technology of nineteenth century science. The revolutionary discoveries of twentieth century science began to influence technology only during and after World War II. For this reason, the rise of electrical science in the nineteenth century ranks as one of the most significant historical movements of its time. Michael Faraday understood this. When Prime Minister Gladstone visited Faraday's laboratory at the Royal Institution, he inquired what use all the electrical gadgetry might have. Faraday replied, "Someday, you will tax it."

*Summary*

*The ancient Greeks knew of the attraction between iron and lodestone, and rubbed amber for dust or other light objects. Our word "electricity" comes from the Greek word for "amber," and "magnet" from the Greek word for stones from Magnesia. Despite the fact that the words "electricity" and "magnetism" have Greek origins, their study did not form a coherent discipline in ancient times. Electrical and magnetic attractions seemed more magical than scientific.*

*William Gilbert placed the study of both electricity and magnetism on a modern footing. Gilbert was acquainted with mariner's compass charts of the earth compiled from several centuries of exploration. He fashioned a spherical lodestone—a* terrella—*and showed that his compass positioned itself on the surface of his "little earth" much as mariner's compasses did over the real earth. The earth was, therefore, a giant magnet whose magnetic influence extended deep into space. The north pole of a magnetized compass needle was attracted to the north geomagnetic pole—the south pole of the magnet Earth, and was repelled from the south pole. Gilbert rubbed other substances besides amber, and showed that many, but not all, attracted a needle made of any sort of metal, not just magnetized iron. Thus, electricity was sufficiently common to warrant scientific attention. The materials that Gilbert found could be electrified, he called electrics; the others, he called non-electrics. Metals were all nonelectrics.*

*A century after Gilbert, Stephen Gray found that electricity could be transmitted through certain conducting materials—the metals.*

## Summary

*The basic distinction was not between Gilbert's electrics and non-electrics but between conductors and insulators—those that allowed electricity to move freely and those that did not. Gilbert's electrics were all insulators; his non-electrics were conductors. Charles DuFay studied and compared the phenomena of electrical attraction and repulsion. Rubbed glass repelled rubbed glass; rubbed amber repelled rubbed amber; but amber and glass attracted each other. Since electricity could move, DuFay thought of it as a fluid that passed easily through conductors but remained fixed on insulators. Since there were both attractive and repulsive electrical forces, DuFay reasoned that there should also be two electrical fluids, vitreous—or glass-like, and resinous—or amberlike. Ordinarily, vitreous and resinous electricity were found together in the same amounts so that ordinary matter never exhibited electrical phenomena. However, friction removed some vitreous or resinous electricity, as the case may be, creating an imbalance. Benjamin Franklin further refined this concept and added a fundamental new concept: the conservation of charge. There was only one kind of fluid, charge. When glass is rubbed with silk, some charge is transferred from the silk to the glass. The glass acquires an excess and is said to be positively charged; the silk loses the same amount the glass gains, and is negatively charged. Charles Coulomb viewed charge in a different way. Regardless of what model one used to aid the imagination, charge was a measurable quantity that determined the forces between electrical objects. Using a sensitive torsion balance, Coulomb measured the forces between charged pith balls. The force was proportional to the charges on the pith balls and inversely proportional to the square of the distance between them. Coulomb's law, which was analogous to Newton's law of universal gravitation, was the first step towards putting electricity in a Newtonian framework, the first step towards a theory that would become, in Maxwell's hand, as logically complete and mathematically exact as Newton's theories of mechanics and gravitation.*

*In 1800, Allessandro Volta invented the battery and created the first continuous electrical current. In 1820, Oersted found that a compass needle was deflected by an electrical current. Charges in place created electrical forces; moving charges created magnetism. Ampère, inspired by Oersted's work, immediately found that the compass lines surrounding a straight current were circular, and that a circular current loop had a straight compass line passing through its center; it, therefore, acted like a bar magnet. Inspired by Ampère, Biot and Savart measured the magnetic force between two current-carrying wires. They found that the force was proportional to the current in the two wires and inversely proportional to the distance between them.* Ampère immediately constructed a Newtonian theory of magnetism based on forces acting at a distance between currents. He postulated that permanent and induced magnetism could be explained if matter were made of electro-dynamic molecules, whose microscopic current loops made them behave like microscopic compasses.

Michael Faraday was as interested in the pattern of electrical and magnetic forces in the space surrounding charges and currents as in the complexities of the electromagnetic properties of matter. He refined the old techniques of mapping magnetic forces, by defining magnetic lines of force. Their direction was the direction a north pole of a compass points. The strength of the magnetic force was proportional to the density per unit area of lines of force. Electrical lines of force could be defined in a similar way: their direction was the direction of an electric force on a positive charge, and the strength of the electrical force was proportional to the density per unit area of electrical lines of force.

Faraday also found that when the number of magnetic lines of force threading a closed wire circuit changed with time, a current flowed through that circuit. Since current could only flow in a wire when electric forces were present, Faraday concluded that electrical lines of force accompanied changing magnetic lines of force.

James Clerk Maxwell made mathematical sense of Faraday's lines of force. He defined electric and magnetic fields, which could be calculated from the forces on test charges or currents. Maxwell also added the complement of Faraday's Law to his equations: a time-changing electric field creates a magnetic field. Maxwell's equations related the fields to their sources. Charge produced an electric field, current a magnetic field. Charges and currents were found only in matter. However, since time-changing electric fields created magnetic fields—and vice versa, *an electromagnetic wave could propagate in empty space.* Maxwell calculated the speed of the wave based upon information derived from electrical and magnetic experiments and found it to be the same as the speed of light. Light, therefore, is an electromagnetic wave.

## Appendix Chapter 8

### The Electroscope

An electroscope is illustrated in Fig. 8-20. It consists of a metal plate on top, a vertical metal rod, and a very light piece of metal, usually gold leaf, hinged onto

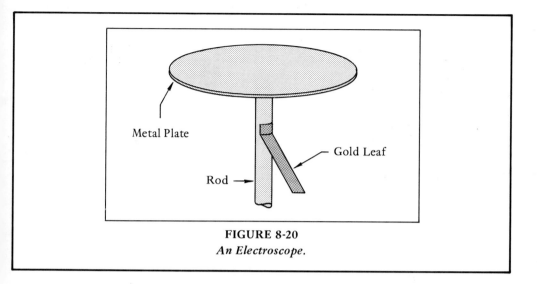

**FIGURE 8-20**
*An Electroscope.*

the vertical rod. In the absence of electrical forces, the leaf falls vertically. However, when an electrified rod (either positive or negative) is brought near the top of the device, the gold leaf moves away, evidently repelled by the vertical rod. When the electrified rod is removed, the gold leaf falls back vertically.

However, if the electrified rod touches the top of the electroscope, a novel thing happens: The leaf moves up, and stays up, even after the rod is removed. There is a simple explanation. The electric fluid, vitreous or resinous, has been transferred to the electroscope; the vertical rod and the gold leaf now have an excess of the same kind of electric fluid, and, therefore, they repel each other.

How can one explain the fact that the gold leaf is repelled by the rod when an electrified object is simply brought near the top of the electroscope? The electroscope, being neutral, has equal amounts of vitreous and resinous fluids. The charged object attracts one kind to the top, and repels the other towards the bottom, where the gold leaf is. The lower part of the vertical rod, and the gold leaf, then have an excess of the same kind of electric fluid, and they repel each other.

We can understand the electroscope in terms of positive and negative charges also. As in Fig. 8-21, there are two ways of "charging" the electroscope, by "induction" and by "contact."

Suppose the rod has a negative charge; the electroscope is neutral. As the rod is brought near the top of the electroscope, it attracts the positive charge and repels the negative charge. Since electricity flows easily through a conductor, the top plate becomes positively charged, while the rod and the gold leaf both become negatively charged and repel each other. This effect is called "induction." If the charged rod is removed, the charges in the electroscope redistribute themselves, all parts of the instrument become neutral, and the gold leaf falls down.

Suppose, however, that the negatively charged rod is placed in contact with the electroscope, then some negative charge is physically transferred to the electroscope, and the gold leaf goes up. If the rod is removed, a net negative charge re-

# Magnetism and Electricity

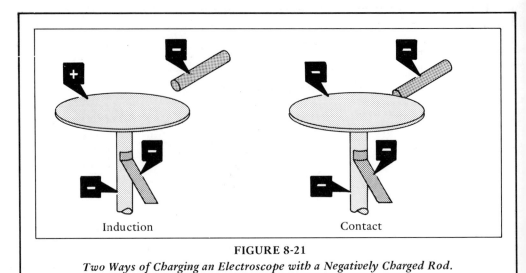

**FIGURE 8-21**
*Two Ways of Charging an Electroscope with a Negatively Charged Rod.*

mains on the electroscope as indicated by the fact that the gold leaf stays up, repelled by the rest of the instrument.

Gradually, the charge leaks off, and the leaf falls down. The charge leaks off faster on a humid day than on a dry day. Humid air is a much better conductor than dry air.

### Franklin and Lightning

Franklin performed many experiments that demonstrated that electricity "leaked" off a conductor most easily if the conductor's surface had a point. A pointed surface also drew electricity from an electrified object more easily than a flat surface. Franklin believed that lightning was just giant sparks, discharges of electricity from cloud to cloud, or from a cloud to the ground. He proposed that a tall pointed conductor inserted into the ground could draw electricity from clouds and keep lightning from striking structures. Franklin is regarded as the inventor of the lightning rod.

Franklin suggested that one could test the idea that lightning was electricity by sitting beneath the spire of a church equipped with standard electrical apparatus, like the electroscope described above, or more complicated instruments used for storing large quantities of electricity. Several Europeans attempted the experiment; most were severely shocked, and at least one Englishman was killed. Franklin himself was saved by delays in the completion of the high spire on Christ Church in Philadelphia. Eventually, he gave up waiting; and instead, he flew a kite during a thunderstorm. The electricity flowed down the wet string (a good conductor) into his apparatus, demonstrating that lightning and terrestrial electricity are the same.

### More on Microscopic Magnetism

We know today that all matter is made up of atoms. Each atom is made up of a heavy, fixed, positively charged nucleus surrounded by light, negatively charged electrons. The electrons circle the atoms, just as the planets circle the sun. These electrons are permanently moving charges. Their orbits are little current loops, indistinguishable from microscopic magnets, whose north and south poles stick out perpendicularly to the current loop.

In most substances, these microscopic current loops point in random directions, cancelling each other out so that there is no large scale magnetic effect. But in iron, some iron alloys, and to some extent in the metals cobalt and nickel, the microscopic current loops can line themselves up; the magnetic forces add, causing a large-scale magnetic effect. This is a permanent magnet.

The magnetic force on a current carrying wire does not really act on the wire, it acts on the charges moving in the wire. These charges are tiny particles called electrons. The wire moves simply because the magnetic force deflects the electrons, which carry the wire along with it, since the electrons are stuck in the wire. Today we can make beams of electrons, (and other charged particles as well) and see them deflected by magnets, in the complete absence of wires. This is the principle of the cyclotron. (It is also the principle of the picture tube of a television set. A beam of electrons moves rapidly back and forth, covering the entire face of the tube thirty times a second. The face of the tube is coated with sensitized material that glows when struck by electrons. The signal from the television station makes the intensity of the electron beam vary very rapidly. The beam is made to sweep over the whole area of the screen by a system of magnets whose strengths change in a systematic way.)

## *Lines of Force and the Inverse-Square Law*

Suppose we want to calculate the force on a small charged particle in the vicinity of some distribution of electric charges. It seems as if there are two different ways of proceeding. One is to draw carefully Faraday's lines of force, being sure each line at every point is in the direction of the electric force. Then, the strength of the force is given by the density of the lines at the place the test charge is. But, we could also have added up the forces due to each charge, using Coulomb's law. A remarkable property of Faraday's idea is that both methods give exactly the same result!

How can this be? Faraday himself probably did not know the general proof; this was left for Maxwell to discover. However, it is easy to see that the rule is correct in a simple case. Consider the field of a single charge as in Fig. 8–13. How does the density of field lines change with distance from the center?

It is obvious from the picture that the lines are farther and farther apart as you go out, so that the force decreases. To get quantitative information, remember that the field is really three-dimensional. We can calculate the density at a distance $R$ from the center as follows: The density of lines in any volume of space is just the number of lines passing through that volume, divided by the volume. Since the picture is symmetric about the center, the density is the same at a distance $R$ in any direction. Let us draw a thin spherical shell at distance $R$ from the charge, with a tiny thickness (call it $d$). The volume of the shell is the area of the surface of the sphere, multiplied by $d$. Since all the lines have to go through the shell, the number of lines is the same for a shell at any distance. (The absolute number depends only on the strength of the charge.) Call this number $N$. If $V$ is the volume of the shell, the density if $N/V$. $N$ does not depend on $R$, but $V$ does.

# Magnetism and Electricity

The volume of our shell is, therefore, $4\pi R^2 d$. (Ignore the fact that the outer shell is slightly larger than the inner one. We take $d$ to be very small.) Therefore, the density of lines is:

$$\frac{N}{4\pi R^2 d} = \frac{N}{4\pi d} \cdot \frac{1}{R^2}$$

This is Coulomb's law! The force diminishes inversely as the square of the distance from the charge. (The number, $N/4\pi d$, simply tells us the overall strength of the electric field. The way it is defined is not important since it depends on how we measure charge. What *is* important is the way the field strength depends on the distance $R$ from the charge.)

In fact, Faraday's rule always works, and even for a complicated set of sources is entirely equivalent to the inverse-square law.

## Some Quantitative Rules for Fields and Forces

In the text, we have not written down many formulas for the forces due to fields, or for the fields of charged particles. These rules follow from Coulomb's law and the law of Biot and Savart (Ampère's Law). Here they are given in the following paragraphs, in their most elementary form.

An electric charge produces an electric field $E$. The value of $E$ at a distance $r$ from the charge is:

$$E = \frac{q_1}{r^2}$$

where $q_1$ is the size of the charge. The field points away from the charge $q_1$ if $q_1$ is positive. The total field at a point is the sum of all the fields to all the charges around. As with forces, we have to add separately the fields in each direction. The force on a charge of size $q_2$ where the field is $E$ is:

$$F = q_2 E$$

The combination of these two rules is just Coulomb's law.

Magnetism is more complicated. Suppose a charge, $q_1$, is moving with a speed $v_1$. We want to know the magnetic field at distance $r$ from the charge. For simplicity, assume that a line drawn from the point where we want to know the field to the moving charge is perpendicular to the direction of the motion. Then, the rule is that the magnetic field (usually denoted $B$) is perpendicular both to the direction the charge is moving and the line drawn from the charge to the point of observation. That is why the magnetic force lines around a wire are circles surrounding the wire. The size of the magnetic field is:

$$B = \frac{v_1}{c} \frac{q_1}{r^2}$$

The letter $c$ stands for the speed of light. Since, ordinarily, $v_1$ is very small, compared to $c$, the magnetic field of a single charged particle like an electron is very small. But large magnetic fields can be built up from the combined fields of many moving charges, even if each one individually is not moving too fast. This happens in a current-carrying wire, or in a permanent magnet.

The force on a particle of charge, $q_2$, in a magnetic field, $B$, also depends on the speed, $v_2$, of the second particle. If it is at rest, there is no force; magnetic fields affect currents, not charges. If it moves in a direction perpendicular to the field, the force is:

$$F = q_2 \frac{v_2}{c} B$$

Again, the speed of light, $c$, occurs in the equation.

These are not the original forms of the laws because the amount of charge moving to produce a measured current was not at first known. It was Maxwell who discovered that the speed of light, $c$, occurs in these equations, which at first sight have nothing to do with light, only with the magnetic effects of electric charges.

Similar formulas hold if all the directions are not perpendicular, as assumed above, but then they involve the angles between the directions of motion and the direction of one particle relative to the other; and the subject becomes complicated.

## *Faraday and Electrochemistry*

Faraday's fame rests not only on the discovery of magnetic induction and the idea of force-fields but also on a large number of other discoveries about electricity, currents, magnetism, and chemistry. We have no time to study them in detail, but it would be unfair to neglect completely to mention his pioneering work showing the close connections between chemistry and electricity. Faraday seems to have been driven by a sure intuition of the unity of science.

Faraday spent a great deal of effort in studying how batteries work, and gradually came to the conclusion that a chemical reaction occurred between the liquid and each of the conducting electrodes. He achieved his most interesting results by inverting the battery: Faraday inserted two different metals into an acid solution, or salt water, and instead of getting an electric current out, he passed a very strong electric current, from a big battery, through the solution. Small bubbles arose from each electrode. Faraday caught the gases in tubes and examined them chemically. One was hydrogen and the other was oxygen. Faraday had decomposed water into its constituent elements by the use of electric currents. Moreover, he discovered that a fixed quantity of electric charge was needed to liberate any particular amount of hydrogen, no matter what chemical reaction was involved. In a salt solution, the same amount of electricity could liberate twenty-three times that weight of sodium at one electrode, and 35.5 times as much chlorine at the other. What was so special about those numbers? They were the relative "atomic weights" of hydrogen, sodium, and chlorine recently proposed by the chemist, John Dalton.

The explanation is simple. Matter is composed of atoms. Water is hydro-

## Magnetism and Electricity

gen and oxygen, salt is sodium and chlorine. The atoms each have a heavy nucleus, positively charged, surrounded by negatively charged electrons. In water, the sodium loses an electron to the chlorine, so that the salt solution is made up of positively charged sodium "ions" (atoms with one or more electrons missing) and negatively charged chlorine ions. An electric current is really a stream of electrons. At one electrode, electrons are added to the sodium ions, making ordinary sodium atoms. At the other electrode, chlorine ions give up an electron to the metal, making ordinary chlorine atoms. Each electron that passes through changes one ion of sodium and one ion of chlorine into an atom; so that the amount of liberated sodium or chlorine is just equal to the weight of a single atom, multiplied by the number of electrons (amount of charge!) passed through.

Thus, Faraday came close to confirming the existence of atoms and understanding their electrical structure. The rest of the story belongs to a chemistry book.

### A Charging Capacitor

We present a simple argument that changing electric fields should produce magnetic fields. A capacitor is two conducting plates, placed a fixed distance apart. It is a device for storing electric charge. Suppose a capacitor is connected with wires to a steady source of current as in Fig. 8-22. A steady current flows in the wire at the top. A magnetic field with circular lines of force surrounds the wire. The current is interrupted by the flat plate of the capacitor. Since the charge carried

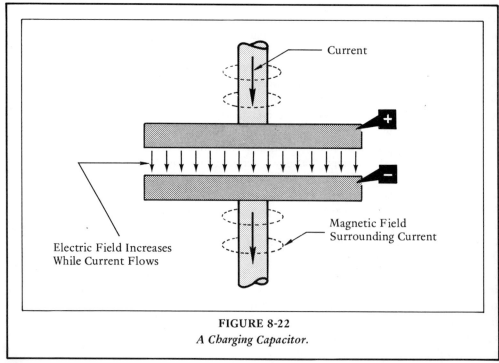

**FIGURE 8-22**
*A Charging Capacitor.*

Appendix Chapter 8

by the current has no place to go, it accumulates on the capacitor plate. A current that flows in continuously causes the charge on the capacitor plate to increase with time. By Coulomb's law, charges create electric fields. In the space between the capacitor plates, therefore, there is an electric field. The electric field increases continuously with time, because of the progressive accumulation of charge. The electric field draws negative charges, which are attracted to the positive charge on the top capacitor plate, to the bottom plate, where they accumulate. The flow of negative charges to the bottom plate constitutes a current in the lower wire, in the same direction as the current in the upper plate. (Negatives flowing upward are the same as positives flowing downward.) The lower wire is, therefore, surrounded by a magnetic field of the same direction and magnitude as the upper wire. According to the old understanding, there would be no magnetic field between the capacitor plates. This "hole" in the magnetic field is unphysical, argued Maxwell. The changing electric field must create a magnetic field. Solving Maxwell's equations, with the new effect included, leads to the conclusion that the changing electric field produces a magnetic field between the capacitor plates equal in magnitude and direction to the magnetic field surrounding the wires. Of course, this is not a "proof" of anything, merely an illustration of one of the consequences of Maxwell's postulate.

## QUESTIONS

1. Imagine a straight wire moving parallel to straight magnetic lines of force. Will a current flow in the wire? Explain your answer.
2. A positively charged pith ball and a negatively charged pith ball are enclosed in a non-conducting box. Imagine an electrical detector made of a small positively charged pith ball at one end of a non-conducting needle and a small negatively charged pith ball at the other. The needle is balanced like a compass needle. Further, imagine that no charge ever leaks off a pith ball. Describe what the electrical detector measures. What about the lines of force?
3. A circular wire has an oscillating current in it—its current flows in one direction and then the other, changing direction at regular intervals. The oscillating current loop is enclosed inside a non-conducting box. What is the best way to detect its presence?
4. Can you guess how a radio works? As a hint, use the brief description of how a telephone works in the Appendix of this chapter. What takes the place, in radio transmission, of the modulated current in the telephone wires?
5. Draw the lines of force for the gravitational field from a point mass. Can a gravitational dipole exist?
6. Two straight wires are tied closely together with a string. The currents in each wire are equal in magnitude and opposite in directions. The currents are steady in time. Draw the lines of force outside the wires.

7. Maxwell showed that electromagnetic waves have the following relationship between their frequency $f$ and their wavelenth $\ell$:

$$\ell f = c$$

where $c$ is the speed of light. ($c$ is about 186,000 miles per second, a mile is 5,280 feet, and one foot equals 12 inches.) Thomas Young showed that the wavelength, $\ell$, of light is about 1/50,000 of an inch. What is the frequency, $f$?

8. The electromagnetic waves used in radio broadcasting have a frequency, $f$, of about 100,000 cycles per second. What is their wavelength, $\ell$?

## REFERENCES

1. *The Dialogues of Plato,* trans. and ed. by B. Jowett (Random House, New York, 1937) vol. 1., p. 281.

2. Christopher Columbus' letter to the King and Queen of Spain, quoted in *Physics, the Pioneer Science,* by Lloyd W. Taylor (Dover, New York, 1959), p. 589.

3. From *Michael Faraday: A Biography,* by L. Pearce Williams (Simon and Schuster, New York, 1971), p. 182.

4. *Ibid.,* p. 99

5. *Ibid.,* p. 27.

6. *Ibid.,* p. 512.

7. Quoted in *Physics, the Pioneer Science,* by Lloyd W. Taylor (Dover, New York, 1959), p. 749.

8. From *Michael Faraday: A Biography,* by L. Pearce Williams (Simon and Schuster, New York, 1971), p. 333.

ALBERT EINSTEIN
*(Photograph by Harold M. Lambert and by permission of the Houghton Library, Harvard University)*

CHAPTER

# 9

# EINSTEIN AND RELATIVITY

### THE AETHER RETURNS

Everything seemed to fit together. Nature obeyed Newton's Laws of Motion. Two forces were completely understood. The first was gravity; Newton's Law of Universal Gravitation accounted for nearly all phenomena due to gravity. The other was electromagnetic force. Maxwell's equations described exactly how it worked and their predictions were confirmed by a great variety of experiments. Their most interesting prediction was that there should be "electromagnetic" waves that travel at the speed of light. Assuming that light itself was an electromagnetic wave of extremely high frequency and small wavelength, one could account for all its properties including reflection, refraction, and diffraction.

Now, Maxwell predicted that light waves travel at a particular speed, $c$, which could be calculated from the known laws governing electric and magnetic fields. In this fact itself lay the seed of a great puzzle. Sound waves, for example, travel in air at about 1000 feet per second. If you are moving through the air in the opposite direction at 100 feet per second, the sound would be moving *relative to you* at 1100 feet per second. But Maxwell predicted that *any* observer would measure the same speed, $c$, for electromagnetic waves or light. The natural assumption was that there is also a medium for the propagation of light waves, which was called by the old name "aether." It was difficult to conceive of waves without a medium for them to propagate in. Maxwell's theory seemed valid only for observers at rest relative to the aether.

But then, which observers were at rest relative to the aether? Without this knowledge, Maxwell's theory would remain incomplete. Could one assume that the aether is at rest relative to the Earth, as Maxwell's equations seemed to say? No one wanted to fall into that trap again, by proclaiming that the Earth was the center of the aether universe. Then, the Earth should be moving through the aether, in different directions at different times as it moves along its orbit around the sun; and the measured speed of light should vary at different points along the Earth's orbit. Measuring this effect became the crucial experiment to do in the late 1800s. The most accurate attempt was made by an American, Albert Michelson.

In 1869, a sixteen year old Polish immigrant set out from Virginia City, Nevada, where his parents had settled after the Gold Rush. He traveled to Washington, by foot, on horseback, and by carriage and by rail. It was well known that each day President Grant walked his dog at a certain hour. Albert Michelson waited for him on the White House steps, to beg for admission to the U.S. Naval Academy, despite the fact that his examination score had not qualified him. A special place was opened for Michelson at the Academy. His four years at the Academy passed without special distinction. His marks in science and engineering were good, but it was clear he was not suited for the seafaring life. It was with some relief on all sides then that after his required tour of duty at sea, Michelson was appointed an instructor in physics at the Naval Academy. He married and settled into the routine of family and teaching. Research then was done in his spare time. Maxwell's revolutionary ideas made it an interesting time for those interested in optics; and like many others, Michelson conceived the need for more accurate measurements of the speed of light. For Michelson, like Tycho Brahe, sheer accuracy of measurement became an obsession. "The fact that the velocity of light is so far beyond the conception of the human intellect, coupled with the extraordinary accuracy with which it may be measured, makes this determination one of the most fascinating problems that fall to the lot of the investigator." Paying for his equipment himself, Michelson proceeded to measure the speed of light, the first of many such improvements in accuracy that he would carry out.

His instrument was a twelve-sided mirror, set up on Mount Wilson in Southern California. A nearby source of light shone on one face of the mirror; the light was reflected off a very distant mirror, on Mount Baldy, accurately placed to reflect the light back to another face of the twelve-sided mirror, and then to an observer, as in Fig. 9-1.

The mirror was rotated rapidly. By the time the light returned,

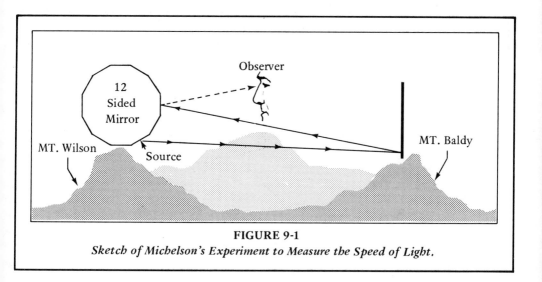

**FIGURE 9-1**
*Sketch of Michelson's Experiment to Measure the Speed of Light.*

the first face had moved slightly so that the light was reflected in a different direction, and was not detected. But if the mirror was rotated fast enough, then when the light returned a second face was in the place the first face had been when the mirror was not rotating, and once more reflected the light to the observer. The speed of rotation at which this happened was measured, and so Michelson knew how long it took for one face to go 1/12 the way around the rotating mirror. Michelson also got the U.S. Coast and Geodetic Survey to measure the distance between Mount Wilson and Mount Baldy with fantastic accuracy (a few inches!). Michelson then knew the distance the light had gone and the time it had taken (1/12 the period of rotation of his mirror), and thus, he could compute the speed of light. An accurate modern value is: $c = 186,282$ *miles per second.*

One might measure a ship's speed through the water by throwing a log overboard. Michelson attempted to measure the earth's speed through the aether using a sophisticated version of this method. His experiment used a device he called an interferometer, sketched in Fig. 9-2. Michelson beamed light at a half-silvered mirror, which was aligned at exactly 45° to the direction the light was coming from. A half-silvered mirror reflects some of the light and transmits the rest. (So-called, one-way glass is half-silvered.) According to the law of reflection, the reflected beam makes a right angle (90°) to the transmitted beam. Mirrors at $A$ and $B$ reflect both halves back to the half-silvered mirror. Again, half of each is transmitted and half is reflected

**Einstein and Relativity**

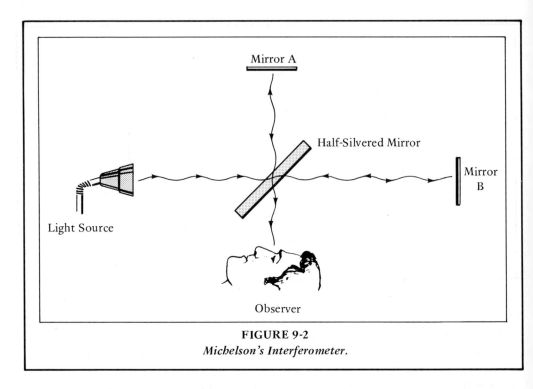

**FIGURE 9-2**
*Michelson's Interferometer.*

so that an observer at C sees a beam made up partly of the light reflected from A, and partly of the light reflected from B.

Now suppose the mirrors A and B are exactly the same distance from the half-silvered mirror. Then the observer at C sees a combination of two beams of light that have traveled exactly the same distance from the source, and, therefore, have taken exactly the same time. The two beams are "in phase;" i.e., the crests and troughs of the waves arrive simultaneously, interfering constructively.

Suppose, however, that the interferometer, along with the earth, moves through the aether in a direction parallel to the line connecting the half-silvered mirror to the mirror B. Then the time it takes light to travel through the aether from the half-silvered mirror to B and back is not the same as the time light takes to travel to A and back. The light reflected off B takes a little longer than the light reflected off A. If $w$ is the distance between the half-silvered mirror and the mirrors A and B, the time difference is given by the formula:

$$\frac{w}{c} \cdot \frac{v^2}{c^2} = \textit{time difference}$$

In this formula, $v$ is the speed of the earth through the aether. The calculation is given in an Appendix section to this chapter. Notice that $v/c$ is very small so that $v^2/c^2$ is an exceedingly small number; e.g., $(1/1,000,000)^2$ is $1/1,000,000,000,000$.

This time delay is not directly measurable. But the two beams, which are recombined before they are observed, interfere. After being split, the light from mirror $A$ should arrive somewhat ahead of that from $B$, as in Fig. 9–3; and the two beams should interfere destructively.

How could Michelson know whether what he saw was due to a real difference in the speed of light in the two directions, or rather to an accidental error in not placing the two mirrors $A$ and $B$ exactly the same distance from the half-silvered one? To check, he rotated the whole apparatus slowly through ninety degrees, so that the beam that was parallel to the earth's motion through the aether was exchanged with the one that had been perpendicular to it. As the interferometer rotated, the phase of the two beams, i.e., whether the troughs and peaks add or cancel, should have changed gradually, so that a continuously changing sequence of light and dark spots should have been observed.

Michelson first tried his experiment in the laboratory of Hermann von Helmholtz at the University of Berlin. However, it was so sensitive that it picked up vibrations from street traffic, and it had to be moved to Potsdam, then in the country. Even there, footsteps a block away disturbed his measurements. Nevertheless, he succeeded in finishing his experiment by working in the middle of the night when there were fewer vibrations. His results were disappointingly negative. There appeared to be no interference pattern; apparently the motion of the Earth did not affect the speed of light.

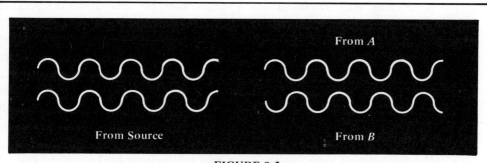

**FIGURE 9-3**
*After Passing through the Interferometer, the Troughs and Crests of the Two Waves Do Not Arrive Simultaneously.*

One possible explanation was that the earth, by coincidence, was at rest in the aether when Michelson performed his experiment. If that were so, it was necessary only to wait a half year and try again. The speed of the earth around the sun was known to be eighteen miles per second. No matter how fast the solar system might move through the aether, the speed of the earth through the aether in any direction must change by 36 miles per second during six months. In particular, if the earth were at rest in the aether the first time Michelson did his experiment, then six months later he should have measured a speed of 36 miles per second.

Once again, the knowledge of just how accurate one's observations were played a crucial role; for Michelson carefully analyzed his apparatus and concluded that his interferometer could easily detect a change of 36 miles per second through the aether, or even changes one-tenth as great. Nevertheless, his result was negative. Michelson never measured a speed through the aether anywhere near the speed of the earth in its orbit around the sun.

Upon his return to America, Michelson resigned his naval commission to become a professor of physics at the new Case Institute in Cleveland. There, beginning in 1887, he collaborated with a chemist, Edward W. Morley, on a more definitive version of the same experiment, designed to solve the difficulties Michelson encountered in Germany. They floated a stone, 5 feet square and 14 inches thick, on a pool of liquid mercury. The mercury was poured into a doughnut-shaped trough, a doughnut-shaped piece of wood was floated on the mercury, and the stone rested on the wood. Michelson's interferometer was mounted on the stone. The entire experiment was supported by a column mounted directly on bedrock. The whole apparatus was so well balanced and vibrationally isolated that a push of the finger would set it rotating. With these elaborate precautions, the measurements became much easier. Even so, it took them several months to align their delicate instrument. For the next several years, they carried out thousands of measurements, at noon and at 6 p.m., for sixteen directions, while the Earth was constantly changing its position. At no time did they find a changing interference pattern. The Michelson-Morley experiment has since been repeated over and over again, but never has a variation in the speed of light been detected.

Here was a great puzzle: on the one hand, how could one explain the Michelson-Morley experiment and still believe in the aether; and on the other hand, how could one understand how light propagated without an aether? Several explanations attempted to patch things up; like Ptolemy with his minor epicycles, to "save the appearances." For example, Sir Oliver Lodge suggested that a layer of aether could be dragged around with the Earth, so that the aether near the Earth's surface would be at rest relative to the Earth. Since the Michelson-Morley experiment would then have been at rest relative to the aether, its results would be trivially explained. Lodge searched for an effect in the laboratory, beaming light near rapidly spinning metal plates that might set the aether near them to spinning, but he never found one.

A Dutch theoretical physicist, Hendrik A. Lorentz, and an Irish mathematician, G. F. Fitzgerald, independently suggested an explanation for the negative result of the Michelson-Morley experiment. Michelson and Morley had assumed that their two light beams traversed paths of precisely identical length. But suppose the distance between the half-silvered mirror and the mirror $B$ shrinks due to the motion through the aether; i.e., suppose there were a law of nature that said things shrink along the direction of their motion through the aether. It is not hard to work out just how much they have to shrink so that the two *times* the split halves of the beam take to pass through the interferometer were exactly the same, thereby "explaining" the negative results no matter how fast one was moving through the aether. But could such a law of nature be consistent with all the other known laws?

## ALBERT EINSTEIN

The year 1895, the year of Lorentz's paper interpreting Michelson's interference experiment in terms of length contraction, also found a sixteen year old boy posing logical paradoxes to himself, after the fashion of adolescents. "Suppose I run away from a mirror at the speed of light; will I still see my reflection?" "If I move beside a light wave at the speed of light, I should see a spatially oscillatory electromagnetic field at rest; is this possible?" The boy's chronically unsuccessful father, who had seen a succession of small electrical busineesses fail in Ulm, Munich, and Milan, a victim of his good nature and naiveté, urged the boy to forget his "philosophical nonsense" and study the "sensible trade" of electrical engineering. Only Albert Einstein's uncle Cäsar in Stuttgart seemed sympathetic; to him Einstein addressed many letters pouring out his thoughts, including one especially interesting essay entitled, "Concerning the Investigation of the State of Aether in Magnetic Fields," indicating an early interest in the problems of electricity and magnetism.

Albert Einstein must have been a trial to his parents. Although the boy read Kant's philosophy at age thirteen, and pored over texts of advanced mathematics, he would not apply himself in school, and consequently did not earn his *Gymnasium* certificate. Without it, it would be difficult to attend a university. When Hermann Einstein's business in Milan failed, Einstein left his wandering family to take the entrance examination for the Technical University of Zurich. He made no attempt whatsoever to prepare himself for the examination and failed. A year's extra study at the Cantonal School in the nearby town of Aarau was arranged for him. Apparently Aarau's gentle atmosphere appealed to him, for there he worked. After a year in Aarau, he passed his entrance examination. The open-minded democracy in Switzerland was such a contrast with German authoritarianism that Einstein renounced his German citizenship in favor of Swiss nationality.

Admission to the University did not change Einstein's way of life. He lived and dressed casually and unconventionally, his mind on other things. Einstein rarely attended class, relying upon notes taken by his friends to pass examinations. His detached, apparently disinterested, air offended his professors. At this time, the tangible direct contact with nature found in the laboratory was the only part of the formal curriculum that appealed to him. While friends went to class for him, he remained in his room to read the works of Maxwell, Helmholtz, Hertz, Poincare, and, above all, Ernest Mach.

Mach had developed a point of view about physics that was very influential in the early twentieth century. Most of his ideas appeared in his "Science of Mechanics," which appeared in German in 1883. His central point was that one should reject from the language of science ideas and constructions that are not observable. Rather, that science only provided a set of simple, economical rules, which, assisted by logical and mathematical reasoning, correctly describe the collection of observed facts about nature. Of course, this idea was at least as old as Galileo, but Mach applied it critically to the state of science in his time. Two examples especially influenced Einstein's thought in different ways.

The first had to do with the existence of atoms and molecules. The idea that matter was made up of tiny, invisible atoms and molecules explained the vast amount that was known about chemical transformations. It was also consistent with the physical behavior of matter; for example, the way gases and fluids behave when compressed, expanded, or heated. However, the physicist did not need to know that matter was made up of tiny particles. All the properties of matter could be explained just as well by assuming matter to be a continuous substance, as Aristotle had, infinitely divisible. No one had ever seen an effect that could be explained only by their existence:[1]

> Chemical, electrical, and optical phenomena are explained by atoms. But the mental artifice atom was not formed by the principle of continuity; on the contrary, it is a product especially devised for the purpose in view. Atoms cannot be perceived by the senses, like all substances; they are things of thought. Furthermore, the atoms are invested with properties that absolutely contradict the attributes hitherto observed in bodies. However well fitted atomic theories may be to reproduce certain groups of facts, the physical inquirer who has laid to heart Newton's rules will only admit those theories as provisional helps, and will strive to attain, in some more natural way, a satisfactory substitute.

Mach also attacked the idea of absolute space, or the existence of an absolute frame of reference, and the related idea of the aether, which had never been seen. Despite the knowledge, since Galileo and Newton, that only relative speeds entered into the laws of mechanics, the same was not true of accelerations and rotations. Newton's famous experiment with a rotating bucket of water seemed to prove that there was absolute meaning to accelerated motion. When the bucket was still, the surface of the water was flat. But when the bucket, and the water in it, was set rotating, the surface was no longer flat, but higher on the sides than in the middle. Newton concluded that there was an absolute meaning to the idea of rotation, and, therefore, a real meaning to the idea of absolute motion and rest. The aether, which was popular in Mach's time, gave physical substance to this idea. Absolute motion was the type of motion Michelson and Morley were trying to measure; motion relative to the aether.

Mach insisted that the whole concept of the aether, and of absolute motion, had no basis in experiment and, therefore, should not have a place in physics. Newton's bucket experiment only showed that the surface of the water becomes curved when it rotates relative to the fixed stars, or better, relative to the average motion of everything else in the universe. To test this idea, one would have to "try to fix Newton's bucket and rotate the heavens of fixed stars." Since we cannot do that experiment, we cannot conclude that we know what would happen. Mach was sure that it was motion relative to the other matter in the universe, not the aether or some abstract absolute space, which caused the physical effects of acceleration and rotation. Of Newton's bucket, he wrote, "No one is competent to say how the experiment would turn out if the sides of the vessel increased in thickness and mass until they were ultimately several leagues thick."

While Einstein was pondering these deep questions, he applied himself to passing the university examinations, and managed to graduate in 1900, at considerable emotional cost:[2]

> The coercion had such a deterring effect upon me that, after I passed my final examination, I found the consideration of any scientific problem distasteful to me for an entire year.

In the normal scheme of things, a young man such as Einstein would have become an assistant in the physics laboratories at Zurich. But not having played the game, he could not reap its rewards. All the available positions went to his friends and fellow-students; and somehow, one could not be found for abstracted, arrogant Einstein. Now the question, what to do, became pressing. Einstein took a job for a few months at the Swiss Federal Observatory; his first scientific paper being recently published, he sent copies of it to various laboratories around Europe asking for a position, uniformly receiving no answer. He taught mathematics for a few months. During periods of unemployment, he worked on his doctoral dissertation. Finally, a more or less permanent solution to the problem of the impractical Einstein was found, much to the relief of friends and family. Einstein's good friend, Marcel Grossmann, induced his father to speak for Einstein to the director of the Swiss patent office in Berne. The job would be good for him; he could settle in for life and forget his ineffectual dreaming about physics. And so, in the summer of 1902, Einstein moved to the Swiss capital, Berne, to take up a job as Technical Expert (Third Class). He would remain there seven years.

Einstein's job was by no means uninteresting. It required examining new inventions (often proposed by unschooled individuals), determining what was really new and describing concisely the essence of the inventor's inventions. There Einstein learned the simplicity and clarity of expression so characteristic of his later work. Einstein, who became quite adept, soon received a number of minor promotions.

If Einstein's job took eight hours a day from his life, they were at least stress-free and not unpleasant, and there were, after all, evenings and weekends to pursue his hobby: the deep speculations upon physics that so pleased him. Some evenings were spent in tutoring physics, some spent in the intellectual company of other young scientists and philosophers, and some spent on his Ph.D. dissertation, which was accepted in 1905 but not accorded any special distinction. He also wrote and published a number of minor scientific papers. While it was not an ordinary record for a patent examiner, still it was not what was expected from one with aspirations to a career in physics.

In 1904 and 1905, Einstein's speculations suddenly matured.

Only once before did one so young discover so much in a single year: in 1666, the year of Newton's exile in Woolsthorpe. Curiously, both Einstein and Newton were twenty-six. During one year's time, Einstein published, in addition to some relatively minor works, three immortal articles, each one on a different subject: one on Brownian motion, one on the photoelectric effect, and a third on space and time, now called the special theory of relativity.

Einstein's work on Brownian motion was his answer to Mach's criticism of the concept of atoms and molecules. Nearly a century before, a Scottish naturalist, Robert Brown, had discovered that pollen dust suspended in water and observed under the microscope was in perpetual irregular motion. "These motions," he wrote, "were such as to satisfy me, after frequently repeated observation, that they arose neither from current in the fluid nor from its gradual evaporation, but belong to the particle itself." The idea that it was in the very nature of certain particles to generate their own motion was a throwback to Aristotle's ideas and an affront to modern physics. Einstein, as always, attracted by paradoxes, set out to resolve the anomaly. His doctoral dissertation had treated the then relatively new kinematic theory of gases developed only a generation earlier by Maxwell and Boltzmann. In it, gases are postulated to be large collections of atoms rushing about rapidly in random directions.

Einstein argued that the cause of Brownian motion was the collision of the rapidly moving atoms with the dust particles. One atom, of course, was not usually heavy enough to push a dust particle so fast that one could see it. But from time to time a particle would be struck by several atoms moving very rapidly in the same direction, producing a motion that could be seen in a microscope. Einstein not only put forth this general idea but, more importantly, demonstrated mathematically that the pattern of motion of the suspended dust particles was *exactly* that predicted by the atomic theory. His paper removed all doubt that atoms actually exist; Brownian motions, after all, can be seen with our own two eyes. Upon seeing Brownian motions, Einstein remarked:[3]

> It is an impressive sight. It seems contradictory to all previous experience. Examination of the position of one suspended particle, say, every thirty seconds, reveals the fantastic form of its path. The amazing thing is the apparently eternal character of its motion. A swinging pendulum placed in water soon comes to rest if not impelled by some external force. The existence of a never diminishing motion seems contrary to all experience. This difficulty was splendidly clarified by the kinetic theory of matter.

Atoms, once a convenient hypothesis for the kinetic theory, could no longer be denied.

Einstein's paper on the photoelectric effect, the fact that one can produce an electric current by shining light on certain substances, was even more revolutionary; for, after over one hundred years of the detailed success of the wave theory of light, Einstein proposed to explain the photoelectric effect by assuming light to have some properties of a stream of particles, which he named "photons." Einstein was awarded the Nobel Prize in 1922 for his explanation of the photoelectric effect.

### THE SPECIAL THEORY OF RELATIVITY

A definite proof of the existence of atoms and a radical new theory of light was a reasonable three-month output for a part-time physicist. But in the spring and summer of 1905, Einstein also resolved the contradictions that were cropping up in attempts to reconcile Maxwell's electromagnetism with the ideas about the nature of space and especially time, which lay behind Newton's mechanics.

He started from the following two observations: first, that Maxwell's laws of electromagnetism were derived from experiment and did not contain any internal, logical contradictions; and second, that despite the popularity of the aether, there was really no evidence, as the Michelson-Morley experiment, among others, had shown, that the idea of absolute motion or rest had any meaning, although the laws of electromagnetism *seemed* to speak of absolute motion. He began with the following example:

We know that electric charges at rest produce only electric fields, whereas charges in motion produce magnetic fields as well; charges at rest are unaffected by magnetic fields, while a magnetic field exerts a force on a moving charge (current). Can we not easily tell whether a charge is at rest or in motion simply by observing whether or not it makes a magnetic field, thereby refuting the principle of relativity?

To answer that question, consider a wire loop at rest, and imagine moving a magnet through it with constant speed. In the vicinity of the wire is a changing magnetic field, which according to Faraday's law produces an electric field; this acts on the charges on the wire and causes a current to flow. Now think of the same experiment from the point of view of the magnet. Imagine the magnet at rest and let the wire loop slide over it. This time there is no electric field. But the charges in the wire move through a magnetic field, and we know from Ampère's experiments that when charges move through a magnetic field there is a force on them (that is why a magnetic field exerts a force on a current-carrying wire). This force causes the charges to move around the wire, causing a current. In fact, if one works out the strengths and directions of the forces from the detailed laws of electromagnetism, one concludes that the current induced in the wire is exactly the same in both cases, provided only that the *relative* speeds of the magnet and the wire loop are the same!

Now, the only experiment one can do is to measure the current in the wire, and we have just seen that from this experiment it is impossible to tell which description is correct. There is no way to tell whether the wire or the magnet is at rest. Nevertheless, our description in terms of electric and magnetic fields was quite different in the two cases. Evidently, electric fields and magnetic fields are different aspects of the same thing. An observer at rest relative to the magnet says that no electric field is present, while an observer at rest to the wire says there is. Electric and magnetic fields are not absolute properties of a point in space at a given time; their description depends on the motion of the observer in just the right way to prohibit one from distinguishing rest from absolute motion.

The situation just described is the subject of the opening paragraph of Einstein's paper on special relativity, published in 1905 with the modest title, "On the Electrodynamics of Moving Bodies."[4]

> It is known that Maxwell's electrodynamics—as usually understood at the present time—when applied to moving bodies, leads to asymmetries which do not appear to be inherent in the phenomena. Take, for example, the reciprocal electrodynamic action of a magnet and a conductor. The observable phenomenon here depends only on the relative motion of the conductor and the magnet, whereas the customary view draws a sharp distinction between the two cases in which either the one or the other of these bodies is in motion. For if the magnet is in motion and the conductor at rest, there arises in the neighbourhood of the magnet an electric field with a certain definite energy, producing a current at the places where parts of the conductor are situated. But if the magnet is stationary and the conductor in motion, no electric field arises in the neighbourhood of the magnet. In the conductor, however, we find an electromotive force which gives rise—assuming equality of relative motion in the two cases discussed—to electric currents of the same path and intensity as those produced by the electric forces in the former case.

Einstein goes on to observe that this kind of accident always happens in electromagnetic phenomena, and then makes an astounding generalization:

> Examples of this sort, together with the unsuccessful attempts to discover any motion of the earth relatively to the "light medium," suggest that the

> phenomena of electrodynamics as well as of mechanics possess no properties corresponding to the idea of absolute rest. They suggest rather that, as has already been shown to the first order of small quantities, the same laws of electrodynamics and optics will be valid for all frames of reference for which the equations of mechanics hold good. We will raise this conjecture (the purport of which will hereafter be called the 'Principle of Relativity') to the status of a postulate, and also introduce another postulate, which is only apparently irreconcilable with the former, namely, that light is always propagated in empty space with a definite velocity $c$ which is independent of the state of motion of the emitting body. These two postulates suffice for the attainment of a simple and consistent theory of the electrodynamics of moving bodies based on Maxwell's theory for stationary bodies. The introduction of a "luminiferous ether" will prove to be superfluous inasmuch as the view here to be developed will not require an "absolutely stationary space" provided with special properties, . . .

Thus, the theory of relativity is based on two simple and "only apparently irreconcilable" postulates. First, that the Principle of Relativity is true for all physical phenomena: That is, *all* the laws of physics are the same to two observers who are moving relative to each other in a straight line at constant speed. It is exactly Galileo's old idea that no experiment can detect "absolute" motion or rest, now generalized to electromagnetism. The second principle is that the value of the speed of light in empty space, $c$, a direct result of Maxwell's theory is, therefore, also the same to any observer. The fact that the Michelson-Morley experiment (of which Einstein seems to have been hardly aware) found no aether drift is entirely consistent with this postulate. As they observed, the speed of light, relative to them and their apparatus, was always the same, in any direction, noon and night, June and December.

What is revolutionary about these two postulates is that they are only "apparently irreconcilable." How can two observers, moving relative to each other, observe the same numerical value for the speed of the same beam of light? The answer, Einstein says in the next portion of his paper, is that we must consider carefully what we mean by time.

For Newton and his successors, time had the absolute meaning that we usually use intuitively. It was as if there were a great, universal clock in the sky that anyone, anywhere, could consult any time to see what time it was. "Absolute, true, and mathematical time," Newton wrote, "of itself, and from its own nature, flows equally and without relation to anything external." This was the sort of mental construction Mach had warned against. Einstein criticized it as follows:

We have to take into account that all our judgments in which time plays a part are always judgments of simultaneous events. If, for instance, I say, "That train arrives here at 7 o'clock," I mean something like this: "The pointing of the small hand of my watch to 7 and the arrival of the train are simultaneous events."

It might appear possible to overcome all the difficulties attending the definition of "time" by substituting "the position of the small hand of my watch" for "time." And in fact such a definition is satisfactory when we are concerned with defining a time exclusively for the place where the watch is located; but it is no longer satisfactory when we have to connect in time series of events occurring at different places, or—what comes to the same thing—to evaluate the times of events occurring at places remote from the watch.

Thus, we have to be careful when we ask what time it is at a distant point. It is natural to proceed as follows. Suppose an event occurs some distance away; for example, a light flashes or a supernova explodes. We want to know at what time did the event occur. It will not do to look at our own clock when we see the flash, for it has taken a certain amount of time for the light to travel from the source to the observer. If we know how far the source is from the observer, we can calculate the extra time the light spent travelling, since, according to the second postulate of the theory of relativity, we know its speed to be exactly $c$.

Suppose an observer were also at the source. Apparently he could synchronize his watch so that both he and the original observer get the same time for the flash of light. Thus, at first sight it seems that time can be given a universal meaning, in the sense that all observers can agree at what time an event takes place.

However, if we repeat the above analysis with the two observers *moving* relative to one another, we will see that an absolute view of time is not tenable. Of course, two observers can always arrange to record the same time for any event, just by setting their watches appropriately. So the question to ask must be whether they always will record the same *time interval* between *two* events. Einstein provided a most striking example of the fact that this cannot be so if one accepts the two postulates of his theory of relativity; that is, there can be no universal, absolute meaning to the statement that two events happen at the same time.

Einstein himself provided the classical example. He was fond of making scientific points using what he called "thought-experiments." They are experiments you do not have to do; you have to imagine what would happen if you did them. Suppose a man in a railroad car determined the exact middle of the car. And from the middle, he simultaneously launched two light beams in exactly opposite directions towards mirrors at each end of the car. Knowing that the two light beams must travel the same distance with the same speed, he would know that the two beams would strike the mirrors simultaneously, as in Fig. 9-4. Consider an outside observer looking in as the railroad car with its optical experiment moves by him at a constant speed, $v$. The outside observer knows the speed of light is constant and that the motion of its source does not affect its speed. Furthermore, he reasons that the light beam traveling opposite to the railroad car's motion must travel a shorter distance than the beam traveling with the car's motion, because during the time the light traverses the car, the car moves. Relative to the observer on the ground, the beam going backward goes a shorter distance because between the instant it is emitted and the instant it strikes the mirror, the train has moved forward a bit. By similar reasoning, the beam moving forward has a longer distance to go. Therefore, he would find that the light beams do *not* arrive simultaneously at the two mirrors. Thus, the judgment of simultaneity, the most fundamental concept in the definition of time, depends upon the state of relative motion between observers.

The intuitive, Newtonian idea of absolute time is incompatible with Einstein's two postulates. Once we give it up, we see why they are "only apparently irreconcilable." Now imagine that two observers—for example, one on the train and one on the ground—move relative to each other (see Fig. 9-5). Suppose they synchronize their watches so that they both read the same time for one flash of light. We have seen that they cannot both read the same time for the other flash of light; the two flashes cannot appear simultaneous for

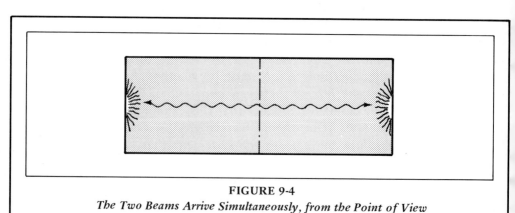

**FIGURE 9-4**
*The Two Beams Arrive Simultaneously, from the Point of View of an Observer at Rest Relative to the Train.*

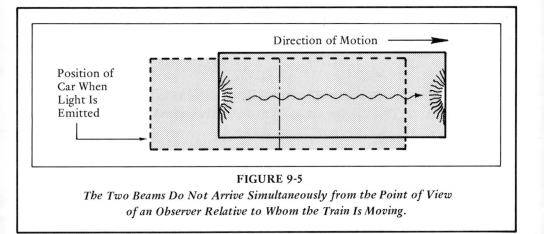

**FIGURE 9-5**
*The Two Beams Do Not Arrive Simultaneously from the Point of View of an Observer Relative to Whom the Train Is Moving.*

both observers. Therefore, one observer's watch, according to the other, must be either fast or slow. Following Einstein, we can calculate the rate of a moving clock. We can actually deduce what that rate must be just from the two postulates of the Theory of Relativity.

To this end, we do another "thought-experiment." Let us imagine that time is measured by an idealized device we can call a light clock. A light clock consists of two mirrors facing each other, held apart by a rigid rod of known length, as in Fig. 9-6. A light pulse is reflected back and forth between the mirrors. Each time the light bounces off a mirror is like a tick of a clock. We know that the speed of light, $c$,

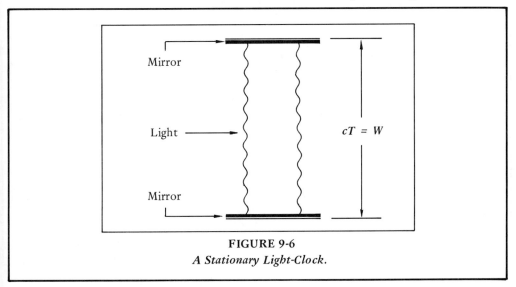

**FIGURE 9-6**
*A Stationary Light-Clock.*

is about 186,000 miles per second. Thus, if for example the mirrors were one mile apart, the time between the ticks of the light clock would be 1/186,000 of a second.

Now, let us imagine two observers, each with a light clock; the two light clocks are identical. Suppose the distance between the mirrors in the light clocks is $W$ (in the example just mentioned, $W$ was one mile). Let $T$ stand for the time between ticks. Then, since speed is distance divided by time, and since the speed of light is always $c$:

$$c = \frac{W}{T}$$

or,

$$T = \frac{W}{c}$$

Now one observer—call him the "stationary observer"—wants to know how fast the moving observer's clock is ticking. For convenience in setting up this thought experiment, let us imagine that the moving clock is moving at some speed, $v$, at right angles to the line connecting its two mirrors, as in Fig. 9-7. This illustration shows the light path as seen by the stationary observer. Between ticks, the light clock has moved, so that the distance the light goes is larger than $W$. Since, according to the second postulate of the theory of relativity, the speed of light is always $c$, we conclude that according to the stationary observer's measurements the time between the moving clock's ticks is longer than $T$; the moving clock is running slow! Of course to an

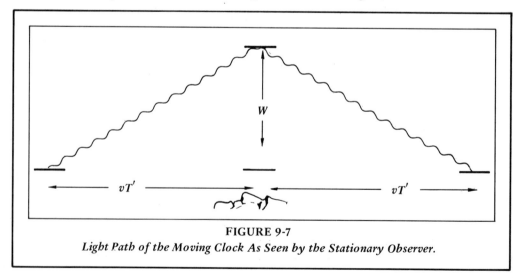

FIGURE 9-7
*Light Path of the Moving Clock As Seen by the Stationary Observer.*

# The Special Theory of Relativity

observer moving with the "moving" clock, the time between ticks is $T$; from his point of view, the "stationary" observer's clock goes slow! This is a startling conclusion, but inescapable if one accepts Einstein's two "apparently irreconcilable" postulates.

We can calculate the rate of the moving clock. The result is so basic that we shall do it here in the text. Call the unknown time between ticks of the moving clock $T'$. The distance the light travels during that time is $cT'$. Suppose, for example, the mirror is moving at 3/5 the speed of light. Then, during the time between ticks (the unknown time $T'$), the mirrors move to the right [in Fig. 9-7, where $v$ is $(3/5)c$], a distance $(3/5)cT'$. The distance between the two mirrors is still $W$. The points $A$, $B$, and $C$ form a right triangle, with the light-path $AC$ the hypotenuse. Its length is $cT'$. The base has length $(3/5)cT'$, and the height is $W$. From Pythagoras' theorem:

$$(cT')^2 = \left(\frac{3}{5}cT'\right)^2 + W^2$$

Solve for $W^2$:

$$W^2 = c^2 T'^2 - \left(\frac{3}{5}\right)^2 c^2 T'^2 = \left[1 - \left(\frac{3}{5}\right)^2\right] c^2 T'^2$$

Since $1 - (3/5)^2 = 1 - 9/25 = 16/25$:

$$W^2 = \frac{16}{25} c^2 T'^2$$

Now we can solve for $T'$:

$$T'^2 = \frac{25}{16} \frac{W^2}{c^2}$$

or, taking the square root:

$$T' = \frac{5}{4} \frac{W}{c}$$

But, $W/c$ is just the time between ticks of the stationary clock: $W/c = T$; therefore:

$$T' = \frac{5}{4} T$$

The interval between the ticks of the moving clock is 5/4 greater than for the stationary clock; i.e., the moving clock is ticking at 4/5 the rate of the stationary clock.

Let us do the calculation in general. Suppose the moving clock is moving at speed $v$. Then, everything is the same as before, except the base of the right triangle in Fig. 9-7, which is $vT'$, since the distance the clock has moved between ticks is $vT'$. Pythagoras' theorem reads:

$$(cT')^2 = (vT')^2 + W^2$$

where $W = cT$ as before; then:

$$W^2 = c^2 T'^2 - v^2 T'^2 = (c^2 - v^2)T'^2$$

so,

$$T'^2 = \frac{W^2}{c^2 - v^2}$$

Divide numerator and denominator on the right by $c$:

$$T'^2 = \frac{W^2/c^2}{1 - (v^2/c^2)} = \frac{(W/c)^2}{1 - (v^2/c^2)} = \frac{T^2}{1 - (v^2/c^2)}$$

And finally:

$$T' = \frac{T}{\sqrt{1 - (v^2/c^2)}}$$

The time between two events measured in a moving system is longer by the above amount. When $v = 0$, the clocks are at rest relative to one another; and the times agree: $T' = T$. When $v/c$ is very small, as in our ordinary experience, the denominator $\sqrt{1 - (v^2/c^2)}$ is so close to one that the time increase is negligible. Even at the speed of an airplane, the correction is so small that no ordinary clock is accurate enough to measure the difference.

On the other hand, should we observe a clock moving relative to us very close to the speed of light, so that $v^2/c^2$ is very near 1, and $\sqrt{1 - (v^2/c^2)}$ is very small, we would find that the time in the moving frame slows down very much relative to the time in the stationary frame; in other words, $T'$ would then be very much larger than $T$. Einstein called this phenomenon time dilation. Time dilation is completely symmetric. For in the above analysis of time dilation, the role of the two observers could be reversed, without changing the conclusions. Each observer sees the other's clock running slow.

We emphasize that time dilation occurs for all clocks, not just light clocks. They were just a mental construction to help us think about the geometry of the problem. All moving clocks run slow, at the rate predicted by the time-dilation formula. Otherwise, the comparison of another clock with a light clock would be different to the two observers, and one could tell which was moving, in violation of the principle of relativity.

Einstein's prediction of time dilation has been verified billions of times. Certain elementary particles decay naturally into other particles after a fixed time. In other words, they have a natural internal clock that is their decay time. Should such particles be at rest, their decay times would be very short; however, they are ordinarily observed either as cosmic rays or in high energy particle accelerators when they are moving with speeds near the speed of light; their decay times in the "stationary" frame of the laboratory appear very much longer, in complete accordance with quantitative expression for time dilation derived above.

Measurements of length, too, are relative, Let us define the unit of length by a rigid yardstick; a stick is a yard in length so long as it is at rest in the observer's reference frame. Imagine a second yard stick in uniform motion relative to the first. Will it still appear to be a yard in length? To decide this, we must first specify how to measure the length of a *moving* stick. We could specify that at a given moment two observers in the stationary frame simultaneously take snapshots, one of each end of the stick. Since the pictures are taken simultaneously in the rest frame, we can compare the length of the stationary yardstick with the moving yardstick. However, since simultaneity itself is a relative concept, it is conceivable that the measurements of stationary and moving yardsticks differ. In fact, if $W$ is the rest length, then $W'$, the moving length, turns out to be:

$$W' = W \sqrt{1 - (v^2/c^2)}$$

The calculation is similar to that for time dilation and is explained in the Appendix.

The Michelson-Morley experiment is "explained" directly by Einstein's two postulates. The speed of light is the same in the two directions of the split light beam, so both light pulses take the same time to travel from the half-silvered mirror to the two mirrors, $A$ and $B$, in Fig. 9-2 and back; and arrive simultaneously, independent of the motion of the interferometer and the whole laboratory relative to anything else. But suppose we were perverse enough to try to understand this fact from the point of view of an observer at rest, say, relative to the sun. Then, the interferometer is moving at some speed, $v$, the speed of the earth around the sun. Imagine that the first arm is perpendicular to the direction of motion, that the second is parallel, and that both arms have length $L$ as measured by the moving observer. According to the observer at rest relative to the sun, the pulse of light that travels along the first (perpendicular) arm and back takes an amount of time:

$$\frac{L}{c} \cdot \frac{1}{\sqrt{1 - (v^2/c^2)}}$$

to complete the round trip. This is just the time dilation effect. To calculate the time the parallel beam takes, we must take into account the contraction: the length of the path is not $L$ but $L\sqrt{1 - (v^2/c^2)}$; then, it is not hard to calculate the time for this round trip, using the method of the Appendix. It is, of course, also:

$$\frac{L}{c} \cdot \frac{1}{\sqrt{1 - (v^2/c^2)}}$$

So, both pulses of light arrive simultaneously. The contraction of lengths is needed for this observer to understand why the two pulses return simultaneously to the half-silvered mirror, a fact that is obvious to an observer moving along with the interferometer. The fact that this all works out shows that Einstein's two postulates are not really "irreconcilable."

Einstein now could attack some of the problems he had been wondering about since his boyhood, such as, "What could happen if you try to catch up with a light beam?" Suppose you were moving nine-tenths the speed of light. Would you see an electromagnetic wave moving at one-tenth the speed of light? No. Because of the second postulate: all observers measure the same speed, $c$, for a beam of light.

How does this work? Suppose a train moves at 90 miles per hour, and a rider in the train walks down the aisle toward the front of the train at five miles per hour (relative to the train). Relative to an observer on the ground, the rider moves at a speed of ninety-five miles per hour. At least this is the "common sense" answer that we developed when we discussed the principle of relativity as developed by Galileo. Speeds simply add. In general, if the walker's speed relative to the train is $u$, and the train's speed is $v$, the walker's speed relative to an observer on the ground is $u + v$. But, in the theory of relativity, the "common-sense" law of addition of speeds must be wrong. Suppose we had considered a pulse of light instead of a person walking at five miles per hour in a train. Relative to the train, its speed is $c$; relative to the observer on the ground, its speed is also $c$, not $c$ plus 90 miles per hour. What is the general rule?

As a "thought-experiment," let us replace the train by a spaceship, moving relative to an observer on earth—whom we will call the "stationary" observer—at three-fifths the speed of light, or, in symbols, $(3/5)c$. Next, suppose the space-ship shoots out a "life-raft," moving at the same speed, $(3/5)c$, relative to the space-ship. How fast is the "life-raft" moving relative to the "stationary" observer? According to the Galilean idea of adding speeds, the answer would be $(3/5)c + (3/5)c$, or $(6/5)c$. The life raft should move faster than light relative to the stationary observer.

What does the theory of relativity say? The life-raft moves at a speed $(3/5)c$ ahead of the space-ship, which itself moves at speed $(3/5)c$ relative to a stationary observer. To calculate the life-raft's speed relative to the stationary observer, we must know how long it takes to go a given distance. We can proceed as follows: Let $L$ be the length of the space-ship, measured by an observer at rest in the ship. Let us ask how long it takes for the life-raft to go from the back to the front of the space-ship. If the time it takes, as measured by a clock at rest in the space-ship, is $T$, then, from the definition of speed:

$$\frac{L}{T} = \frac{3}{5} \cdot c$$

We need to know another time, the length of the time interval as measured by a clock at rest in the life-raft. Call this time $T_0$ ($T_0$ is sometimes called the *proper time* between the two events; it is the time interval measured by an observer for whom both events happen at the same place). Repeating our earlier calculation, the time ($T$) measured in the space-ship is related to $T_0$ by:

$$T = \frac{T_0}{\sqrt{1 - (3/5)c^2/c^2}} = \frac{T_0}{\sqrt{1 - 9/25}} = \frac{T_0}{\sqrt{16/25}} = \frac{T_0}{4/5} = \frac{5}{4}T_0$$

As expected, $T$ is longer than $T_0$. Solving for $T_0$, we get $T_0 = (4/5)T$.

Let us call the unknown speed of the life-raft, as measured by the stationary observer, $v$. Let $d$ be the distance the life-raft moves, and $t$ the time it takes, all as measured by the stationary observer; then $v = d/t$. Now $d$ is not just the length of the space-ship, since during the $t$ the space-ship has moved forward a distance $(3/5)ct$ (speed times time). Therefore, $d$ is the length of the space-ship plus $(3/5)ct$. The length of the space-ship is not $L$, because the moving space-ship appears to be contracted by the factor $\sqrt{1 - [(3/5)c]^2/c^2} = 4/5$; therefore;

$$d = \frac{4}{5}L + \frac{3}{5}ct$$

and,

$$v = \frac{d}{t} = \frac{4}{5} \cdot \frac{L}{t} + \frac{3}{5}c$$

Now, $t$ can also be related to the proper time, $T_0$, by the time dilation formula:

$$t = \frac{T_0}{\sqrt{1 - (v^2/c^2)}}$$

Notice that the unknown speed, $v$, appears in the formula for $t$. Since we know that $T_0 = (4/5)T$:

$$t = \frac{(4/5)T}{\sqrt{1 - (v^2/c^2)}}$$

We substitute this formula for $t$ in the formula above for $v$, where $t$ appears in the denominator of the first term. The factor 4/5 cancels, and we get:

$$v = \frac{L}{T}\sqrt{1 - (v^2/c^2)} + \frac{3}{5}c$$

But $L/T$ is just the life-raft's speed relative to the space-ship, namely, $(3/5)c$, so:

$$v = \frac{3}{5}c\sqrt{1 - (v^2/c^2)} + \frac{3}{5}c = \frac{3}{5}c\left[\sqrt{1 - (v^2/c^2)} + 1\right]$$

This is an equation for $v$ that is not hard to solve. See the Appendix for the method. The answer is $v = 15/17\ c$. You may check easily that the equation is indeed true if $v$ is replaced by $15/17\ c$ both places it occurs.

So, relative to the earth, the life-raft is not going at speed $6/5c$, faster than light, but rather at speed $15/17c$, slightly slower than light. Common sense addition of speeds does not hold in relativity.

We obtain the general rule by repeating the same thought-experiment with the space-ship and the life-raft traveling at arbitrary speeds. Let the speed of the space-ship (relative to the "stationary" observer) be $v_1$, and the speed of the life-raft relative to the space-ship be $v_2$. According to Galileo and Newton, the speed of the life-raft relative to the stationary observer would be $v_1 + v_2$. According to Einstein's relativity, the answer is (see Appendix):

$$v = \frac{v_1 + v_2}{1 + (v_1 v_2/c^2)}$$

This relativistic rule for addition of speeds is very important. First of all, notice that if $v_1$ and $v_2$ are both $3/5c$, the rule says $v = 15/17c$; our example was a special case. Secondly, notice that if both speeds, $v_1$ and $v_2$, are very small compared to the speed of light, then $v_1/c$

and $v_2/c$ are very small numbers, so $v_1 v_2/c^2$ is tiny. Not much error will be made if we neglect $v_1 v_2/c^2$ compared to one in the denominator of the formula for $v$; and the Galilean approximation, $v = v_1 + v_2$, is a good one. For example, suppose both $v_1$ and $v_2$ are 25,000 miles per hour, which is about the fastest speed anyone has ever traveled (on real moon-bound space-ships). Even this speed is only 1/27,000 the speed of light, and the error in using the common sense rule, $v = v_1 + v_2$, is slightly greater than one part in a trillion. In ordinary usage, the relativistic correction to the addition of velocities is too small to be measured. However, relativistic speeds are obtainable in cyclotrons and particle accelerators, and occur naturally in the motions of atomic electrons, in the radiation emitted by the sun and by radioactive elements, in cosmic rays, and in the relative speeds of distant stars. In all these phenomena, the relativistic addition of speeds, as well as time dilation and length contraction, are confirmed in detail.

No matter what $v_1$ and $v_2$ are, if both are less than $c$ so is $v$. You cannot attain speeds faster than $c$ simply by ordinary accelerations. By this we mean the following: Suppose, relative to some observer, an object travels very fast, almost at the speed of light; and suppose there is a force on it. Imagine a second observer moving with the object; to him it is at rest. The force accelerates it; so after some time, it has acquired some speed $v$ that, at least if we wait a short enough time, must be less than $c$, since it starts at zero. The speed according to the first observer is obtained by the relativistic formula for adding speeds, and so is always less than $c$. It is impossible to accelerate something up to speed larger than the speed of light using ordinary forces. Nothing may go faster than light.

Suppose the space-ship, moving at speed $v_1$, sends out a beam of light instead of a life-raft. The speed of the beam of light relative to the space-ship is $c$, according to the second postulate. What is its speed relative to the stationary observer? We can find out by using the formula with $v_2$ replaced by $c$:

$$v = \frac{v_1 + c}{1 + (v_1 c/c^2)} = \frac{v_1 + c}{1 + (v_1/c)} = \frac{c[(v_1/c) + 1]}{1 + (v_1/c)} = c$$

If the speed of the beam relative to the space-ship is $c$, its speed relative to the stationary observer is also $c$, no matter what the speed $v_1$ of the ship is; $v_1$ simply cancels out. The speed of light is always $c$, relative to everybody. The relativistic rule for adding speeds is consistent with the second postulate: Light travels at speed $c$ relative to any observer.

The logic is very tight here. On the face of it, the two postulates of the theory of relativity seem irreconcilable. Yet, as we follow through their consequences carefully, we find no contradiction. From the postulates, we found the rule for time dilation and length contraction. Proceeding, we deduced a formula for the addition of speeds. A particular consequence is that whatever moves at speed $c$ relative to one observer, also moves at speed $c$ relative to any other observer. As Einstein indicated, the second postulate does not lead to any logical contradictions.

The fact that it is impossible to accelerate an object to a speed above the speed of light means that Newton's laws must in some way be modified. Since Newton's laws work well for almost all applications, the new laws should differ from them only for speeds close to the speed of light.

As an object's speed approaches $c$, it becomes harder and harder to accelerate. The same force results in a smaller and smaller change in speed. In other words, an object's resistance to acceleration, its inertia, increases. Einstein found that the form of Newton's second law—"Force equals the rate of change of momentum"—could be saved only if the meaning of momentum were modified. Specifically, instead of defining momentum as simply $mv$, mass times speed, the relativistic version of momentum must be:

$$\frac{mv}{\sqrt{1 - (v^2/c^2)}}$$

Or, to put it differently, one may preserve the old definition of momentum, but the mass of an object then depends on its speed! The relativistic mass, or inertia, is:

$$\frac{m}{\sqrt{1 - (v^2/c^2)}}$$

For slow speeds, these definitions agree with Newton's, so relativistic mechanics does not change the description of ordinary phenomena. This is because the ubiquitous factor, $\sqrt{1 - (v^2/c^2)}$, is indistinguishable from one if $v/c$ is very small. As $v$ approaches $c$, $v^2/c^2$ becomes almost one; and $\sqrt{1 - (v^2/c^2)}$ becomes close to zero. Since it is in the denominator of the formulas for relativistic momentum and inertia, these quantities become very large so that the resistance to acceleration is very large also. If its speed were to become equal to $c$, its inertia (mass) and momentum become infinite.

Einstein showed that with this definition, Newton's third law still holds for collisions between massive objects. The total momentum of any system of colliding objects is unchanged by collisions; and further, only with this relativistic definition of momentum is it conserved according to any observer, no matter with what speed he is moving relative to the objects. For objects that act on each other at a distance, the situation is a little more complicated. Electric and magnetic forces, for example, propagate by means of the electro-magnetic field. We started this section by observing that the exact description of the electromagnetic force, for example, how much is electric and how much is magnetic, depends on the relative motion of the observer and the moving charged objects, and can get very complicated. However, it all works out. Maxwell's complete theory of electromagnetic forces, together with the waves that can propagate with the field, make complete and consistent sense only with the ideas about space and time described by the theory of relativity.

Newton's theory of gravity, however, was not so easily accommodated.

## A NEW VIEW OF GRAVITATION

Special relativity slowly embedded itself in the consciousness of the physics community. At no time was there much serious contention that it might be wrong, yet, nevertheless, the new philosophy of physics it called for required getting used to. As the theory of relativity became better known, so also did its creator. Einstein, the obscure patent officer, developed an extensive correspondence, and often personal acquaintance, with any who cared to discuss the deep questions of physics. The few brief years before the first world war were idyllic for physics. Physics was a truly free international community whose members corresponded and traveled freely across international boundaries. No longer did a visit like Sir Humphry Davy's to France seem remarkable. The mails and railroads were so good that people from different countries now were doing research together day by day. Of course, nothing suggested that physics might have any more political significance than, say, Einstein's other love, music. All this was to end abruptly, never to return, in 1914.

Offers of academic positions, once all but closed to Einstein, became plentiful. Still unknown to the general public, he had become a full-fledged member of the scientific community. In 1908, he lectured at the University of Berne without relinquishing his position at the patent office. He was to be notable for lecturing with complete lucidity without notes and for his willingness, remarkable in the formal university atmosphere of the time, to promote discussion with his students during class. The year 1909 found him giving his first invited paper before the German Physical Society; he later remarked that up to the age of thirty he had "never met a real physicist." In that year, he forsook his peaceful patent office once and for all to become an associate professor at Zurich. In short order, he left Zurich for a professorship at the German University at Prague, only to return soon thereafter to Zurich. He was determined to attain as soon as possible a position that would permit the same freedom he had to forgo when he embarked upon an academic career.

Einstein left the nurturing of his brain-child, special relativity, to others. He was preoccupied with the paradox of photons, and the restriction of relativity theory to inertial "frames of reference," i.e., to observers in uniform motion relative to one in which the laws of physics held. No one but Einstein showed concern for this problem. (After all, was not Newtonian mechanics limited in the same way?) This, despite the doubts, which Mach had raised, that an inertial frame of reference, i.e., one in which there were no forces, existed even in principle. Special relativity could only be preserved if the relativity principle could be extended to *all* frames of reference, even those accelerating or rotating relative to each other. The world is full of forces, so Einstein tried to extend the relativity principle to *all* systems, including accelerating ones. Yet, it appeared that an accelerating observer could *feel* that he was accelerating. Relative to the observer, things did not move uniformly in straight lines. How could relativity apply to such a situation?

Einstein found the key in Galileo's famous experiment in the Leaning Tower of Pisa. All objects fall with the same acceleration. In Newton's mechanics, this fact was a bit artificial. The gravitational force on an object depended on its "gravitational" mass, which happens to be the same as the "inertial" mass that enters into the second law. Gravitational mass, which determines the strength of the attractive gravitational force, is equal to a body's inertial mass, which measures its resistance to acceleration according to Newton's second law of motion. From Newton's point of view, this is an interesting coincidence. For example, there is no universal rule relating the electrical force on an object to its mass. If the earth attracted an apple through electric forces, all apples would not fall at the same rate; those more highly charged would fall faster. Newton himself, impressed by the equality of gravitational and inertial mass, had performed many experiments to verify it, by constructing pendulums of the same size out of different materials and checking that they all swing back and forth at the same rate.

Einstein perceived that coincidences like this do not just happen;

rather that the Tower of Pisa experiment must express a deeper underlying natural principle. Because of the equality of gravitational and inertial mass, gravity, unlike other forces, has something special to do with the structure of space and time.

Suppose, when Galileo threw the two balls off the Tower of Pisa, he had jumped off with them. What would he have observed? Relative to his astonished companions who remained behind, both balls and Galileo would have fallen with the same unchanging acceleration, 32 feet per second per second, accelerated by the force of gravity. According to Newton's law of universal gravitation, this force is stronger on the more massive objects. But, the acceleration of the falling objects depends not only on the force, but is larger for smaller masses than for larger ones ($a = F/m$). Since the mass in Newton's second law—inertial mass—equals the mass in the law of universal gravitation—gravitational mass—the acceleration works out to be the same for all objects independent of the mass.

What would Galileo have seen? The two balls would have remained at rest relative to him, as if there were no gravitational force. Suppose Galileo, his two balls, and anything else he had were enclosed in a falling elevator, a "thought-experiment" Einstein was fond of when asked to explain his theory of gravity. This makes it clear Galileo would have had no way to decide whether he was in empty space, where there is no gravity, or was falling in a gravitational field.

We may turn the experiment around. Imagine the elevator really to be in empty space, far from any source of gravitational attraction, and let us supply it with rockets so that it is a space ship. If the space ship is "at rest" or moving with a constant speed relative to someone "at rest" (by which we mean only that he feels no forces of unknown origin), the relative positions of "Galileo" inside and all his equipment never changes. There are no forces on anything, so nothing can move. We have all seen this happen in television reports from real space ships. To an observer on the ground, the events in the space ship also accord with the laws of nature, as they must according to the principle of relativity. If the space ship is moving relative to the observer on the ground, all the objects in it continue to move with the same speed according to Newton's second law, because there are no forces on them.

Let us imagine that the elevator-spaceship is at rest relative to the stationary observer; in it "Galileo" and all his equipment are floating free. "Gallileo" presses the "up" button. The elevator-spaceship starts accelerating; "Galileo's" feet are planted firmly on the floor. What happens to the objects floating freely in space? Relative to the stationary observer, they remain at rest. Relative to "Galileo," accelerating in one direction as the spaceship speeds up, the objects accelerate in the other direction until they hit the floor. He concludes that they are falling because of a gravitational force! Relative to him, all objects are falling at the same rate—the rate the spaceship is "really" accelerating.

If gravitational and inertial mass were even slightly unequal so that different objects fell with different accelerations, then "Galileo" could tell the difference between a uniform gravitational field and a uniformly accelerating elevator; for then only the uniformly accelerating elevator would lead to the same apparent motions for all objects. But the facts are otherwise. Years of experimentation had proven the equality of gravitational and inertial masses. Consequently, there can be no difference between physical systems in a uniform gravitational field and in accelerating reference frames.

Einstein suggested that the equivalence between constant, uniform gravitational fields, such as the gravitational field at the surface of the earth, was not just a property of gravity but an example of a general law of nature, which he called the "Principle of Equivalence." This principle says that all the laws of nature are the same in a constant gravitational field as they are to a constantly accelerating observer in the absence of a gravitational field. In other words, if the spaceship is accelerating at 32 feet per second per second, all objects in it would fall toward the floor with that same acceleration. "Gallileo," inside the elevator-spaceship could conclude either that he was accelerating at 32 ft/sec/sec or that there was a gravitational force causing all objects to accelerate toward the floor at the same rate. Einstein's Principle of Equivalence asserts that this idea applies not only to gravitational experiments such as dropping balls but to all phenomena, including those of electricity, magnetism, and light. You cannot do an experiment of any kind to tell whether you are in a constant gravitational field or, instead, uniformly accelerating.

The Principle of Equivalence provides the answer to the question: "Does light fall?" Suppose a beam of light is sent out exactly horizontally from one side of a room. Does it arrive at the wall across at exactly the same height; or does it, like material objects, fall under the influence of gravity? Posed in this fashion, the question is difficult to answer experimentally. Why? Because light, traveling at 186,000 miles per second, travels one mile in 1/186,000 of a second. If light falls like material objects, we may ask how far it falls while traveling one mile horizontally. In any time, $t$, it falls a distance $(\frac{1}{2})gt^2$, according to Galileo's law. If $t$, measured in seconds, is 1/186,000 (the fraction of a second light takes to go one mile) and g is 32 ft/sec/sec, over one mile light will fall only about:

$$\frac{1}{2} \times 32 \times \left(\frac{1}{186,000}\right)^2 \text{ ft} = .0000000006 \ (6 \times 10^{-10}) \text{ ft}$$

This distance is far too small to be measurable, much less the distance light might fall across an ordinary size room.

Nevertheless, the fact that light falls this tiny distance is precisely what the principle of equivalence predicts. Again consider "Galileo" in his spaceship-elevator, accelerating in empty space at 32 ft/sec/sec. A light beam, according to an observer at rest, shoots across the spaceship elevator in a straight line. (Light always travels in straight lines; this has been believed since the time of Euclid, and is still the only experimental definition we have of "straight line.") But, from "Galileo's" point of view in the accelerating vehicle, the light seems to fall, accelerating downward at 32 ft/sec/sec. According to the Principle of Equivalence, all phenomena in the accelerating vehicle are the same as if the vehicle were at rest in a gravitational field. We conclude that in a gravitational field, for example here at the surface of the earth, light, like everything else, falls at the same acceleration, g, as in Fig. 9-8.

Thus far, Einstein had logically analyzed the foundations of

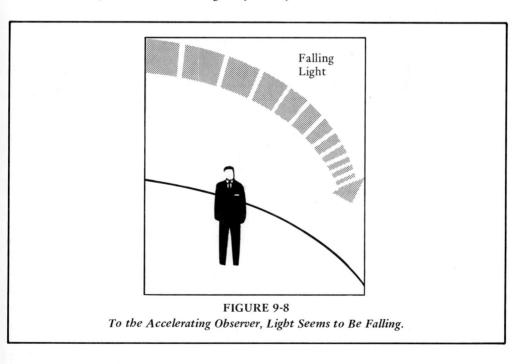

FIGURE 9-8
*To the Accelerating Observer, Light Seems to Be Falling.*

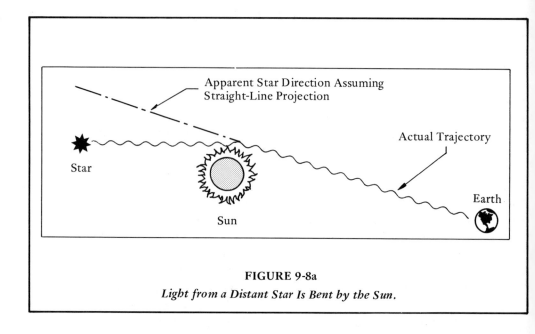

**FIGURE 9-8a**
*Light from a Distant Star Is Bent by the Sun.*

mechanics. Since all bodies are experimentally observed to accelerate equally under gravity, gravity must be equivalent to an accelerating reference frame. Relative to an accelerating reference frame, light will appear to fall. But now, Einstein could make a daring physical speculation. Since an accelerating frame is equivalent to a gravitational field, light must actually be deflected in passing near massive gravitational bodies. In 1911, Einstein calculated the slight deflection suffered by the light ray from a distant star, which passes near the sun, to be 0.83 seconds of arc, tiny but perhaps not beyond measurement during a total eclipse of the sun. Indeed, a Scotch-German physicist, Erwin Findlay-Freudlich, started a correspondence with Einstein in the hopes of measuring the gravitational deflection of starlight, only to have his elaborate plans dashed by the outbreak of World War I.

Despite the fact that he had renounced his German citizenship, in 1914 the German Academy of Sciences offered Einstein the position of "Professor Extraordinary" at its Kaiser Wilhelm Institute in Berlin. It carried the astounding salary of 12,000 marks per year. The opportunity to work with Max Planck, Walter Nernst, Fritz Haber, Otto Stern, Max Born, and the rest of the outstanding faculty and students at the Institute was highly attractive. Its facilities were the best in all Germany, if not Europe. Best of all, his position would leave him free of distraction, for it carried no specific responsibilities. Einstein could do what he pleased, so long as he did it in Berlin. The free time was the main consideration; for this, he would even return to the Germany he renounced. This is obvious; for, in his letter of acceptance, he wrote:[5]

> When I consider that each working day weakens my thinking powers, I can accept this post only with a certain degree of awe. Accepting the post, however, has encouraged me to think that one man cannot ask of another more than that he should devote his whole strength to a good cause, and in this respect, I feel myself truly competent.

Einstein knew he needed the time; for he perceived looming ahead of him a struggle with mathematics. He had formulated a principle of relativity appropriate to "inertial" frames of reference, i.e., those in which Newton's second law holds. His doubts about the existence of inertial frames drove him to extend the principle to uniformly accelerating frames of reference. In transforming to such a frame of reference, one had only to include the effects of an equivalent uniform gravitational field. The trouble was: nature does not ordinarily produce constant, uniform gravitational fields. For example, we know that the earth's gravitational attraction diminishes as the square of the distance from the center of the earth, and can be considered "constant" only as an approximation over a distance of a few miles.

Let us return to the elevator falling under gravity, only now let it fall in a realistic gravitational field; say, the earth's field. Newton had shown correctly that the acceleration of gravity diminished as the square of the distance from the center of the Earth. The objects at the top of the elevator, being farther from the Earth, would accelerate slightly more slowly than those at the bottom, which are closer to the Earth. If the elevator is small, the difference in accelerations is very small, but the observer in the elevator eventually would perceive the objects to separate from another. This example, and others, indicated that the principle of equivalence between uniform accelerated frames and uniform gravitation could itself be only an approximation that is valid only when you could consider distances "small" compared with the distance over which the gravitational field varied appreciably. By the same reasoning, we usually treat the earth's gravitational acceleration as uniform for most ordinary circumstances because we only consider objects falling over distances small compared to the distance over which the earth's field changes noticeably.

Einstein had single-handedly shown that Newtonian mechanics, so apparently complete and powerful, in fact is valid only for speeds much smaller than that of light. Special relativity solves this problem, but it too is an approximation since it is limited to inertial, non-accelerating, frames of reference. Using the equivalence of inertial and gravitational masses, the relativity principle may be extended to uniform gravitational fields and uniformly accelerating reference frames; but this too is only an approximation, since, in general, gravitational fields may be considered approximately uniform only over a small region. No longer would the simple logical analysis he had used with such devastating power suffice. Einstein had to face the fact that a completely general theory of relativity, involving non-uniform fields, like the earth's, would require a formidable mathematical apparatus. Only with deep mathematics could Einstein satisfy Faraday's requirement that gravitation, like electricity and magnetism, be describable as a continuous field rather than as action at a distance.

The outbreak of the first World War awakened Einstein briefly from his involvement with physics. Some weeks after the war started, Einstein wrote to Ehrenfest in Holland: [6]

> Europe, in her insanity, has started something unbelievable. In such times one realizes to what a sad species of animal one belongs. I quietly pursue my peaceful studies and contemplations and feel only pity and disgust.

To Einstein, war was simply not logical. From the safety of his Swiss citizenship, Einstein signed a number of pacifist manifestos, to the great annoyance of his colleagues, most of whom had enthusiastically joined the German war effort. Einstein's life as a public figure had commenced. Still, physics came first. The combined stresses of war and his mental abstraction caused Einstein and his wife to separate, she returning with their two sons to the safety of neutral Switzerland. Things proceeded in this fashion for two long years, years of complete mental absorption, years comparable in intensity with the two spent by Newton on the *Principia Mathematica,* years of horror for Europe, and years of sublime struggle for Einstein. His labors, and his last great work in physics, were over in 1916. He had created a truly general theory of relativity.

Einstein developed a whole mathematical theory of gravitation that was consistent with special relativity and included the Principle of Equivalence; the idea that at each point in empty space the laws were such that an observer falling freely in whatever gravitational field was there, like Galileo in his "elevator," could not detect any effects of gravity at all provided he did experiments on a small enough scale. This meant that Einstein had to rewrite the laws of gravity and of the motion of objects under the influence of gravity in such a way that they could be used not just by observers in intertial frames of reference, i.e., those moving with constant, straight line motions relative to each other, but even to observers *accelerating* relative to one another. Thus, one can use Einstein's theory to find out how things will move under gravity no matter how one is moving relative to anything else; and all observers will agree on questions whose answers they can compare, e.g., will two balls flying through the air in some way collide? In this way Einstein did away completely with any idea of absolute space. The way things moved under the influence of gravity depended simply on the relative positions and motions of everything else in the universe. An inertial frame of reference, one in which Newton's second law holds, is indeed just one that is at rest relative to the average motions of everything else in the universe. Thus, Einstein's theory answered Mach's criticism that Newtonian mechanics seemed to need the idea of absolute acceleration, while no experimental evidence existed for this idea.

Because Einstein's theory of gravitation was written to hold for *all* observers, he called it the General Theory of Relativity, in contrast to the Special, or Restricted Theory of Relativity, which he had developed in 1905. The mathematical details of his General Theory are beyond the level we are restricting ourselves to here. However, contrary to a popular misconception, they are not all that difficult or arcane; anyone with a scientific education can understand them.

Einstein's theory has a startling consequence: when large distances are considered, the laws of ordinary Euclidean geometry, the kind we have been using all along, are not exactly right. The basic arguments are simple. Our conception of space is defined by the procedures we undertake to locate the objects in it. To locate an object, we must send a signal to the object and receive a return signal. The best such signal is that which travels fastest over the most direct route. Therefore, the best we can do is to use light waves. In an inertial frame, the light will travel in a straight line. In this case, the geometry of space is the normal Euclidean geometry. On the other hand, in general, gravitational fields will curve the path of the light rays, as for example, in the case already mentioned of light from a star passing close to the sun. There is a theorem of geometry that says that if you add up the three angles in any triangle, the sum is always 180°. Now imagine three observers on three different stars, far apart; and suppose each one measures the angle subtended by the other two at his own star. We have observed that light is bent when passing through a strong gravitational field. Therefore, the direction a star appears to be in is not the direction it "really" is, and the sum of the angles will not come out quite 180° if the light from one of the stars has to pass near a massive object (like another star). It can be objected that here we are playing with words. The light does not really go in a straight line; so its sides do not really form a triangle. It is possible to talk this way, but it is not a useful viewpoint.

Non-Euclidean geometry is not all that mysterious either. For example, if you draw on the surface of a sphere a triangle whose three sides are great circles, and add up the three angles, the sum is always greater than 180°. Suppose the ancient Greeks had tried to test the theorem about the sum of angles by measuring the three angles of a very large triangle, with one corner at the North Pole and the other two on the equator, far apart. The two angles on the equator would be exactly 90° each, so the sum of the three would be greater than 180°. What would they have concluded? That Euclid's geometry was wrong? No, simply that it does not apply to the earth because the earth is not flat. The same thing happens in Einstein's theory, except that we are in three instead of two dimensions and it is difficult to visualize the analog of a sphere.

Einstein preferred another example. Suppose one measured, with a ruler, both the circumference of a disk and its diameter. According to geometry, their ratio will be exactly the number $\pi$. Next, suppose the disk is rotating, and one makes the same measurements. Whatever ruler one uses to measure the circumference will now be contracted, due to the contraction of lengths predicted by Special Relativity, so the measured circumference will be greater (more lengths of the ruler will be needed); and the ratio, as defined by the results of physical measurements, will be *greater* than $\pi$. But in General Relativity, the rotating reference frame is supposed to be just as good as the first one to do physics in. To the rotating observer, the rule that the ratio is $\pi$ just does not hold; his geometry has different rules. The fact that geometries other than the traditional one of Euclid could make logical sense had been discovered only about a generation before by the mathematicians Riemann and Lobachevski.

With the General Theory of Relativity, Einstein had a logically consistent description not only of electricity and magnetism, but now of gravity also. Although in principle the General Theory's predictions for gravitational effects differed profoundly from Newton's Universal Gravitation, the predictions of the two theories agree for ordinary, everyday phenomena. Only when you ask what happens over large, cosmological distances, or to objects moving at speeds comparable to the speed of light, or near very massive objects, like the sun, creating strong gravitational fields, are there measurable discrepancies predicted. Was there no experimental evidence? Did Einstein expect the world to accept the General Theory solely on the basis of the simplicity and clarity of its logical foundations? The answer is probably yes. The inner coherence of the General Theory, the way it combines the Principle of Relativity and the Principle of Equivalence, must have been enough for Einstein.

For the doubters, there was one fact, out of the millions of experiments and observations that were known, which Newtonian mechanics had failed to explain. Recall that almost one hundred years earlier, Leverrier and Adams had studied the motions of the planet Uranus. After all the effects of the other known planets, especially Jupiter and Saturn, had been taken into account, as well as the attraction of the sun, Uranus' motion still could not be explained by the Law of Universal Gravitation. Leverrier and Adams did not conclude that Newton's Laws were wrong. Rather, they guessed that an undiscovered planet was responsible for the discrepancy. They calculated where it should be, and directed astronomers where to point their telescopes in order to discover Neptune.

Leverrier found a similar problem with the planet Mercury. Mercury's orbit is an ellipse, like all the planets, but is more elongated, more cigar-shaped, than any of the others. Centuries of observation had shown that Mercury did not quite repeat its orbit exactly each revolution. Rather, while the shape of the orbit hardly changed, the orientation of Mercury's ellipse gradually rotated. Imagine drawing a line from the sun to the point on Mercury's orbit where it is closest to the sun, Mercury's "perihelion." This line, from the sun through the perihelion of Mercury's orbit, moves slightly each revolution. Over one century, it rotates almost ten minutes of arc. This phenomenon is called the precession of the perihelion of Mercury.

Leverrier tried to explain the precession of the perihelion of Mercury by the gravitational attraction of the planets, and was able to account for most of it. But, a tiny discrepancy remained between the calculated and the observed rate of the precession of the perihelion of Mercury. It amounted to 43 *seconds* of arc per *century*. It was possible, of course, that such a tiny effect was an observational error. Leverrier, after his success with Neptune, thought it more likely that Mercury's motion would be explained by another new, undiscovered planet, this time closer to the sun than Mercury. Lost in the sun's glare, it was not unreasonable that such a planet might exist undetected. Leverrier had some idea of its mass and orbit, and even named it Vulcan. For decades astronomers searched for the planet Vulcan, occasionally even found it, but always lost it again.

Einstein calculated the orbit of a planet around a massive object like the sun, according to the General Theory of Relativity. The orbits were ellipses, closed on themselves, to a very good approximation. But not quite. Very close to the sun, objects do not move quite as Newton's inverse square law predicts. Here is the modest end of Einstein's 1916 paper introducing the General Theory of Relativity:[7]

> According to this, a ray of light going past the sun undergoes a deflexion of 1.7"; and a ray going past the planet Jupiter a deflexion of about .02".
>
> If we calculate the gravitational field to a higher degree of approximation, and likewise with corresponding accuracy the orbital motion of a material point of relatively infinitely small mass, we find a deviation of the following kind from the Kepler–Newton laws of planetary motion.
>
> Calculation gives for the planet Mercury a rotation of the orbit of 43" per century, corresponding exactly to astronomical observation (Leverrier); for the astronomers have discovered in the motion of the perihelion of this planet, after allowing for disturbances by other planets, an inexplicable remainder of this magnitude.

And so, the new theory resolved a century-old puzzle. Once again, extreme accuracy of a scientific observation proved decisive. And once again, the motion of the planets was decisive in teaching us the laws of physics.

Einstein proposed a second test of the General Theory, but he had no evidence. Recall that in 1911 he had suggested that the Principle of Equivalence required that light from a distant star be bent as it went near the sun, and had estimated 0.83" for a ray of light grazing the sun's surface. Now, in 1916, with the General Theory completed, he realized that he had made an error in using Newton's inverse-square law for the strength of gravity close to the sun. Doing the calculation correctly, he now predicted that light would be bent twice as much, about 1.7 seconds of arc.

In England, a colorful physicist and astronomer, Arthur Stanley Eddington, received a copy of Einstein's 1916 paper through mutual friends in neutral Holland. Eddington was a Quaker and a pacifist. He refused to be inducted into the army, and insisted on being exempted as a conscientious objector; to his disappointment he was deferred because the draft board felt that he was more useful as an astronomer than as a soldier. He must have felt affinity for the Swiss Jew in Berlin, who also hated war.

Eddington, an immediate convert to the new ideas, remained an enthusiastic supporter and popularizer of Einstein's theory for many years. The English government was interested in investigating the feasibility of any new German scientific program. Eddington called its attention to the "unique opportunities afforded by the eclipse of

1919," for testing Einstein's theory. Secret plans were made for British expeditions to Sobral in Brazil and Principe Island off the coast of Guinea in West Africa in order to test the theories of a scientist working in the enemy capital. The war was ended in November, 1918, but the eclipse expeditions went ahead.

The deflection of starlight near the sun was measured on May 29, 1919. Eddington waited until the next meeting of the Royal Society in November to announce the dramatic result. The deflection of starlight by the sun had been observed during the eclipse, and agreed with the predictions of the General Theory of Relativity. The philosopher, A. N. Whitehead, described the scene at the Royal Society meeting:[8]

> The whole atmosphere of tense interest was exactly that of the Greek drama; we were the chorus commenting on the decree of destiny as disclosed in the development of a supreme incident. There was dramatic quality on the very staging:—the traditional ceremonial, and in the background the picture of Newton to remind us that the greatest of scientific generalisations was now, after more than two centuries, to receive its first modification.

A war-weary world was impressed by the universality of science, by a British expedition confirming a German theory. It was dazzled by a man, a pacifist, who passed the agonizing war years contemplating the mysteries of the universe, successfully replacing Newton's laws by his own more profound analysis of space, time, and gravity.

Einstein was catapulted to instant fame upon a tidal wave of publicity. The only comparable event in the history of science was the reception accorded Galileo's *Starry Messenger.* Like Galileo, he and his life were permanently altered. His views on religion, philosophy, and politics were insistently probed and incessantly discussed. Like Galileo, he used the fame of his discoveries and name to advocate fundamental changes in society. This time, it was not to establish the position of rational thought and methodology within science, but to advocate the use of the rationality by then so characteristic of science to solve the world's deep social problems. Einstein lent his name to a plethora of pacifist causes. Having suddenly become the world's leading Jew, his advocacy of Zionism was very influential. At the same time, the first demonstrations against "Jewish physics" began in Germany, demonstrations that would grow in bitterness and intensity as Hitler was finally swept into power.

Like Newton, Einstein saw his scientific creativity diminish after the age of forty. Both seem to have been exhausted by their prodigious efforts, their dedication to the austere life of a creator undermined by the civilized pleasures and distractions of fame. Einstein gradually drifted out of the mainstream of the science he did so much to create.

Einstein was away from Germany on a speaking tour when Hitler was elected Chancellor of the Third Reich. Einstein did not return, announcing his decision publicly in protest. He eventually settled in America, where so many of his European colleagues, hounded by Hitler's persecution, would soon follow, to change the face of American science and society. The Institute of Advanced Study in Princeton was his refuge. When, in 1939, the ominous possibility that German experiments on nuclear fission contained the key to a new and most horrible weapon dominated all scientific discussion, a number of Einstein's European colleagues prevailed upon him to use his prestige to awaken America to the danger. He did so. In a letter to President Roosevelt, Einstein, the world's most famous pacifist, advocated the construction of the atomic bomb. Fate, in the form of Hitler's barbaric persecutions, had forced upon this man of reason and peace a heart-rending turnabout.

Between the first two world wars, Einstein's General Theory of Relativity was developed by Einstein and others to include a cosmology, the study of the universe as a whole. What emerged was a picture of an expanding universe; its components are huge groups of stars, called galaxies (from the Greek word for "milky way"), separating from each other like dots on the surface of a balloon that is being blown up.

The evidence for this expansion is the "red shift" of light from distant galaxies. Stars are composed mainly of hydrogen and helium, which emit light of characteristic colors. If a star is moving rapidly away, the wavelength of the light it emits, as observed by us, will be longer than it would be to an observer at rest relative to the star. It will be redder. This is because a star has moved between the instants when the star emits two successive crests of a lightwave so that the spacing between the crests is larger than it would be if the star were at rest. This effect, called the Doppler shift, provides a tool modern astronomers have used to observe that the distant galaxies are moving away from us. The farther away they are, the faster the galaxies are receding. The most distant ones recede at a good fraction of the speed of light.

This expanding universe seems to be well described by Einstein's theory of gravitation, and completes the change that began with Copernicus in our vision of man's place in the universe. Indeed, just as Greek science had put man at the center of an intimate universe, so modern cosmology is creating a new vision, whose outlines are being completed in the language of Einstein's general theory. We are alone in the solar system, on the single fertile planet circulating the sun, which is itself a common nondescript star far from the center of our galaxy, the Milky Way, which contains tens of billions of stars. The known galaxies themselves number in the billions, all rushing apart from one another, probably the aftermath of a primeval explosion in the beginning.

## Summary

*Maxwell's electrodynamics seemed to need a medium to transmit electric and magnetic forces: something to vibrate when electromagnetic waves were transmitted. Furthermore, it seemed to speak of absolute motion: only charges in motion exert, and are acted upon, by magnetic forces. The medium was called the aether; and the old idea of absolute motion was resurrected in the form of motion relative to the aether.*

*However, no experiment was able to detect absolute motion. The most accurate attempt was the experiment of Michelson and Morley. Their instrument, called an interferometer, could detect the interference of two beams of light, one parallel to, and one perpendicular to, one's motion through the aether. No effect was ever found; the earth seemed always at rest in the aether.*

*In 1905, Albert Einstein proposed that absolute motion was a meaningless concept, and that Galileo's principle of relativity included all physical phenomena. His "theory of relativity" was founded on two postulates. First, all the laws of physics are equally valid to all observers in relative, uniform motion (the principle of relativity); and, second, Maxwell's electrodynamics, including the conclusion that electromagnetic waves travel at the speed c, is one of these laws.*

*Einstein explained that these postulates seem paradoxical only if we believe in an absolute definition of time intervals and lengths. He carefully examined how one measures times and distances, and concluded that his postulates were not logically irreconcilable; however, if they are correct, one must abandon the idea of absolute time. If two events appear simultaneous to one observer, they will not be simultaneous to another observer who is moving relative to the first one. Quantitatively, if the time interval between two ticks of a clock (or any two events) is $T_0$ according to an observer at rest relative to the clock, the time interval T according to an observer moving relative to the clock at speed v is:*

$$T = \frac{T_0}{\sqrt{1 - (v^2/c^2)}}$$

$T$ is longer than $T_0$, and the effect is called time dilation.

Lengths of moving objects are contracted. An object of length $L_0$, measured at rest, has a shorter length, $L$, measured by a moving observer. If he is moving at speed $v$ parallel to the object, then:

$$L = L_0\sqrt{1 - v^2/c^2}$$

Einstein showed that time dilation and length contraction do not lead to any contradictions, and have many interesting consequences, including the relativistic addition of speeds. If an object moves at speed $v_1$ relative to some observer, and a second object moves at speed $v_2$ relative to the first object; the second object moves at speed $v$ relative to the first observer where:

$$v = \frac{v_1 + v_2}{1 + (v_1 v_2 / c^2)}$$

If $v_1$ and $v_2$ are both less than $c$, so is $v$. If either is equal to $c$, then $v = c$.

The momentum of an object of mass $m$, moving at speed $v$, is $mv/\sqrt{1 - (v^2/c^2)}$. For speeds close to $c$, the momentum becomes very large. Therefore, a force on an object, while increasing its momentum, can never increase its speed beyond $c$.

Ten years later, Einstein completed the General Theory of Relativity, which included a new theory of gravity. Its basic postulate, called the Principle of Equivalence, states that motion under the influence of gravity, relative to an observer at rest in the gravitational field, is indistinguishable from motion in the absence of gravity relative to an accelerating observer, provided one considers a region of space small enough so that within it the strength of the gravitational field is effectively constant. For most phenomena, Einstein's theory made the same predictions as Newton's. But, Einstein's theory also explained a tiny discrepancy in the orbit of Mercury, which Newton's theory had been unable to account for, and predicted the deflection of starlight by the sun, which was confirmed during a total eclipse in 1919. The General Theory of Relativity is expressed by mathematical equations that are true for all observers, even accelerating ones. No longer is there anything special about those observers, all of whom are moving uniformly relative to one another, for whom Newton's laws correctly describe motions that are slow, relative to the speed of light. These special "inertial frames of reference", in which Newton's laws hold for objects moving slowly compared to the speed of light, are at rest relative to the average motion of all the matter in the universe (the "fixed stars").

# Appendix Chapter 9

*Time-Difference in the Michelson-Morley Experiment*

Michelson based his experiment on a remarkably simple idea. Imagine the aether to be like a flowing stream of water, and light to be like a fish swimming at a constant speed relative to the water. Then it takes less time to swim *across* a moving stream and back than to swim the same distance upstream and back. Suppose the fish swims at speed $c$ relative to the water, which is flowing with speed $v$, across a stream of width $w$. Since the water carries the fish downstream, the fish must point a bit upstream in order to arrive precisely at the opposite bank, as in Fig. 9-9. (Figure 9-9 is drawn by an observer at rest relative to the water. Relative to the shore, the line labeled $ct$ is actually horizontal.) In the time $t$, the fish swims a total distance $ct$, the stream carries him downstream a distance $vt$. Using Pythagoras' theorem, the time $t$ to cross the stream may be computed:

$$c^2 t^2 = v^2 t^2 + w^2$$

or,

$$t^2(c^2 - v^2) = w^2$$

or,

$$c^2 t^2 \left(1 - \frac{v^2}{c^2}\right) = w^2$$

or,

$$t^2 = \frac{w^2/c^2}{1 - (v^2/c^2)}$$

By taking the square root, we arrive at:

$$t = \frac{w/c}{\sqrt{1 - (v^2/c^2)}}$$

If the stream is not moving, $v = 0$, then $t$ is just the width of the stream divided by the speed with respect to the water, $w/c$. If $v$ is non-zero, the time is larger. If the fish can only swim at the water speed, so that $v/c = 1$, it will never reach the opposite bank; for then $\sqrt{1 - (v^2/c^2)} = 0$ and $t$ is infinite. To go back and forth, the time is clearly twice as great, so over all:

$$t = \frac{2w/c}{\sqrt{1 - (v^2/c^2)}}$$

Let us now imagine that the fish swims upstream the same distance $w$, and then back downstream to its starting point. Swimming upstream, its speed relative to the banks of the stream is $c - v$; swimming downstream, its speed is $c + v$. The time to go upstream is $w/(c - v)$, and to go downstream is $w/(c + v)$. To go back

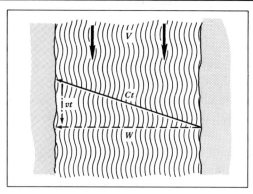

**FIGURE 9-9**
*Calculation of Time for Fish to Swim Across River.*

and forth then, the total time elapsed $t'$, is:

$$t' = \frac{w}{c-v} + \frac{w}{c+v} = \frac{2w}{c[1-(v^2/c^2)]}$$

When $v = 0$, the two times $t$ and $t'$ are the same. However, when the stream is moving $v \neq 0$, the two expressions differ in that one has a square root in the denominator and the other does not. Since $1 - (v^2/c^2)$ is larger than $\sqrt{1-(v^2/c^2)}$, it takes longer to swim upstream and back than across and back. Let us denote the time difference between the two propagation paths by $T = t' - t$; then (the symbol $\approx$ is often used to mean "roughly equal to"):

$$T = \frac{2w}{c}\left[\frac{1}{1-(v^2/c^2)} - \frac{1}{\sqrt{1-(v^2/c^2)}}\right] \approx \frac{w}{c}\left(\frac{v^2}{c^2}\right)$$

where the last form is approximately true when $v/c$ is very much less than one.

### Contraction of Moving Yardsticks

In discussing time dilation, we assumed that if the two light clocks were of identical construction, the length of the moving clock, as viewed from the stationary clock, was also $W$. We can prove that this is always true, provided that the relative motion of the two clocks is at right-angles to the line connecting the two mirrors. Suppose we arrange it that the two clocks leave marks on each other as they pass by each other, and suppose the bottoms of the two clocks coincide. Do the tops coincide? If the top of the moving clock left a mark on the stationary clock below the top, then a stationary observer could conclude that moving clocks are shorter. But to the "moving" observer, the "stationary" clock, moving relative to him, was longer. The "moving" observer concludes that moving clocks are longer. The principle of relativity assures us that there can be no difference between the laws of nature as known to the two observers. The clocks do not change lengths.

Now suppose instead that the moving clock is moving parallel to the line connecting the two mirrors at some speed $v$. Now the argument just given will not work. One cannot just compare the marks on the two clocks, but also has to know *when* the marks were made. Let us analyze this situation more carefully.

Suppose the moving clock emits a flash of light from the rear mirror. How long does it take to reach the front mirror? According to an observer moving along with the moving clock, this time interval is, just as before:

$$T = \frac{W}{c}$$

So is the time for the return journey, $T$. The total time for the pulse of light to leave the rear mirror and return is:

$$2T = 2\frac{W}{c}$$

Now consider the same events from the viewpoint of the "stationary" observer. Let $T_1$ be the time for the light to go from the rear mirror to the front mirror, and let us admit the possibility that from his measurements the stationary observer determines the length of the moving clock not to be $W$, but some other number, $W'$. Now the distance the light goes during time interval $T_1$ is not just $W'$ because during that time the front mirror has moved a bit. Since its speed is $v$, it has moved the distance $vT_1$; and, therefore, the total distance the light goes, according to the stationary observer, during the time $T_1$ is:

$$W' + vT_1$$

Its speed is this distance divided by the time, namely:

$$\frac{W' + vT_1}{T_1} = \frac{W'}{T_1} + v$$

According to the second postulate of the Theory of Relativity, this speed is $c$:

$$c = \frac{W'}{T_1} + v$$

Let $T_2$ be the time the light takes, according to the stationary observer, to make the return journey from the front mirror back to the rear mirror. By an argument similar to the one we made for the forward trip, the distance traveled is shorter than $W'$ by the distance the rear mirror goes during that time, namely, $vT_2$. The distance is $W' - vT_2$, and the speed is:

$$\frac{W' - vT_2}{T_2}$$

Again, this must be $c$:

$$c = \frac{W' - vT_2}{T_2} = \frac{W'}{T_2} - v$$

### Einstein and Relativity

Now we have two equations for $c$:

$$c = \frac{W' + vT_1}{T_1} \qquad c = \frac{W' - vT_2}{T_2}$$

We solve them both for $W'$, and so get two equations for $W'$:

$$W' = T_1(c - v) \qquad W' = T_2(c + v)$$

Subtract the second from the first:

$$T_1(c - v) = T_2(c + v)$$

Now we can express $T_1$ in terms of $T_2$, $c$, and $v$, by dividing the equation above by $c - v$:

$$T_1 = T_2 \frac{c + v}{c - v}$$

Now $T_1 + T_2$ is the total time for the light's round trip, which to a moving observer is $2T$. According to the time dilation rule for an imaginary clock located at the rear mirror, this time is:

$$T_1 + T_2 = 2T/\sqrt{1 - (v^2/c^2)}$$

Since $T = W/c$, this equation becomes:

$$T_1 + T_2 = \frac{2W}{c} \frac{1}{\sqrt{1 - (v^2/c^2)}}$$

or,

$$T_1 = \frac{2W}{c} \frac{1}{\sqrt{1 - (v^2/c^2)}} - T_2$$

Next we write $T_1$ in terms of $T_2$:

$$T_2 \frac{c + v}{c - v} = \frac{2W}{c} \frac{1}{\sqrt{1 - (v^2/c^2)}} - T_2$$

Add $T_2$ to both sides:

$$T_2 + T_2 \frac{c + v}{c - v} = \frac{2W}{c} \frac{1}{\sqrt{1 - (v^2/c^2)}}$$

or,

$$T_2 \left[1 - \left(\frac{c + v}{c - v}\right)\right] = \frac{2W}{c} \frac{1}{\sqrt{1 - (v^2/c^2)}}$$

For the bracket on the left-hand side, we may write:

$$\left[1 + \left(\frac{c + v}{cv}\right)\right] = \left[\left(\frac{c - v}{c - v}\right) + \left(\frac{c + v}{c - v}\right)\right] = \left[\frac{(c - v + c + v)}{(c - v)}\right] = \frac{2c}{(c - v)}$$

Thus,

$$T_2 \left[ \frac{2c}{(c-v)} \right] = \frac{2W}{c} \cdot \frac{1}{\sqrt{1-(v^2/c^2)}}$$

or,

$$T_2 = \frac{2W}{c} \cdot \frac{c-v}{2c} \cdot \frac{1}{\sqrt{1-(v^2/c^2)}} = \frac{W}{c^2} \cdot \frac{(c-v)}{\sqrt{1-(v^2/c^2)}}$$

Now we can find the length $W'$ from the rule, $W' = T_2(c+v)$:

$$W' = \frac{W}{c^2} \frac{(c-v)}{\sqrt{1-(v^2/c^2)}} (c+v)$$

Since $(c-v)(c+v) = c^2 - v^2$:

$$W' = \frac{W}{c^2} \cdot \frac{c^2 - v^2}{\sqrt{1-(v^2/c^2)}} = \frac{W[1-(v^2/c^2)]}{\sqrt{1-(v^2/c^2)}} = W\sqrt{1-(v^2/c^2)}$$

This is our result. Moving rigid bodies must be shortened, in the direction of the motion, by a factor $\sqrt{1-(v^2/c^2)}$. This effect is often called the Lorentz contraction, since it is similar to the proposal made, in a different context by Lorentz and Fitzgerald, to explain the Michelson-Morley experiment.

## *Relativistic Addition of Speeds*

Consider two objects moving in the same direction. The first, the "space-ship," is moving at speed $v_1$ relative to the "ground." The second, the "life-raft," is moving with speed $v_2$ relative to the first. What is the speed, call it $v$, of the life-raft relative to the stationary ground?

In the example we worked out, $v_1$ and $v_2$ were both $(3/5)c$. Then, $v$ satisfied the following equation:

$$v = \frac{3}{5} c \left[ \sqrt{1 - \frac{v^2}{c^2}} + 1 \right]$$

Multiply both sides by $(5/3)c$ and subtract 1:

$$\frac{5v}{3c} - 1 = \sqrt{1 - \frac{v^2}{c^2}}$$

Now square both sides of this equation:

$$\frac{25}{9} \cdot \frac{v^2}{c^2} - \frac{10}{3} \cdot \frac{v}{c} + 1 = 1 - \frac{v^2}{c^2}$$

Subtract 1 from both sides, and add $v^2/c^2$:

$$\frac{v^2}{c^2} + \frac{25}{9} \cdot \frac{v^2}{c^2} - \frac{10}{3} \cdot \frac{v}{c} = 0$$

now,

$$\frac{v^2}{c^2} + \frac{25}{9} \cdot \frac{v^2}{c^2} = \left(1 + \frac{25}{9}\right) \frac{v^2}{c^2} = \frac{34}{9} \cdot \frac{v^2}{c^2}$$

so,

$$\frac{34}{9} \cdot \frac{v^2}{c^2} - \frac{10}{3} \cdot \frac{v}{c} = 0$$

Divide each term by $v/c$, and add $10/3$:

$$\frac{34}{9} \cdot \frac{v}{c} = \frac{10}{3}$$

The solution is:

$$\frac{v}{c} = \frac{10}{3} \cdot \frac{9}{34} = \frac{15}{17}$$

We will only outline the method for the general problem here. It is elementary, but a bit long. As in the text, let $L$ be the length of the space-ship and $T$ the time it takes for the life-raft to go from the rear of the space-ship to the front, both as measured in the space-ship. Again, let $T_0$ be the proper time for the trip, i.e., the time as measured in the life-raft. Then, from the time-dilation rule:

$$T = \frac{T_0}{\sqrt{1 - (v_2^2/c^2)}}$$

or,

$$T_0 = T\sqrt{1 - \frac{v_2^2}{c^2}}$$

Again, let $d$ and $t$ be the distance and time interval as measured by the stationary observer; then:

$$v = \frac{d}{t}$$

and,

$$t = T_0 \bigg/ \sqrt{1 - \frac{v^2}{c^2}}$$

Now $d$ is the length of the space-ship, multiplied by the Lorentz contraction factor, plus the distance the space-ship travels during the time $t$. Since its speed is $v_1$:

$$d = L\sqrt{1 - \frac{v_1^2}{c^2}} + v_1 t$$

# Appendix Chapter 9

and,

$$v = \frac{d}{t} = \frac{L}{t}\sqrt{1 - \frac{v_1^2}{c^2}} + v_1$$

Substitute for $t$:

$$v = \frac{L}{T_0}\sqrt{1 - \frac{v_1^2}{c^2}}\sqrt{1 - \frac{v^2}{c^2}} + v_1$$

and substitute for $T_0$ in terms of $T$:

$$v = \frac{L}{T}\frac{\sqrt{1 - (v_1^2/c^2)}\sqrt{1 - (v^2/c^2)}}{\sqrt{1 - (v_2^2/c^2)}} + v_1$$

Since $L/T$ is $v_2$:

$$v = v_2\frac{\sqrt{1 - (v_1^2/c^2)}\sqrt{1 - (v^2/c^2)}}{\sqrt{1 - (v_2^2/c^2)}} + v_1$$

This is an equation for $v$ in terms of $v_1$, $v_2$, and $c$. The way to solve it is to subtract $v_1$ from both sides and then square both sides:

$$v^2 - 2vv_1 + v_1^2 = v_2^2\frac{[1 - (v_1^2/c^2)][1 - (v^2/c^2)]}{1 - (v_2^2/c^2)}$$

Multiply both sides by $1 - (v_2^2/c^2)$, and multiply everything out:

$$v^2 - 2vv_1 + v_1^2 - \frac{v^2 v_2^2}{c^2} + \frac{2vv_1 v_2^2}{c^2} - \frac{v_1^2 v_2^2}{c^2}$$

$$= v_2^2 - v_2^2\frac{v_1^2}{c^2} - v_2^2\frac{v^2}{c^2} + v_2^2\frac{v_1^2}{c^2}\frac{v^2}{c^2}$$

Next, collect the coefficients of $v^2$, the coefficients of $v_1$ and the terms independent of $v$ separately, all on the left-hand side:

405

$$v^2\left(1 - \frac{v_2^2}{c^2} + \frac{v_2^2}{c^2} - \frac{v_1^2 v_2^2}{c^2 c^2}\right) + 2v\left(\frac{v_1 v_2^2}{c^2} - v_1\right)$$

$$+ \left[v_1^2 - v_1^2 \frac{v_2^2}{c^2} - v_2^2 + \frac{v_1^2 v_2^2}{c^2}\right] = 0$$

There are some cancellations:

$$v^2\left(1 - \frac{v_1^2 v_2^2}{c^4}\right) + 2vv_1\left(\frac{v_2^2}{c^2} - 1\right) + (v_1^2 - v_2^2) = 0$$

This is a "quadratic equation" for $v$. In general, if:

$$ax^2 + bx + c = 0$$

the solutions are:

$$x = \frac{-b + \sqrt{b^2 - 4ac}}{2a}$$

and,

$$x = \frac{-b - \sqrt{b^2 - 4ac}}{2a}$$

In our case, we have:

$$a = 1 - \frac{v_1^2 v_2^2}{c^4}$$

$$b = 2v_1\left(\frac{v_2^2}{c^2} - 1\right)$$

$$c = v_1^2 - v_2^2$$

The solution is:

$$v = \frac{v_1 + v_2}{1 + (v_1 v_2/c^2)}$$

(The other solution to the quadratic equation is $(v_1 - v_2)/[1 - (v_1 v_2/c^2)]$. But, this is not a solution to the original equation. Any solution to the original equation must be a solution to the final one; however, the logic does not go backward.)

# REFERENCES

1. Mach, Ernest *The Science of Mechanics,* trans. by Thomas J. McCormack (LaSalle, Open Court Publishing Co., 1960), pp. 588–589.
2. Clark, Ronald W. *Einstein, the Life and Times* (World Publishing Co., New York and Cleveland, 1971), p. 39.
3. Ibid., p. 62.
4. Einstein, Albert "On the Electrodynamics of Moving Bodies," *Annalen der Physik 17,* 1905, trans. by W. Perret and G. B. Jeffery in *The Principle of Relativity* (Dover, 1923), pp. 37–38.
5. Clark, Ronald W. *Einstein, the Life and Times,* op. cit, p. 170.
6. Ibid., p. 183.
7. Einstein, Albert "The Foundation of the General Theory of Relativity," *Annalen der Physik 49,* 1916, trans. by W. Perret and G. B. Jeffery, op. cit., pp. 163–164.
8. Clark, Ronald W. *Einstein, the Life and Times,* op. cit., p. 232.

# Index

Absolute motion, 175ff, 370, 391, 397
Absolute space, 231, 365, 391
Absolute time, 168, 231ff, 370
*Academia dei Lincei,* 152, 192
Academy of Plato, 26, 39
Academy of Sciences (France), 192, 285, 309, 315
Acceleration, 161ff, 166, 180, 193, 206, 231, 336, 382, 384; absolute, 391; average, 166; definition, 165; due to gravity, 172ff, 195, 197, 208; instantaneous, 166, 199; of objects in uniform circular motion, 199, 235, 273
Action at a distance, 47, 315, 330
Adams, John Couch, 226, 393
Aether, 47, 286, 357–363, 370, 397, 399
Alhazen, 253–255, 260, 287
*Almagest,* 63, 106
Alphonsine tables, 73, 106
Ampère, André Marie, 315ff, 325, 326, 334, 336, 345–346, 368
Ampère's law, 320, 351
Angle of incidence, 251
Angle of refraction, 251
Annual motion, 13, 20, 39
Aphelion, 126, 136
Archimedes, 26, 49
Aristarchus of Samos, 35, 48–51, 52, 53, 55–58, 93, 95
Aristotle, 26, 29, 41, 43–48, 52, 63, 73, 76, 95, 109, 118, 148, 156, 157, 159, 160, 161, 172, 175, 179, 258, 272, 286, 297, 342, 365, 367; biography, 43ff; on effect of a resisting medium, 46; elements, 45, 52, 192, 247, 295; law of falling bodies, 46, 144, 211, 336; Lyceum, 26, 43; physics of, 43ff, 51, 143; on light, 245, 249, 287; on the possibility of a vacuum, 46
*Astronomia Nova* (*see* "The New Astronomy")
Astronomical unit, 86
Astronomy: ancient, 3–6, 25, 110; Babylonian, 4, 25, 29, 32, 110; Chinese, 4, 109; Egyptian, 4, 25; Greek, 4ff, 25ff, 110; Korean, 109; Mayan, 4, 25
Atoms, 45, 46, 53, 233, 247, 271, 295, 314, 364, 367
Attraction, electric, 303, 305, 345

Barrow, Isaac, 194, 203
Batteries, 313, 343, 345
Bellarmine, Cardinal Robert, 153, 156
Bentley, Richard, 230
Boyle, Robert, 192
Brahe, Tycho, 49, 106–112, 153, 156, 192, 224, 225, 256, 260, 267, 299, 351, 358; biography, 106ff, 118
Brownian motion, 367
Bruno, Giordano, 93
Buridan, Jean, 159

Calculus, 191, 194, 205
Callipas, 40, 64
*Camera Obscura,* 254–260
Cavendish, Henry, 220, 239, 309
Celestial poles, 10, 13, 21, 29, 30, 35, 77, 79, 97, 221
Celestial sphere, 10, 12, 20, 21, 30, 37, 39, 41, 47, 49, 51, 52, 64, 81, 93, 95, 108, 111, 221
Charge, electric, 306, 337; conservation of, 306, 345
Chromatic aberration, 262, 266
Color, 248, 261–266, 269
Columbus, Christopher, 29, 75, 298
Comet, 214, 225ff; of 1577, 109; Halley's, 203, 226
Compass, 297, 309, 315, 344; dip angle of, 298
Compass lines, 316, 335
Compound motion, 172–175, 176
Conic sections, 125, 175, 224
Constellations, 6, 7
Copernicus, Nicholas, 58, 63, 71, 74–103, 105, 113, 114, 116, 118, 122, 128, 143, 145, 175, 178, 180, 192, 230, 301, 302, 305, 396; biography, 74ff; "On the Revolutions of the Heavenly Spheres," 74, 75–90, 106, 156
"Cosmographic Mystery" (Kepler), 113ff, 120, 145
Coulomb, Charles de, 308ff, 345; biography, 308ff; Coulomb's law, 308–312, 315, 322, 333, 345, 350, 351
Currents, electric, 312ff

Dante, 51
Davy, Humphry, 314, 318, 324
*De Revolutionibus Orbium Caelestium* (*see* "On the Revolutions of the Heavenly Spheres")
Deferent, 64ff, 82, 91, 94, 128
Degrees (definition), 4
Democritus, 45, 53
Descartes, René, 191, 193, 205, 211, 214
"Dialogue Concerning the Two New Sciences" (Galileo), 158, 181, 267

409

# Index

"Dialogue on the Two Great World Systems" (Galileo), 157, 181
Diffraction, 261, 269, 272, 277, 280, 284, 287, 291, 340, 357
Dip angle, 298
Diurnal motion, 12, 13, 15, 16, 19, 20, 41, 77
Donne, John, 94
Doppler shift, 396
Dufay, Charles, 305, 306, 308, 345

Eccentric, 71, 76, 95, 119
Eclipses, 3, 12, 31, 33, 48, 106, 254, 256, 611
Ecliptic, 13, 15, 19, 20, 21, 33, 64, 79, 98, 119, 230
Eddington, Arthur S., 394
Einstein, Albert, 157, 168, 221, 227, 234, 305, 330, 339, 357–397; biography, 363ff, 383ff, 390, 396
Electricity, 302–344; electrical attraction and repulsion, 304, 305, 345; in motion, 312ff; one-fluid model, 305, 345; two-fluid model, 305, 345; static electricity, 302–312
Electromagnetic molecules, 321, 345
Electromagnetic spectrum, 341
Electromagnetic waves, 338, 339ff, 346, 357
Electromagnetism, 316–322, 392
Electroscope, 347–348
Elements (of Aristotle), 45, 52, 192, 247, 295
Ellipse, 125, 167, 175, 202, 224; focus of, 125–126
Epicycle, 64, 76, 83, 91, 94–95, 119, 120, 128, 286, 362
Equal areas law, 124, 127, 197, 222
Equant, 71, 76, 95, 120, 128, 136
Eratosthenes, 34, 35–36, 52
Euclid, 26, 144, 194, 246, 247, 248, 254, 387, 392
Eudoxus, 38–42, 52, 63, 64, 76, 95, 156, 256–257

Falling bodies, law of, 158, 167–172, 180, 183
Faraday, Michael, 305, 314, 322–331, 343–344, 346, 352, 368

Fields, electric and magnetic, 332ff, 337, 346, 368–369; gravitational, 336, 383–397
Fitzgerald, G. F., 363
Force, according to Kepler, 117, 119, 192; according to Newton, 206, 207, 211, 382
Franklin, Benjamin, 305ff, 345, 348–349
Frequency of a wave, 279
Fresnel, Augustin, 285, 320
Friction, 160, 161, 180

Galilei, Galileo, 6, 27, 34, 118, 143–189, 191, 192, 193–193, 201, 208, 211, 231, 235, 245, 267, 287, 297, 303, 305, 364, 365, 370, 378, 380, 384–385, 397; adjuration, 157–158; biography, 143ff, 156ff; Dialogues Concerning the Two New Sciences, 158, 181, 267; Dialogues on the Two Great World Systems, 157, 181; Letters on Sunspots, 156; Letter to the Grand Duchess Christina, 156, 157; *Siderius Nuncius* or the Starry Messenger, 118, 146ff, 152, 181, 192, 258, 260, 395; trial, 157
Galilei, Vincenzo, 143, 157
Geomagnetic poles, 299ff
Gilbert, William, 299ff, 317, 344
Gold standard, 232
Gray, Stephen, 304, 344
Gravity, 161, 214, 383–397; acceleration due to, 172ff, 195, 197, 208; according to Einstein, 383ff, 398; gravitational field, 336, 383ff, 398; and lines of force, 330; inverse square law, 201, 203, 216, 228 (*see also* Universal gravitation, law of)
Greatest elongation, 16
Grimaldi, Francisco, 261, 277, 287
Guericke, Otto von, 303, 308, 310

Halley, Edmond, 203–205, 221, 224, 234
Halley's Comet, 203, 226
Hammurabi, 4
*Harmonice Mundi* (*see* "The Harmonies of the World")

"Harmonies of the World, The" (Kepler), 129
Henry, Joseph, 317
Heraklides, 48, 53, 112
Herschel, William, 226
Hipparchus, 21, 64, 73, 108
Hooke, Robert, 192, 203, 205, 211, 224, 234
Huygens, Christian, 191, 205, 211, 215, 224, 273–278; "Treatise on Light," 273; wave model of light, 273–278; wavelets, 274, 287
Hveen, 107

Impetus, 159, 172
Inclined planes, 167
Inertia, 159ff, 235, 297; law or principle of inertia, 160, 166, 175, 180, 193, 194, 196, 206, 311, 382
Interference, 279ff, 287, 340, 359
Interferometer, 359–360

Julian calendar, 76, 99
Jupiter, 10, 14, 18–19, 42, 49, 67, 78, 80, 81, 88, 89, 114, 193, 222, 224, 225, 227, 393; moons of, 146, 147, 150–152, 180, 183, 200, 236, 267ff, 287
Kepler, Johann, 113–132, 143, 167, 179, 191, 192, 214, 222, 225, 226, 245, 255–261, 299, 302, 329, 351; *Mysterium Cosmographicum* or "The Cosmographic Mystery," 113ff, 120, 145; *Astronomia Nova* or "The New Astronomy," 119ff; *Harmonice Mundi* or "The Harmonies of the World," 129; *Dioptrice,* 260; biography, 113ff, 118, 132; First Law, 128, 222, 267; Second Law, 128, 194, 197, 222; Third Law, 129–132, 183, 195, 200, 203, 235; on light, 192, 255–261
Kidinnu, 4
Kuhn, T. S., 35, 75

Lagrange, Louis de, 225, 234
Laplace, Pierre-Simon, 225, 234
Law of falling bodies, 158, 167–172, 180, 183

Leibnitz, Gottfried, 191, 194, 233
Length contraction, 377, 378, 381, 398, 400
"Letter to the Grand Duchess Christina" (Galileo), 156, 157
"Letters on Sunspots" (Galileo), 156
Leverrier, Urbain Jean Joseph, 226, 393, 394
Light, 26, 29, 245–293, 295, 339; ancient models for, 245–248; particle theory of, 247, 270–273, 368; speed of, 192, 248, 266–270, 286, 339, 359, 370; wave theory of, 348, 273–286, 368; wavelength of, 279, 284
Light clock, 373, 376
Lightning, 348-349
Lines of force, 325, 328, 330, 346
Lodestones, 295ff, 316, 344
Lorentz, Hendrik A., 363
Lowell, Percival, 227, 247
Luther, Martin, 75, 94
Lyceum of Aristotle, 26, 43

Mach, Ernest, 364ff, 370, 384
Maestlin, Michael, 113
Magellanic clouds, 228
Magnets, 295-302
Magnetic induction, 327ff, 337, 343, 346
Magnetic poles, 300
Magnetism, 295-344; early knowledge of, 295ff; electromagnetism, 316-332, 392; microscopic, 349; induced, 327
Mars, 10, 14, 18ff, 42, 49, 67, 78, 80, 82, 83, 88, 118-128
Mass, 207, 208-211; inertial and gravitational, 220-221; 384-386; of the earth, 213, 219, 238-240
Maxwell, James Clerk, 305, 329, 330, 331-342, 346, 358, 367, 369, 397; biography, 331; Maxwell's equations, 337, 357, 368; unification of electricity, magnetism, and light, 339
Mechanistic philosophy, 193, 205, 272
Mercury, 10, 14, 18-19, 42, 49, 70, 78, 80-82, 87, 95, 96-97, 111, 152, 227, 393, 398; greatest elongation of, 18; synodic period, 18; precession of the perihelion, 393
Michelangelo, 75, 143, 158, 260
Michelson, A. A., 107, 286, 358-363, 397; biography, 358
Michelson-Morley experiment, 359-363, 368, 377, 397, 399-400, 403
Milky Way, 146, 180
Milton, John, 73
Minute of arc, definition, 4
Momentum: definition, 207; conservation of, 212
Moon, 5, 9, 20, 32, 41, 42, 47, 147; distance to, 35, 55-58; mountains on, 34, 146, 147, 180, 181-183, 193; motion of, 14-16, 20, 21, 39, 65, 197, 235, 237-238; parallax of, 108; phases of, 15, 32, 42, 52
Museum of Alexandria, 64
*Mysterium Cosmographicum* (see "The Cosmographic Mystery")

Natural and unnatural motions, 44-46, 159, 173, 231
Neptune, 227, 228, 393
"New Astronomy, The" (Kepler), 119, 145, 146
Newton, Isaac, 46, 71, 119, 167, 178, 191-242, 273, 278, 285, 295, 305, 310, 329, 336, 337, 365, 367, 370, 380, 395, 398; biography, 194-195, 203-205, 232, 234; on color, 261-266; the "crucial experiment," 265; the falling apple, 195; the laws of motion, 205-213, 357, 382; first law of motion, 206; second law of motion, 206-207, 384, 385, 391; third law of motion, 211-213, 318; and the Mint, 232; the "miraculous year," 194-195, 367; on light and optics, 213, 245, 270-272, 277, 278, 280, 287; "Opticks," 230, 233; on the role hypotheses in science, 229; *Philosophiae Naturalis Principia Mathematica* or the "Mathematical Principles of Natural Philosophy," 75, 194, 195, 203-205, 214-229, 232, 233, 234, 286, 390; on theology, 230-231, 233; universal gravitation, law of, 34, 129, 215-220, 225, 226, 235, 236, 311, 319, 336, 339, 345, 357, 389
Non-Euclidean geometry, 391-392
North Pole Star (Polaris), 7, 8, 10, 37
Norman, Robert, 298
Nova: of 1572, 108, 113, 132; of 1054, 109; of 1604, 145

Oersted, Hans Christian, 315, 325, 326, 336, 343, 345
Opposition, 19, 20, 67, 78, 83, 87, 91, 95, 110, 268
"Opticks" (Newton), 230, 233
Oresme, Nicole, 159

Parabola, 125, 175, 187, 224, 231
Parallax, 5, 49-50, 53, 58-59, 77, 82, 93, 108, 110, 111, 113, 132, 153
Particle theory of light, 247, 270-272, 287, 368
Pendulum, law of, 144, 273
Perihelion, 126, 136-137, 393
Period of a wave, 279
Periods of revolution of the planets, 87-90, 100-103
Permanent magnets, 301
Photoelectric effect, 367-368
Photons, 368, 384
Planets, motions of, 6, 9, 10, 16-19, 20, 39
Plato, 26, 37-42, 43, 44, 45, 49, 52, 71, 114, 116, 119, 128, 246
Pluto, 227-228
Pope, Alexander, 234
Precession of the Equinoxes, 21, 76, 79, 81, 97-99, 214, 221
Precision of measurement, 110, 122, 132, 224, 267-268, 358
*Principia* (Newton), 75, 194, 195, 203-205, 214-229, 232-234, 286, 390
Principle of Equivalence, 386, 387, 390, 398
Printing, 75
Projectile motion, 143, 145, 172, 180, 184-187
Proper time, 379, 404
Prutenic tables, 106

# Index

Ptolemy of Alexandria, 26, 51, 64–73, 106, 120, 144, 154, 194, 245, 252, 254, 256, 258, 286
Pythagoras, 27–28, 29, 45, 52, 114, 128, 134, 168; theorem of, 27–28, 34, 52, 53, 198, 199, 271, 290, 375, 376, 399
Raphael, 256
Reflection, law of, 249, 255, 269, 271, 275–276, 286, 357
Refraction, law of, 191, 249–253, 258, 260–261, 262, 269, 271, 276–277, 286, 357
Regular solids (polyhedra), 114–116, 133–136
Relativity: general theory, 221, 227, 319, 330, 383–397, 398; principle of, 157, 175–177, 178, 180, 187, 206, 231, 370, 377, 378, 385, 397; special theory of, 367, 368–383, 390, 397
Repulsion, electric, 303, 305, 345
Retrograde motion, 16–20, 41, 42, 65–67, 69, 78–84, 90, 91, 95
"Revolutions of the Heavenly Spheres, On the," 74, 75–90, 106, 156
Rheticus, 74
Roemer, Olaus, 267
Royal Society of London, 192, 203, 232, 324, 331, 395

Saturn, 10, 14, 18–20, 42, 49, 65–67, 78–83, 88, 89, 95, 100, 114, 215, 222, 224, 225, 227, 236, 273, 332, 339
Scientific method, 144–145, 156, 157, 178–179, 181, 229, 265, 303
Seasons, 3, 7, 11, 64, 79
Second of arc, definition, 4

Shakespeare, William, 143
Sidereal period, 87–90, 98, 100–103
*Siderius Nuncius* (*see* "The Starry Messenger")
Simultaneity, 371, 372
Size of the earth, 34–36, 54–55
Sizes of the planets' orbits, 84–87, 99–100
Shape of the earth, 28–33
Snell, Willebrord, 191, 260; Snell's law, 260, 271, 277, 287
Socrates, 26
Solar system, 91
Solenoid, 317, 326
Speed, 47, 124, 161ff; addition of (in relativity), 380–381, 398, 403–406; average, 164; instantaneous, 165
Stars: distance to, 58–59, 150; motion of, 6–11; number of, 147, 148; sizes of, 7, 154
"Starry Messenger, The" (Galileo), 118, 146ff, 152, 181, 192, 258, 260, 395
Statistical mechanics, 332, 367
Sun, 3, 5, 9, 42, 78, 82, 119, 126, 128, 152, 246; motion of, 11–13, 39, 64; gravitational force of, 222–225
Sunspots, 156
Supernova, 109
Synodic period, 18, 19, 67, 87, 95, 103

Telescope, 6, 145, 146, 152, 180, 192, 228, 262, 266, 287; reflecting, 233, 266
Television, 340
Terella, 299, 317, 344
Thales of Miletus, 45
Thought experiment, 371, 378
Time dilation, 376, 377, 381, 382, 398

Toricelli, 192
*Traite de la Lumiere* (*see* "Treatise on Light")
"Treatise on Light" (Huygens), 273
Two-sphere model of the universe, 36–38

Universal gravitation, law of, 34, 129, 215–220, 225, 226, 235, 236, 311, 319, 336, 339, 345, 357, 389
Uraniborg Castle, 107
Uranus, 226–228, 393

Vaccum, 46–47, 52, 193, 205
Venus, 10, 14, 16, 42, 49, 67–69, 78, 80, 82, 87, 91–92, 95–97, 99–100, 111, 152; greatest elongation, 16–17, 69, 154; phases of, 91, 146, 154, 180; retrograde motion of, 16–18, 154
Vinci, Leonardo da, 256
Vision, 245
Volta, Allesandro, 312, 345
Voltaic cells, 313
Voltaire, 233
Vulcan, 227

Wave theory of light, 248, 273–286, 287, 368
Wavelength, 279, 290, 341
Weight, 208–211; of the earth, 240
Wren, Christopher, 203

Young, Thomas, 278–286

Zodiac, 15–19, 42, 65, 69, 94, 151